Lecture Notes in Earth Sciences

Edited by Somdev Bhattacharji, Gerald M. Friedman,
Horst J. Neugebauer and Adolf Seilacher

20

Peter Baccini (Ed.)

The Landfill

Reactor and Final Storage
Swiss Workshop on Land Disposal of Solid Wastes
Gerzensee, March 14–17, 1988

Springer-Verlag

Berlin Heidelberg New York London Paris Tokyo

Editor

Prof. Dr. Peter Baccini
Swiss Federal Institute for Water Resources
and Water Pollution Control (EAWAG)
Federal Institute of Technology
CH-8600 Dübendorf, Switzerland

Scientific Organizing Committee

Peter Baccini (Chairman)	Swiss Federal Institute for Water Resources and Water Pollution Control, Switzerland
Michel Aragno	Institute of Biology Université de Neuchâtel, Switzerland
Robert K. Ham	Environmental Engineering University of Wisconsin, USA
Adrian Pfiffner	Institute of Geology Universität Bern, Switzerland
Rainer Stegmann	Environmental Engineering Technische Universität Hamburg-Harburg, FRG

Director of Administration

Walter Ryser	AG für Abfallverwertung AVAG Uttigen, Switzerland

ISBN 3-540-50694-2 Springer-Verlag Berlin Heidelberg New York
ISBN 0-387-50694-2 Springer-Verlag New York Berlin Heidelberg

© Springer-Verlag Berlin Heidelberg 1989
Printed in Germany

Printing and binding: Druckhaus Beltz, Hemsbach/Bergstr.
2132/3140-543210 – Printed on acid-free paper

PREFACE

In a densily populated industrialized country, waste disposal must be compatible with the requirements of the environment. This is one of the indispensable requirements to guarantee an effective protection of the environment.

In the past years the waste disposal industry has been given increasing attention by the general public as well as the authorities. This confirms the necessity of adapting the quality of waste disposal to the technological standard of the production. While in the past, waste disposal performance was more or less evaluated in terms of short-term costs, there is at present a reorientation in the direction of a science-based waste disposal industry. These new tendencies are taking into account ecological factors as well as the long-term consequences - i.e., for decades and centuries to come - of waste disposal methods. In this light, particular attention is given to the depositing of residues whose utilization does not appear meaningful from an ecological point of view, or would require disproportionate ressources.

It is an important concern of the Federal Authorities to encourage the rapid materialization of disposal solutions which can function as ultimate deposits, and which will therefore cause neither water pollution nor gaseous emissions. In view of this goal it is necessary to establish criteria and regulations for the wastes to be deposited as well as for the characteristics of the deposits. This field confronts science with an urgent but rewarding challenge and calls for close collaboration between many different specialized disciplines.

I wish to thank the participants in the Swiss Workshop on Land Disposal of Solid Wastes for their contribution to the solution of this urgent problem.

Federal Office of Environmental
Protection

B. Böhlen

B. Böhlen

TABLE OF CONTENTS

WORKING GROUPS

	A) THE LANDFILL AS A REACTOR		B) THE LANDFILL AS A FINAL STORAGE	
	A1: Biological and chemical processes	A2: Material transport and properties of reactor envelopes	B1: Scientific and technical criteria for the final storage quality	B2: Methodology for the evaluation of the final storage quality
Chairman	Ham, Robert K. (USA)	Stegmann, Rainer (FRG)	Pfiffner, Adrian (CH)	Brunner, Paul (CH)
Lecturers	Aragno, Michel (CH) Ehrig, Hans-Jürgen (FRG) Förstner, Ulrich (FRG)	Farquhar, Grahame J. (CAN) Gray, Donald H. (USA) Madsen, Fritz T. (CH)	Bülow, Eckehard (FRG) Huggenberger, Peter (CH) Stief, Klaus (FRG) Zeyer, Josef (CH)	Fahrni, Hans-Peter (CH) Francis, Chester W. (USA) Milde, Gerald (FRG)
Troublemaker	Schwarzenbach, René (CH)	Zingg, André (CH)	Lichtensteiger, Thomas (CH)	Müller, Hans-Peter (CH)
Rapporteur	Belevi, Hasan (CH)	Höhn, Eduard (CH)	Stuijvenberg, Jan van (CH)	Krebs, Jürg (CH)
Discussion participants	Colin, François (F) Matter, Christine (CH) Salkinoja, Mirja (SF) Strub, Walter (CH)	Kanz, Werner (CH) Oggier, Peter (CH) Piepke, Gabriel (CH) Ryser, Walter (CH) Schlüchter, Christian (CH)	Egli, Markus (CH) Lagerkvist, Anders (SW) Zurkinden, Auguste (CH)	Billard, Hervé (F) Filippini, Francino (CH) Lechner, Peter (A) Messmer, Jörg (CH) Stämpfli, Dominique (CH)

Chairman: Baccini, Peter (CH)
Secretary: Lichtensteiger, Thomas (CH)

Director of Administration: Ryser, Walter (CH)
Secretary: Peiser, Evelyne (CH)

INTRODUCTION

A SWISS WORKSHOP ON LAND DISPOSAL

MOTIVATIONS - OBJECTIVES - STRUCTURE

Peter Baccini

Swiss Federal Institute for Water Resources and Water Pollution Control

1. Motivations

During the last decades most countries regulated the land disposal of
solid wastes on the basis of their laws for water protection. Pollution
of surface and ground waters by leachates had to be prevented, mainly
by controlling the output fluxes. The physical and chemical properties
of the input were neither well defined nor well controlled. The
long-term implications of such disposals seemed to be of minor importance.   Two events induced a critical revision of contemporary disposal
practice:

1) Costly restorations of landfills polluting or threatening pollution
of their environment raised questions about the "ecological and economical balance" of solid waste disposal with unpredictable reactions.

2) Environmental protection laws with a "holistic" concept, i.e. to
protect man and biosphere from any hazardous anthropogenic residual
fluxes of matter and energy, ask consequently for quality standards of
solid residuals "compatible with short and long-term geogenic processes".   In other words the catalogue of "quality objectives" for air,
water and soil has to be extended for solids in and from landfills.

Today it is widely accepted that any waste management system, an integral part of a national economy, is not just a function of the scientific concepts and the technically feasible processes. It is as much
depending on the general political climate . Each country develops its
own disposal strategy due to its characteristic boundary conditions.
There might be even several "optimal solutions" for a certain region to
reach its quality objectives. Therefore the best "universal solution"

cannot exist by definition. Among scientists however the rationale behind the technical and regulatory processes chosen should be subject of
permanent and critical revision.

The Swiss Federal Government has decided to proceed as follows [EKA
1986]:

A) Each generation handles its residual waste (the not reusable fraction) to a status of "final storage quality". Landfills with solids of
final storage quality need no further treatment of emissions into air
and water.

B) The solid residue with final storage quality should have properties very similar to the earth crust (natural sediments, stones, ores,
soils).
It follows that principally <u>two types of landfills</u> are to be realized:

<u>"Reactor Landfills"</u> producing predictable outputs with respect to
quantity, quality and time. Within about 30 years the emissions should
meet the air and water quality standards without any further treatment.
The residues should meet criteria B). They have become "Final Storage Landfills".

<u>"Final Storage Landfills"</u> are deposits with solids meeting criteria B)
from the beginning. These properties can be achieved, if necessary, by
appropriate pretreatments (e.g. by incineration).

For any expert in waste management it is obvious that such a concept
cannot be realized within a few years. Beside the preponderant political constraints there are many open questions with respect to the
scientific methodology to characterize "final storage quality". Furthermore the term "final storage" is not virginal any more.

The "Swiss Workshop on Land Disposal of Solid Wastes" is a first attempt to discuss - among experts from various scientific disciplines
and in different functions - the landfill as "Reactor" <u>and</u> "Final Storage". Therefore the adjective "Swiss" designates not only the geographical site of the meeting but also one of the main motives, namely
to invite an international group to look critically at one possible
strategy in handling the process of landfill.

2. Objectives

The workshop members, on the basis of the papers prepared in advance by invited "Lecturers",

- discuss and select the criteria which are essential to qualify a "final storage quality" of anthropogenic solid waste in landfills
- discuss strategies to "transform" reactor landfills to final storage landfills
- write a report of comments to the basic topics of the workshop and to the group's specific questions

Although each group had different specific questions, the four introductions and the four group reports show repetitions with respect to basics. By purpose this redundancy was not changed by the editor, in order not to interfere with the character of the individual working group.

3. Structure

Four parallel groups of about 12 persons each treated simultaneously the topic "reactor and final storage" from a somewhat different point of view, according to their expertise in landfills. Plenary sessions were held daily to pass informations and to discuss first results. It was necessary to combine briefly two groups. However it was most important that each group tried to find its own way. The results were distributed among all workshop participants.

The following roles were given:

Group Chairman
the moderator, organizes the group to fix the agenda and to write the reports

Lecturer
explains the methods, results and hypotheses presented in his review paper, acts as expert with respect to the paper's topic

Troublemaker
is responsible for stimulating disturbances if group dynamic processes
tend to false compromises

Rapporteur
summarizes highlights, controversies and open problems of his group in
written form, to be presented in plenary session by the chairman or a
speaker of the group

Discussion Participant
prepares in advance (by reading the review papers) questions or antith-
eses for the lecturers

Reference

EKA 1986, Eidg.Kommission für Abfallwirtschaft, "Leitbild für die
schweizerische Abfallwirtschaft" (Guidelines for the Waste Management
in Switzerland), Schriftenreihe Umweltschutz Nr.51, BUS, 3003 Bern
(1986)

ACKNOWLEDGMENTS

I am greatly indebted to

- Walter Ryser, not only an entrepreneur and engineer of great practical experience in constructing and running landfills, but also a phi-
  losopher of waste management. He is the man who made this workshop
  possible.

- My colleagues of the scientific organizing committee who helped me a
  great deal with their advice

- Paul H. Brunner who had the courage and the skill to replace the illness stricken chairman Wallace Fuller in the last minute

- <u>Evelyne Peiser and Thomas Lichtensteiger</u> who made the hard daily
  work, patiently correcting the errors of their bosses

- the <u>sponsors</u> who were ready to support a new experiment

- to <u>Ulrich Förstner</u> and <u>Hanspeter Müller</u> who made an additional effort at the end of the workshop to present their conclusions in
  public.

THE LANDFILL AS A REACTOR

BIOLOGICAL AND CHEMICAL PROCESSES

Chairman: Robert K. Ham (USA)
Lecturers: Michel Aragno (CH)
                           Hans-Jürgen Ehrig (FRG)
                           Ulrich Förstner (FRG)
                           Walter Ryser (CH)*
Troublemaker: René Schwarzenbach (CH)
Rapporteur: Hasan Belevi (CH)
Discussion participants: François Colin (F)
                           Christine Matter (CH)
                           Mirja Salkinoia (SF)
                           Walter Strub (CH)

SPECIFIC QUESTIONS:

1) Which measures can accelerate the decomposition processes of organic material and by this shorten the time of the "reactor period" of a landfill?

2) How can one define the transfer from "biologically mediated processes" to "geochemically mediated processes" (i.e. practically only abiotic processes) in landfills?

3) Which compounds should be excluded from landfills due to their potential to
           - prolong the reactor period
           - increase emissions
           - produce new xenobiotic and/or hazardous compounds?

4) What is the lifetime of landfill internal installations (e.g. drainage systems) to secure a controlled water household?

5) Which landfills (types, examples) are well documented so that they could serve as objects for multidisciplinary studies?

6) Which "indicator parameters" should be selected to describe the state of evolution of a landfill?

INTRODUCTION

THE LANDFILL AS A REACTOR: BIOLOGICAL AND CHEMICAL PROCESSES

Robert K. Ham

University of Wisconsin-Madison

The typical concept of a sanitary landfill for disposal of solid wastes is that this is a method to dispose or get rid of the waste. In other words, the landfill has been considered in terms of a waste receptor by the public as well as by the professionals involved with planning or operating landfills. In more recent years, we are realizing more and more that materials leave the landfill, either carried by water or as a gas, and have an impact on the environment. This has caused the emerging concept of the landfill as a reactor or treatment device, at least amongst some of the professional community. Under this concept, a landfill will be used to treat waste materials, venting or collecting treatment products to minimize environmental impacts, resulting in a stabilized mass which will not be of environmental concern at some defined time in the future.

In considering the total impact of a landfill on the environment, one must look at the decomposition processes which control the degradation of waste to stable matter and the by-products of the degradation processes. Considerable research and monitoring of landfills has occurred over the last 20 years or so, but such information has, until now, not been incorporated into an overview of the state-of-the-art followed by an evaluation of the future of landfilling. Group Al was given the opportunity to do just that. The objective was to compile and summarize what we know about landfill decomposition and assess this information and apply it to the concept of the landfill as a reactor.

There are two aspects of the landfill reactor which had to be discussed. The first is what happens during the period of active decomposition. What processes occur, how are they affected by different variables, and what are their products of decomposition? Products of decomposition leave the landfill either water-borne in leachate or as a gas. These must be controlled to minimize environmental impact and, in the case of gas, collected for productive use whenever feasible. The

second aspect of the landfill as a reactor is the final approach and
attainment of stabilization or compatibility with the environment, if
in fact this occurs. At such a point, there would be no more products
of decomposition with observable environmental impacts, and the waste
mass would provide a stable land form not subject to further settling.

Four comprehensive papers were prepared to treat the most important aspects in considering the landfill as a reactor. The decomposition processes taking place in a landfill can be broadly classified as biological or chemical, so one paper is presented on each. There is some
overlap as decomposition is, in fact, very complex with a multitude of
biological, chemical, and physical processes occurring simultaneously
throughout the decompositional life of the landfill. However, the basic
processes which control the rate, extent, and degradation pathways are
primarily biologically or chemically mediated. The third paper considers the landfill as a reactor by looking at what goes into and out of
the landfill; in other words, by looking at mass and element balances.
The final paper gives an overview of the technical concepts applied
during the last two decades to control the reactor landfill.

The paper titled The Landfill Ecosystem: A Microbiologist's Look Inside
a "Black Box" provides a review of biological processes occurring inside a landfill. The paper begins with a summary of the physiological
properties of bacteria in general, including nutritional, energy, and
other growth requirements of the different types of bacteria. Following
is a discussion of associations of different bacteria which occur in a
complex environment and with a complex array of substrates and
chemicals, as in a landfill. The role of xenobiotics, defined in the
paper as organic compounds created by man that resist degradation by
naturally occurring organisms, is considered. Finally, this information
is applied to the landfill environment, emphasizing the evolution of
the degradation processes over time and the interactions between bacteria and the environment within a landfill. Products of decomposition
are discussed, as well as mechanisms by which they can escape the landfill and potentially impact the surrounding area. The long-term stability of the landfill mass is considered, and it is pointed out that,
conceptually at least and depending on the natural surroundings, the
landfill will gradually become oxic with largely unknown consequences.

The second paper of this section titled Geochemical Processes in Landfills, considers the chemical processes which occur as waste decom-
poses. Such processes are overshadowed by biological processes during

initial decomposition periods when chemical effects are present but controlled to a large degree by the biological processes. Later, however, as biologically decomposable matter is lost, chemical processes become more important and, in fact, may be expected to dominate processes affecting the final approach to waste mass stability. Chemical processes are therefore critical in
considering the landfill as a long-term storage device.

The paper begins with a discussion of mobilization of pollutants, which summarizes the changes in the chemical environment which result in release of contaminants from solid materials. Factors causing these changes are externally caused pH changes as from acidic
rain water, changes in redox conditions, chemical effects of biological decomposition products, increasing salt concentration, and the presence or formation of complexing agents. Tools for assessing the mobility of different contaminants include column and lysimeter studies which simulate the field conditions, batch equilibrium studies, thin layer chromatography, and single or multiple batch elution tests, all of which can provide input to various modeling techniques. These procedures have varying degrees of success in modeling full-scale landfill decomposition and parameter release patterns.

The paper concludes with a discussion of waste composition during the period late in the decomposition sequence when the waste approaches a stabilized condition. The effects of long-term changes in pH and redox conditions on an otherwise "stabilized" waste is considered with regard to purely inorganic wastes. Pollutant release results from wastes of low organic content provide the best source of information regarding the approach to a stabilized waste, given the lack of long-term monitoring data from municipal waste landfills. Landfill results from coal, sewage sludge, and municipal refuse ash disposal are discussed, and comparison is made to naturally occurring inorganic minerals which are basically stabilized and compatible with their surroundings.

The third paper is titled Water and Element Balances of Landfills. The approach here is not to consider the decomposition processes taking place within a landfill but to consider the landfill as a treatment device or reactor, evolving a fraction of the various materials entering it in the form of leachate or gas. If the fraction of each element or waste component of concern released by these routes can be determined, for example, information is attained about processes taking place within the landfill, and the approach to final stabilization can be observed.

In order to do this, data on the cumulative releases from various landfills of water and different elements is presented. Carbon transfer,
leaving the landfill in the form of COD and TOC in the leachate, plus
gas, is presented because of its importance and as an example of
biologically mediated release. Chloride is discussed as an example of a
largely chemically mediated release, and nitrogen as a combination of
biological and chemical transport processes. Other elements are discussed but to a lesser degree.

The paper concludes that insufficient information is available, and is
unlikely to be available, to reliably predict final storage quality by
the mass balance approach. Mobilization will continue for a very long
time at some low rate, and insufficient information is available or
likely to be available to allow the mass balance approach to predict
final storage quality and hence the state of "stabilized" wastes.

The last paper of this section is titled <u>Control of Reactor Landfills
by Barriers</u>. Although the properties of envelopes are discussed in detail in group A2, the term "multi-barrier strategy" has been used by
different authors to describe a combination of measures to control material fluxes in landfills (see also K. Stief's paper in group B1). The
last paper extends the set of well known barriers, namely the natural
<u>site</u> and the geotechnical <u>envelope</u>, with the controlled <u>reaction</u> in the
landfill (barrier 2) and the <u>technical process</u> to adapt the input (barrier 1). For each type of barrier a list of criteria is given to
evaluate its functioning. Experimental results are given for the behavior of barrier 2. The paper closes with a concept for the "Barrier
Model 2000" in which, according to the Swiss guidelines, only a combination of the barriers 1 and 4 will be necessary.

THE LANDFILL ECOSYSTEM:
A MICROBIOLOGIST'S LOOK INSIDE A "BLACK BOX"

Michel Aragno
Laboratoire de Microbiologie
Université de Neuchâtel (Switzerland)

1. INTRODUCTION

A landfill, when it contains substrates promoting microbial growth and at the same time offers physicochemical parameters (temperature, pH, osmotic pressure, water content, etc.) compatible with microbial life, can be regarded as an ecosystem, or as a complex bioreactor. I prefer the term ecosystem, since bioreactors are in general considered as physically defined environments performing a precise biotransformation. With that meaning, a landfill would rather resemble a mosaic of closely imbricated bioreactors of different physicochemical properties, and thus harbouring different microbial populations interacting systemically.

In this paper, we will first summarize certain physiological properties of bacteria related to their way of life in such an ecosystem. We will then present some principles of the close relationships of bacteria in an ecosystem, with the example of the methanogenic syntrophy, the dominating microbial system in a normal, compacted landfill which has reached its climax. We will then discuss the properties of the landfill environment, from the angle of its microbial life. This paper must be regarded as reflections of a microbiologist, based on his own experience and research, giving a frame for discussions and remarks; it is not intended as a review of the literature on the subject.

2. PHYSIOLOGICAL PROPERTIES OF BACTERIA

2.1 Nutritional requirements

Like every living being, a bacterium must find in its environment the
matter (chemical elements) necessary to build its own biomass , as well
as the energy required for the biosynthesis of cellular components, or
to perform the tasks implied in growth, reproduction and maintenance of
cellular structure

2.1.1 Requirements for assimilation

Microbial biomass contains a number of elements. Some of them, i.e. C,
H, O, N, S, P are constituents of its main components, the macromol-
ecules (proteins, nucleic acids, polysaccharides, lipids); other (metallic ions) participate in the osmotic equilibrium of the cell, in
transport processes, and as cofactors in enzymatic reactions, either in
a ionic or organometallic form. The bacteria must find all these elements in their habitat, and they use them in definite proportions. The
first nutrient to be exhausted will act as a limiting factor for
growth. There are three major chemical elements responsible for this
limitation: carbon, nitrogen and phosphorus.

2.1.1.1 Carbon

Carbon is the main element in the organic matter. The carbon source is
an inorganic one (CO_2 or its derivatives, sometimes CO) in autotrophs,
and an organic one in heterotrophs. In order to reduce CO_2 to the
oxidation level of organic matter, the autotrophs require an inorganic
electron source (see below). Heterotrophs can synthesize all, or most,
of their own organic materials from simple organic compounds such as
sugars, aliphatic aromatic or amino acids, alcohols etc. In addition,
some require already synthesized chemical structures as growth factors
(vitamins, aminoacids, purine or pyrimidine bases, etc.). The small
molecules (monomers and dimers) are transported through the cell wall
by specific permeation systems, whereas polymers must be hydrolysed by
exoenzymes (depolymerases) before the mono-or dimers resulting from the
hydrolysis can be transported into the cell. In most cases, the
depolymerases are released in the medium. Therefore, the hydrolysis

products profit, not only to the enzyme producer itself, but also to other members of the microbial community which are able to transport and metabolize the mono- or dimers, but not to produce the depolymerases. In other cases, such as the cellulolytic aerobes Cytophaga and Sporocytophaga and many anaerobes, the enzymes are strictly located at the cell surface, so that hydrolysis occurs only during close contact between the cell and the polymer fiber. In that case, no hydrolysis product will be released in the medium and these "selfish" organisms will benefit alone from the enzyme activity.

There are big differences with respect to the amount and variety of organic compounds that heterotrophs can assimilate. Some are narrow specialists able to utilize only one or a few compounds, like the methanogens, which can assimilate only acetate, formate, and sometimes methanol, CO_2 and CO, or the strict methanotrophs, which are only able to utilize methane and a few other Cl^- compounds. Other are very versatile organisms, able to use more than a hundred organic compounds as sole carbon and energy source.

Virtually all organic compounds of biological origin can be degraded by naturally occuring microorganisms, although several only under particular conditions (e.g. aerobically or at a neutral pH). This explains that during the geological times, huge amounts (even if these represent a low proportion of the biomass actually synthesized) of organic materials were not completely mineralized, but accumulated, after complex transformations, into the "natural final storage landfills" that the peat, coal and petroleum deposits are. In fact, several compounds, such as aliphatic hydrocarbons, cannot be degraded by anaerobes. Their assimilation requires an attack of the relatively inert molecule by molecular oxygen, catalysed by oxygenases.

The situation is quite different where the organic compounds created by man are concerned. One must understand first that in the case of biological organic substances, evolution progressed in small steps. In this way, the systems able to degrade this matter could follow this evolution, as only one or a few mutations were required for each adaptation. But, when it came to transporting, degrading and metabolizing compounds differing completely from any pre-existent structure, adaptations were much less likely to occur, since it would have required a great number of mutations conferring individually no advantage in order to obtain enzymes with a catalytic structure adapted to the new compound. Such compounds, differing so much from natural ones that they

will resist degradation by naturally occuring organisms, are termed
xenobiotics. In a wider sense, this term is sometimes applied to all
synthetic organic compounds which do not exist in the nature, even if
their resemblance to existing compounds allows them to be easily degraded by existing enzymatic systems.

Actually, the situation is more complex than this simple "all or nothing" picture: one can describe several degrees of "xenobioticity". For
example:

- Certain compounds will be degraded by microorganisms after a mutation
has spontaneously occurred. This in principle will require a long exposure to relatively important concentrations of the compound.

- Others can be easily, but only partly degraded in natural environments. They will be finally transformed into not degradable products,
which will accumulate. This is the case with DDT, and with
tetrachlorethylene: this latter is dechlorinated anaerobically to vinyl
chloride, in such conditions a more stable but at the same time more
toxic compound !

- Some compounds can only be degraded in laboratory conditions, in general in continuous cultures on synthetic media after a long process in
which step by step adaptations have been provoked, through mutagenic
treatments for instance. Such artifical evolution can lead to organisms
adapted to degrade xenobiotics in controlled conditions only, but that
are not likely to survive and/or to perform this task in natural, uncontrolled environments.

- Some compounds can only be degraded if non-xenobiotic substances of
similar structure are present. Then, the enzymes responsible for metabolizing the latter will be induced, and, in some cases, will be able
to deal with the xenobiotic compounds as well, either totally or partially. This is termed co-metabolism.

- Finally, certain polymers cannot be degraded biologically,due to the
lack of an adapted depolymerase. But abiotic degradations of this polymer may occur, leading to biodegradable products.

2.1.1.2 Nitrogen

Nitrogen is often a limiting factor for the development of a microbial
community. A lack of nitrogen in decomposing waste mass will lead to an

arrest of physiological growth (mainly synthesis of proteins and nucleic acids), whereas substrate utilization will continue for maintenance. This may lead to the accumulation of metabolites, such as organic acids, which would decrease the pH and retard certain processes, like methanogenesis.

Many microorganisms utilise inorganic nitrogen, either ammonium salts or nitrate. Some require an organic source of nitrogen, e.g. urea or an amino-acid. Also xenobiotic nitrogen compounds could serve as nitrogen source. This needs further investigation. Many bacterial species can reduce dinitrogen (N_2) into ammonia before assimilating it: they are the so-called "dinitrogen fixers". The presence of dinitrogen fixers in a microbial association could prevent nitrogen limitation.

2.1.1.3 *Phosphorus*

Phosphorus is mainly used as phosphate ions. Because of the low solubility of many inorganic phosphates, it is often a limiting growth factor for microbial communities. Its availability is often even more restricted by its occurrence in a precipitate form. This can be then controlled by the solubilizing effect of metabolites, like organic acids, which are secreted by bacteria.

2.1.2 *Energy requirements*

Energy is required for biosynthesis and for the maintenance of life. Two forms are externally available to organisms: light (the source for phototrophic organisms) or an exothermic chemical reaction (for chemotrophic organisms).

2.1.2.1 *Light*

Light is the main energy source for primary production on earth, mainly for plant production. Several microorganisms can also utilize light energy, but they are of little significance in landfills. Phototrophic (photosynthetic) bacteria can be found associated with algae in waste lagoons where they may immobilize several compounds, including heavy metals. Some also fix dinitrogen.

2.1.2.2 Chemical energy

Chemotrophic organisms depend on an exergonic chemical reaction based, directly or indirectly, upon compounds found in their environment. This reaction will be coupled to the formation of metabolically available energy, such as pyrophosphates bonds (ATP) or trans-membrane cationic and electronic gradients. Two main types of coupling processes can be involved here: respiration and fermentation.

2.1.2.2.1 Respiratory processes

Respiratory processes involve an electron transport through redox intermediates, from an electron donor to an electron acceptor. Two cases can be considered here: if the external electron donor is an inorganic substance, the organism will be called a lithotroph. If the external electron donor is an organic substrate, the organism is then an organotroph.

2.1.2.2.1.1 Lithotrophic metabolism

Lithotrophic organisms are usually autotrophs too. The energy and reducing power from the inorganic substrate are used to reduce CO_2 to synthesize the cell material. This metabolism can occur in aerobic (when the external electron acceptor is molecular oxygen), as well as in anaerobic conditions (with an external electron acceptor different from oxygen, in landfill conditions usually sulfates, protons or CO_2).

In aerobic conditions, in general, this lithotrophic metabolism
is expressed when there is starvation in organic nutrients. The substrates which can be used as electron donors as well as their oxidation products, are listed in table 1 (next page).

In anaerobic conditions, practically only H_2 can be utilised as electron donor. However, in the presence of nitrate acting as electron acceptor (nitrate respiration, or denitrification), reduced sulfur compounds (elemental sulfur, sulfide, thiosulfate) be used as electron donors, too.

Electron donor	Oxidation product	Type of bacteria
Ammonia (NH_3, NH_4^+)	Nitrite (NO_2^-)	Nitrous bacteria*
Nitrite (NO_2^-)	Nitrate (NO_3^-)	Nitric bacteria*
Hydrogen sulfide (H_2S)	Sulfur (S)	| Sulfur-oxidizing
Sulfur (S)	Sulfate (SO_4^{--})	| bacteria
Hydrogen (H_2)	Water (H_2O)	
Hydrogen-oxidizing		bacteria
Ferrous iron (Fe^{+2})	Ferric iron (Fe^{+3})	Iron bacteria

Nitrous and nitric bacteria perform together the oxidation of ammonia to nitrate. As a group, thea are called the "nitrifying bacteria".

Table 1: External electron donors for lithotrophic, aerobic respiration, their oxidation products and the types of organisms responsible for it

The main electron acceptors used by anaerobic chemolitho-autotrophs are sulfate, elemental sulfur and carbon dioxide (or its derivatives). Some species of sulfate- reducing bacteria perform an autotrophic metabolism using hydrogen as electron source and sulfate as an electron acceptor for respiration, according to the following overall equation:

$$4\ H_2 + SO_4^{--} + 2\ H^+ \longrightarrow H_2S + 4\ H_2O + energy$$

Elemental sulfur is mainly used as an electron acceptor by extremely thermophilic Archaeobacteria living in geothermal environments, This has no significance in landfill systems. The use of CO_2 as electron acceptor is ecologically more important: coupled to the oxidation of H_2, it is one of the key steps of the microbial anaerobic degradation of organic wastes, i.e. methanogenic digestion. The global reaction is:

$$4\ H_2 + CO_2 \longrightarrow CH_4 + 2\ H_2O + energy$$

This is performed by the autotrophic methanogens. We will discuss further on the place of these methanogens in the whole microbial syntrophy.

2.1.2.2.1.2 Organotrophic, respiratory metabolism

The utilization of organic substrates as external electron donors for energy production is called organotrophy. In fact, the organic substrates are most often utilized indirectly as electron donors; they are generally submitted to endogenous dehydrogenation reactions leading most often to CO_2 and to reduced endogenous electron donors like NAD(P)H and succinic acid, who transfer electrons directly to the respiratory chain
which is coupled with the formation of trans-membrane gradients.

The same substrates serve in general as carbon source and as electron donors: part of the organic substrate is assimilated to cell matter, and part is oxidized for respiratory purposes. So, most organotrophs are at the same time heterotrophs.

The same oxidants as in lithotrophic metabolisms serve as electron acceptors: oxygen for aerobic respiration, nitrate, sulfate and elemental sulfur for anaerobic ones.

Aerobic respiration gives the highest energy yield per mole of substrate consumed, but also the highest loss of energy as heat. Thus, if heat diffusion is limited, aerobic activities will be accompanied by a strong elevation of local temperature, like in composting processes, for example.

Sulfate reduction is accomplished by specialized bacteria which can utilize, along with molecular hydrogen (see above) a rather limited amount of organic donors, particularly short chain fatty acids. Sulfate reduction leads to the formation of hydrogen sulfide.

The organotrophic, acetoclastic methanogens form a special category: they convert acetate into formyl and methyl groups (respectively electron donor and acceptor). An electron transfer then leads to the formation of methane and CO_2. This, together with the autotrophic methanogenesis, is the most important way of methane production during the anaerobic degradation of organic wastes.

2.1.2.2.2 Fermentative processes

Another way of obtaining energy is also very widespread among anaerobic organotrophs: the fermentations. They are in fact exothermic transformations of the organic substrate, often dismutation reactions, which are coupled by some way to the synthesis of energy-rich pyrophosphate bonds of adenosine- triphosphate (ATP). This mode of ATP formation (substrate level phosphorylation) differs from the synthesis of ATP through respiratory processes (oxidative phosphorylation). The amount of energy available from fermentation is always lower than from respiratory chains.

In most pathways leading to substrate level phosphorylations, there is formation of an excess of reducing power (as NADH or reduced ferredoxin). This excess will then be transferred to a "final electron storage compound", which will then be excreted by the cell. These products (mainly alcohols and volatile fatty acids) will accumulate in the medium, unless they are consumed by other organisms, such as bacteria with anaerobic respiration. High amounts of organic acids can be produced by fermentations, leading to an acidification of the surrounding medium if they are not simultaneously consumed. Some fermentations are also accompanied by the formation of H_2 and CO_2.

In some types of bacteria, the only way to reoxidize NADH and ferredoxin is by transferring the electrons to protons, releasing them then as molecular hydrogen. For thermodynamical reasons, such a reaction is only likely to occur at very low H_2 partial pressure. These bacteria must then live in close contact with a H_2-consumer (see below).

2.2 Some ecological requirements and properties of bacteria

2.2.1 Physicochemical parameters

Bacteria often have precise requirements with regard to physicochemical parameters like temperature, osmotic pressure and pH. In the case of an anaerobic degradation of organic waste, the methanogens are strictly neutrophilic, whereas the fermentative bacteria are mostly acid tolerant. When an important temperature increase occurs, for instance as a consequence of an intense aerobic respiratory activity, many organisms will be killed, but in general they will be replaced by thermophilic or thermotolerant organisms with similar capabilities. In very saline conditions (and consequently very high osmotic pressure), the number of bacterial species will be drastically reduced. Salt and other highly concentrated solutes are in general good preservatives (salted fish, jam).

2.2.2 Attachment of bacteria

In order to remain on the surface of solid particles, bacteria must stick to them by some way. This adherence may be of different nature: by van der Waals or electrostatic forces, or secretion of exopolymers, which act as a glue, fixing the bacteria to their substrate. This is the case, for instance, of Cytophaga cells, which can thus feed directly on the cellulose fiber they hydrolyse with their wall-bound cellulases. The
attachment will prevent the bacteria from wash-off by percolating liquids. The organisms often associate together physically by this way into mixed microbial communities.

3. FUNCTIONAL AND NUTRITIONAL ASSOCIATIONS OF BACTERIA: THE EXAMPLE OF THE METHANOGENIC SYNTROPHY

In the above presentation, the bacteria were more or less considered as separate objects performing their individual tasks independently, without interaction with each other. In a complex habitat like a landfill, the bacteria must interact. Thus, such an ecosystem must be regarded not only as the sum of individual capabilities, but mostly like a true system, in which the activity output is more than the the simple addition of the individual ones. This is particularly true with the methanogenic
association which performs most of the degradation of organic matter, leading to landfill stabilization. Prior to a closer view on the landfill microbial ecosystem, we will discuss here in some detail the main characteristics of such an association (see fig. 1).

In waste, most of the organic matter is composed of biopolymers, polysaccharides, lignin, proteins, nucleic acids, and lipids. As stated above, these polymers cannot be transported into the cell through the membrane, and must first be hydrolysed into their oligomeric subunits. This hydrolysis is the first step of the methanogenic digestion, the so-called liquefaction. The enzymes are in fact synthesized by the organisms responsible for the second step, the fermentative bacteria, which convert these mono- and dimers into fermentation products, mainly C_1 - C_5 acids and alcohols, H_2 and CO_2. H_2 + CO_2, methanol, formic and acetic acids can be converted directly into CO_2 + methane by the autotrophic and acetoclastic methanogens. The other alcohols and acids are no substrates for the methanogens. Two other groups can convert them into methanogenic substrates: the sulfate reducers and the obligate proton reducers.

Several sulfate reducers convert the organic fatty acids to acetate, a methanogenic substrate; but others can oxidize the substrates directly into CO_2. This metabolism is limited by the available sulfate, and produces hydrogen sulfide.

As sulfate is often scarce in waste, the most important organisms performing such a transformation are the obligate proton reducers. As stated above, the only way for them to perform their task (and then to live at all) is to have hydrogen-uptaking organisms close by. Then, the above mentioned substrates will be transformed into methanisable substrates such as acetate, hydrogen and CO_2. The two main groups which are likely to consume hydrogen in these conditions are the autotrophic methanogens and the sulfate reducers. In the presence of high sulfate concentrations, like in sea sediments, the methanogenesis will thus be damped, whereas in normal waste degradation, autotrophic methanogenesis will be the main pathway for the electrons from hydrogen.

Hydrogen uptake will also increase the energy yield of fermentation reactions. Thus, the activity of the methanogens will favour both the obligate proton reducers, which transform the higher acids and alcohols into methanizable substrates, and the "normal" fermentative bacteria too, which will at the same time increase their energy yield and produce more directly methanisable substrates. Increasing the yield of fermentative bacteria will in turn favour the production of exoenzymes for the hydrolysis of biopolymers. It is then easy to understand that such an association, linked by bidirectionnal, mutualistic interactions, is a very tight and stable one, and that the behaviour of the organisms operating in the association will be quite different from the behaviour of the same organisms separate. For these relationships to be narrow enough to increase their efficiency, several conditions must be fulfilled:
homogeneity, sufficient water content, proper and stable temperature, neutral pH, and a well-balanced composition in minerals.

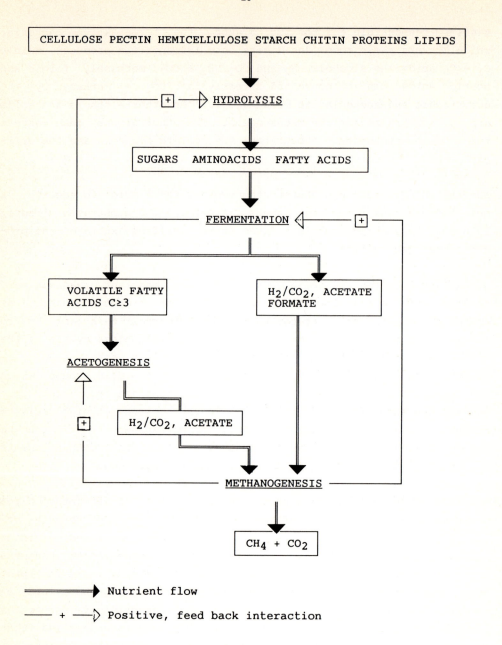

Figure 1:   Scheme of carbon flow and interactions in the methanogenic
syntrophic association.
(From Glauser et al., 1987)

4. THE LANDFILL AS A HABITAT FOR BACTERIA

An important question to be asked in the course of a discussion on
landfills and their final storage quality is whether the biological activity should be activated or repressed. We will keep this question in
mind and try to answer it at the end of this paper.

A landfill for municipal solid waste may be considered in terms of its
general structure, of the way of filling it, or of the topological,
physical and chemical nature of the waste in it. In general, a landfill
consists of several layers of waste dumped over a period of time. The
bottom of a modern landfill should be covered with a liner to collect
the leachate and to avoid its dispersal in the ground waters. This
leachate would then be treated separately, even biologically. A soil
cover can be spread over the surface of the freshly deposited waste, in
order to protect against rodents, birds and wind, and halt combustion
and emission of odours. A gas collection system is necessary, to drain
the biogas produced with a methane content of about 50. This gas may
then be burned or used for heating or for electricity production..

Biologically, a reactor landfill comprizes three main compartments:

- the inner compacted waste mass, which undergoes an
  anaerobic decomposition;

- the superficial layer, materializing the interface between
  the landfill and the air, with aerobic conditions;

- the leachate, which is potentially a biologically active
  liquor which could be treated in aerobic or in anaerobic
  bioreactors, or even recirculated.

4.1 Biological evolution of the waste mass

4.1.1 Sequence of events in a normal compacted landfill

After deposition where the access of oxygen is limited, the waste mass
undergoes a series of phenomena leading to deep anaerobiosis and to the
appearance of the methanogenic syntrophy. This evolution is characterized by the composition and amount of the gas produced (fig. 2).

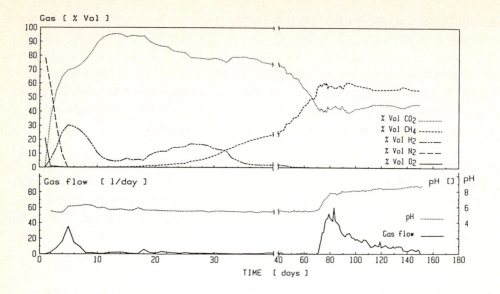

Figure 2: Evolution of gases from a mixture of organic wastes deposited under anoxic conditions. A 1600-l hard polyethylene vessel without external air access was filled with a mixture of 443 kg urban waste (65 % dry matter content) ground and sorted at diameter < 8 cm and 200 l digested sludge (5 % dry matter content). The percolating liquid was recirculated from the bottom to the surface. The temperature was maintained at 33-35 °C. Total gas production was recorded continuously and gas composition was analysed at regular intervals.
(From Dugnani et al., 1986)

First, oxygen is rapidly used up by aerobic organisms, obligate and facultative ones. The facultative anaerobes will then continue to develop and shift to a fermentative metabolism. Easily degradable substrates, such as sugars, will be utilized and converted to fermentation products, among which hydrogen and CO_2. At that time, no significant methanogenic activity is present. This means that, not only hydrogen, but also volatile fatty acids (VFAs) and eventually other fermentation products will accumulate, leading to a decrease in pH. This decrease will delay still more the development of the strictly neutrophilic methanogenic flora. A long, apparently quiescent phase (almost no gas production) then follows, during which methane slowly replaces hydrogen. After a long delay, there is a burst in methanogenic activity, pH increases to neutrality and the concentration of volatile fatty acids in the leachate (not shown) decreases dramatically. Then, while the other parameters remain almost constant, the gas (CH_4 + CO_2) production decreases progressively, due to the exhaustion of the substrate.

4.1.2 Environmental impact of the different phases

How could one define the "final storage quality" that should be the result of this process? In fact, several phases should be considered here:

4.1.2.1 Transitory, "hydrogenic" and "quiet" phases

The "hydrogenic" and "quiet" phases lead to an important environmental charge: volatile, smelly compounds are produced along with a potentially explosive mixture of gases, and the leachate is heavily charged with VFAs and other easily degradable compounds, and with heavy metals solubilized from the waste mass by the low pH and complexing substances. The first environmental goal must be to reach the actively methanogenic phase as fast as possible and to decrease the emissions during the "hydrogenic" and "quiet" phases to a minimum.

4.1.2.2 Methanogenic phase

During the methanogenic phase, the pH increases to neutrality, whereas the concentration of volatile organic compounds diminishes, due to the normal functionning of the methanogenic flora which transforms the intermediate carbon-metabolites into CH_4 and CO_2. This will strongly decrease the odorous emissions, as well as the VFA and heavy metals leachate content. This phase is nevertheless no "final storage", because of the strong gas production, which could become a hazard: it may form explosive mixtures with air, and may kill the vegetation in the vicinity, too. Increasing quantities of biogas must be drained from the land fill. Researches are at present undertaken to purify the gas, by removing the CO_2.

4.1.2.3 Post-methanogenic phase(s)

In a normal compacted landfill, the methanogenic phase will probably last 10 - 30 years. After this time, the landfill has become a relatively inert soil-like mass, whose gas production should no more require control. This "soil status" represents the "final storage quality" expected from such a landfill. In fact, we probably need more information on the post-methanogenic evolution of a landfill. Recently, houses built on former landfills, in the Netherlands, had to be abandonned, because harmful emissions from the underground were detected.
Theoretically we could imagine the following situation:

-At the interface between oxic and anoxic layers at the surface and sides of the landfill, the rate of production of substrates that can be rapidly oxidized aerobically decreases.

- The oxic-anoxic interface will then move toward the interior of the waste mass, when the speed of oxygen diffusion exceeds its consumption rate.

- The humigenic materials, such as lignin, which were not decomposed anaerobically, will then undergo a slow transformation into humic substances. Most humic compounds are very stable, even in aerobic conditions, and will show a very slow mineralization rate

- Certain reduced minerals (ammonia, sulfur or sulfides) which could have accumulated in anaerobic conditions, could possibly be biologically oxidized by aerobic chemolithoautotrophs, leading to the formations of nitric and sulfuric acids. To what extent this could bring further leaching of metals depends of the buffering capacity of the mass.

- Other degradations, of hydrocarbons for example, which require aerobic conditions, could take place at this time. Thus, biotests designed to assess the behaviour of defined substances in a landfill should be performed in aerobic, as well as in anaerobic conditions.

- The aerobic mass could then behave as an organic, peat soil, undergoing a slow mineralization. This mineralization would be accompanied by a release of minerals into the leachate, that would need controlling.

This situation supposes that the waste mass will evolve toward oxic conditions. The rate at which such an evolution can progress, and even the possibility that such an evolution should occur at all, depends on the particular conditions of the landfill: water content, water turn-over, insulation, etc. If anaerobiosis could be maintained in-definitely, it would perhaps lead more rapidly to what could be consid-ered as a "final storage quality" in the narrow sense. Returning to aerobic conditions, the waste mass would then be more like a soil, that is an open system tending toward a dynamic equilibrium, and not an in-ert, unchanging mass.

4.1.3 *Factors affecting the rate of biological degradation in a landfill, and possible improvements*

Anyhow, the biological processes in a conventional landfill are slow, and any improvement leading to acceleration of biological processes would reduce the time needed to reach the successive equilibria evoked here, along with the emission of organic compounds (Gandolla and Grabner, 1983, Glauser et al., 1987), whereas biogas production would be higher and the quality of the gas better, with higher methane con-tent and lower contamination by O_2 and N_2. But, how could one acceler-ate the biological processes in a landfill ?

An organic fraction of urban waste, shredded and ground into particles of millimeter size, amended with sewage sludge in order to optimize its C/N ratio (< 30), in aqueous suspension (e.g. 15% dry weight) can be digested in a digestor run semicontinuously with a retention time of 12 days at 35°C and 9 days at 60°C (Glauser et al, 1987). In these condi-tions, 35 - 40 % of the organic matter is converted to CH_4 + CO_2, and the residue is stable in anaerobic conditions. The time needed for anaerobic stabilization of the waste in a landfill is at least two or-ders of magnitude higher. This can be due to the conjunction of several causes:

- the high C/N content of the waste, which causes nitrogen, rather than available carbon, to be the growth limiting factor; this leads to acidification and inhibition of methanogens.

- the macroheterogeneity of the mass, which counteracts the interactions between the members of the methanogenic syntrophy (see above).

- the low water content, which also contribute to decrease the intensity of the exchanges of the members of the syntrophy, and could lead to local accumulation of high concentrations of substances, with an inhibitory effect.

- the lack of a proper initial microbial flora, particularly of the slow-growing autotrophic and acetoclastic methanogens. The much faster growth of fermentative microorganisms would then lead to an accumulation of acids and a decrease of pH, inhibiting the methanogens.

The conjunction of these factors incited Gandolla and co-workers (Gandolla et al., 1982, 1983; Glauser et al, 1987) to test the following improvements in 100 mm^3 concrete lysimeters lined with polyethylene. They were built in a landfill in Croglio (canton Tessin) and left open at the top. The waste was shredded, iron was removed and the mixture was coarsely sieved (60 mm). It was then mixed with freshly digested sewage sludge so that the final water content was 55% in the mixture. The result was after 3 years a well decomposed, humus-smelling residue, whereas the "hydrogenic" phase was both shorter and characterized by lower concentrations of hydrogen and VFAs and a by a lower COD in the leachate. In fact, shredding increased the homogeneity, while the sludge addition had multiple effects: increasing the water content, optimizing the elemental composition for biological activity (particularly increasing the nitrogen and phosphorus content), and inoculating the waste mass with an actively growing, complex methanogenic flora.

4.2 The air - landfill interface

While the garbage mass in a compacted landfill evolves towards anaerobiosis, aerobic respiratory processes will develop in the superficial layers directly in contact with air. Such processes are based on the utilisation of either inorganic or organic compounds as external electron donors. Organic, volatile compounds will accumulate in anaerobic conditions, mainly during the transient phases, and will reach the surface with the gas phase. Among these, several, like C_4 - C_6 volatile fatty acids and mercaptans, are malodorous. When reaching an oxic environment, these compounds will be oxidized by versatile heterotrophs like Pseudomonads, to CO_2 and H_2O. Hydrocarbons, such as methane, will be oxidized by aerobic microorganisms too, thanks to oxygenases which catalyse the primary attack on these compounds. Similarly, compounds like ammonia, hydrogen sulfide and molecular hydrogen could be oxidized by chemolitho- autotrophs, respectively to nitrate, sulfate and water.

When choosing the suitable materials for the composition of the oxic-anoxic interface of a landfill, the optimization of aerobic biological activities should be considered of primary importance, in order to decrease the emissions of malodorous compounds and eventually of explosive gas mixtures. Due to its heterogeneity, its low surface/volume ratio and to a probably inadequate preexisting microbial flora, the waste material itself, at least the freshly deposited one, would not be ideal. Moreover, it might contain light materials, such as paper and plastic, which could be dispersed by the wind, and it might attract birds and rodents. A better material would be soil, or fresh, aerobically composted organic waste, which already contains a rich and versatile aerobic bacterial flora, acting thus as a biofilter. Heavy soils rich in clay should be avoided, because, in case of drought, cracks form, creating preferential canals for the gases escaping from the anaerobic mass. The contact with the aerobic flora will be reduced in time and surface, and consequently the oxidizing activity too. A sandy, light soil would be best: it provides a great surface to which an important bacterial flora can adhere.

We tested such a system experimentally with a column filled with a sandy soil and gassed with a mixture of "anaerobic gas" (i.e. 20% H_2 in CO_2) and air (Dugnani et al, 1986). After ten days of incubation, all the hydrogen was oxidized by an H_2-oxidizig bacterial flora which developed in the soil. Cell numbers of H_2- oxidizing, chemolithoautotrophic bacteria as high as $3.5 \cdot 10^8$ per g were measured in the soil. Such a high cell density is probably connected with the fact that most of the hydrogen bacteria isolated from this soil were also dinitrogen fixers: the limitation by combined nitrogen was then overcome. Interestingly, these hydrogen-oxidizing bacteria were facultative, highly versatile heterotrophs. With relatively high H_2 concentrations (e.g. 5 - 20 %) in the gas phase, most of the biomass produced will be of chemolithoautotrophic origin, whereas this biomass could oxidize simultaneously the organic compounds, even at low concentration.

Thinking in terms of passive treatment of the emissions after the main decomposition period, landfill could be designed in order to drive the remaining gaseous emissions through soil layers acting as biofilters. More research is needed here to assess the efficiency of such biofilters in eliminating volatile, xenobiotic and non-xenobiotic compounds.

4.3 The leachate

The leachate must be considered as waste water, and as such must not reach natural ecosystems and ground waters. For a study on the migration of landfill leachate into groundwater, see Baedecker and Back (1979). Depending on its content in COD and BOD, as well as in heavy metals and in toxic, organic compounds, one must be decide if the leachate has to be treated in a normal sewage treatment plant, or submitted to a separate pretreatment. In all cases, one should avoid dumping material which could give the leachate the character of a toxic, special waste.

The leachate contains:

> - inorganic salts, part of which originate from the biological mineralization of the waste. Ammonia, sulfides and ferrous iron are potential electron sources for chemolithotrophic bacteria.

> - organic substances easily degradable biologically (e.g. VFAs), either anaerobically in a digester, or aerobically.

> - "humic" compounds in the broad sense, i.e. complex organic molecules of natural origin which are very slowly biodegradable (Gandolla et al., 1982).

> - xenobiotic, artificial compounds.

The simplest pretreatment would be to recirculate the leachate through the landfill. It will help completing the removal of the degradable compounds by the methanogens active in the landfill, and also give this flora a complement of biogenic chemical elements. Recirculation will increase the duration of the contact between leachate and the bacterial flora attached to the landfill solids. In some cases, it can also lead to acidification: further reasearch is then needed towards an optimal use of leachate recirculation.

Another system would consist, on the contrary, in driving the leachate out of the landfill to an anaerobic filter or digestor with pelletized biomass. The biomass is retained in this type of digestors, and hydraulic retention times as low as 12 - 24 hrs are then allowed. Once the most active period of the landfill is over, the emissions of heavy metals and of biogenic elements in the leachate should be reduced by treating the effluent in a more passive manner, i.e. not requiring a heavy maintenance. This could be realized for instance by driving the leachate in a pond. In such conditions, the photosynthetic biomass would act as a concentration agent for most of these pollutants. The heavy metals would then be trapped in the anaerobic sediment. Such "lagooning" systems, if conveniently designed and dimensioned, would require a minimum of maintenance. Some experience has been acquired in the treatment of sewage plants effluents in ponds, but this experience should be extended to the treatment of stabilized landfill effluents.

5. DISCUSSION AND CONCLUSIONS

A modern, compacted landfill is a complex ecosystem dominated by a stable, anaerobic microflora, i.e. the methanogenic syntrophic association, which governs its main behaviour. Except for biogas production, which can lead to energy recuperation, the methanogenic phase has a low environmental impact, compared to the transient, hydrogenic and acidogenic phases which preceed it. Accelerating the biological evolution would have the double advantage of reducing the gaseous and leachate emissions and simultaneously of shortening the delay required to reach stable methanogenesis (Gandolla and Grabner, 1983). Trying to inhibit the biological activity would lead to a landfill burst sooner or later, accompanied by heavy emissions of metabolites, making it a "time bomb".

Conveniently managed and optimized, a reactor landfill could still remain a valuable solution for disposal of urban wastes. Microbiological systems are much more predictable, i.e. produce much less unexpected reactions and are much more controllable than many chemists think generally. An impact evaluation should be made as globally as possible, and compared with the global impact of other systems, e.g. incineration. Incineration requires more and more sophisticated processes to control its direct or indirect emissions. Even if at present supportable economically by a country with a rich economy, it is not so actually with developping countries, and could not be so in the future in developped countries too, where the maintenance of such "heavy" waste management systems could be no more considered as prioritary, due to political changes or economic depression. For that reason, research on improvement of reactor landfills, and developments at pilot and real scale should be promoted, even in countries, like Switzerland, which incinerate most of the urban wastes.

Not only the landfill sectorially taken, but also the whole waste management strategy, before and after the landfill, should be optimized. In that sense, not all the fractions of wastes should be deposited in a re actor landfill containing organic waste. The occurence of xenobiotic compounds and of heavy metals should be minimized, and thus the widespread utilization of such compounds in the household severely repressed, whereas the toxic industrial wastes should be treated more specifically. On the other hand, such few contaminated organic fractions as household sorted kitchen and garden wastes, should rather be composted or methanized in a biodigestor. The residue would then be best used as an improvement for agricultural soils. Among the research needed toward the optimization of the landfill itself and of the system which surrounds it, let us mention, from a biologist's point of view:

- For the landfill itself:

    - Means of shortening the acidic phase and the overall time needed for biological stabilization.

    - Understanding the secondary evolution of a landfill, particularly the latest shift towards oxic conditions and its consequences.

    - Whereas the fate of naturally occuring organic compounds during biological degradation is well known and predictable, research on the evolution of xenobiotic compounds in a landfill ecosystem (complete or partial degradation, sub- products, interactions with the actual microbial flora) in the presence of a complex, aerobic, acidogenic or methanogenic microbial floras is still urgently needed.

-For the waste management system comprizing a landfill:

- Means (political, economical, technical) of limiting the widespread use of potentially dangerous xenobiotic compounds in the household.

- Choice of what may be dumped into an organic landfill, in order to avoid toxic emissions and biotransformation of xenobiotics into toxic end-products.
- Means of controlling in a more passive manner the emissions (leachate and gases) from the landfill after completion of the most active phase of decomposition (e.g. ponds, soil biofilters...)

We are convinced that an optimized, controlled organic landfill has in the future, an important role to play in a global, systemic strategy of waste management. In that sense, the motto "final storage quality" must not be taken in the narrow, dogmatic sense of a "dead" deposit, but rather as a goal: reaching, thanks to a suitable optimization and management, a dynamic status compatible with the environment.

ACKNOWLEDGEMENTS

Most of the ideas developped here originate from research on solid waste methanisation performed in co-working with the CER (Consorzio per l'Eliminazione dei Rifiuti del Luganese) in Bioggio (canton Tessin, Switzerland) and funded by grant 3.262-82 of the Swiss National Science Foundation. The author thanks Mirja Salkinoja, University of Helsinki, and Mauro Gandolla, director of the CER, for their helpful suggestions, and Catherine Fischer, who corrected the manuscript.

LITERATURE CITED

Baedecker, M.J., and Back, W. 1979. Hydrogeological processes and chemical reactions at a landfill. Groundwater 17: 429-437

Dugnani, L., Wyrsch, I., Gandolla, M., Aragno, M. 1986. Biological oxidation of hydrogen in soils flushed with a mixture of H2, CO2, O2 and N2. FEMS Microbiol. Lett. 38: 347-351

Gandolla, M., Grabner, E., Leoni, R. 1982. Ergebnisse von Lysimetern auf der Deponie Croglio, Schweiz. Veröffentl. Inst. Stadtbauwesen, Techn. Univ. Braunschweig 33: 163-182

Gandolla, M., Grabner, E. 1983. Proposte per future ottimizzazioni di discariche controllate. Wasser, Energie, Luft 75: 241-244

Glauser, M., Gandolla, M., Aragno, M. 1987. Anaerobic digestion of urban wastes: sewage sludge and organic fraction of garbage. In: Bioenvironmental systems, 3: 143-225

GEOCHEMICAL PROCESSES IN LANDFILLS

Ulrich Förstner, Michael Kersten and Reinhard Wienberg
Technical University Hamburg-Harburg, D-2100 Hamburg 90, FRG

Abstract

The present review focusses on the qualitative long-term perspectives
of processes and mechanisms controlling the interactions of critical
pollutants with organic and inorganic substrates both in "reactor landfills" and in deposits, which already consist of rock-like material
("final storage quality"). The behavior of pollutants in landfills is
determined by the chemistry of interstitial solutions, i.e. by pH and
redox conditions, and concentration of inorganic and organic ligands;
in "reactor landfills" these conditions are widely variable as a result
of biochemical reactions, while "final storage quality" implies less
variations of chemical interactions. In both alternatives, however,
prediction of short- and long-term effects on groundwater quality
should be based on the proportion of "active species" of compounds
("mobility concept"). Qualitative assessment of potentially mobile
pollutants may involve a controlled significant intensivation of
important parameters such as pH-values. Using sequential extraction
rearrangements of specific solid "phases" can be evaluated prior to the
actual remobilisation of the pollutant into the dissolved phase.

From a geochemical point of view the "reactor landfill" is characterized by labile conditions during the initial aerobic and acid anaerobic
phases, the former mainly due to uncontrolled interactions with organic
solutes. On the other hand, final storage quality, which is defined by
the composition of earth crust material, in most cases is not attained
by simple incineration of municipal waste, i.e. by reduction of organic
fractions only. There is, in particular, the problem of easily soluble
minerals, such as chlorides. Nonetheless the type of inorganic residue
deposits will increasingly receive prevalence as a method of final storage for municipal wastes in the future.

The possible implications with toxic chemicals are presented for the
various types of landfills such as deposits of sewage sludge, municipal
solid waste ("reactor") landfills, sub-sediment deposition of dredged
materials, borrow pit storage of metal oxide residues from industrial

processes, deposit of coal combustion ashes, and finally, residues from incineration of municipal waste materials.

The concept of "geochemical engineering" emphasizes the increasing efforts of using natural resources available at the disposal site for reducing negative environmental effects of all types of waste material. Future efforts should not only be aimed for chemically stabilizing critical compounds in waste materials but - in particular - for recycling valuable components such as strategic metals, e.g., by application of hydrometallurgical extraction procedures.

1. INTRODUCTION

Specific *concerns for land disposal* of contaminated materials include the transport of toxic organics and heavy metals to both surface and ground waters, transport of pathogens to man through such pathways as crops grown in waste-amended soils and contamination of groundwater and surface water systems, as well as the export of nutrients to non-target ecological systems (Capuzzo et al., 1986).

With regard to *loss of groundwater resources*, ascribable to contamination, a United States Library of Congress Report lists 1360 well closings in a 30 year time span (Anon., 1980), broken down as follows: metal contamination: 619 wells; organic chemical contamination: 242 wells, including 170 from trichloroethylene used to emulsify septic tank grease; pesticide contamination: 201 wells; industrial contamination: 185 wells; landfill leachate contamination: 64 wells; chlorides: 26 wells; nitrates: 23 wells.

As for the mechanisms of *metal toxicity*, the most relevant is certainly the chemical inactivation of enzymes. Soil biochemical processes considered especially sensitive to heavy metals are mineralization of N and P, cellulose degradation and possibly N_2-fixation (Domsch et al. 1983). Among the *synthetic organic compounds* polychlorinated biphenyls (PCBs) and polycyclic aromatic hydrocarbons (PAHs) deserve the greatest attention, both because of their quantities present and of their persistence, and especially due to their long-term toxicity (Harms & Sauerbeck, 1983).

With respect to groundwater pollution from landfills *volatile organic compounds* are of particular concern, not only due to the relatively

high mobility of these compounds but also because of the possible formation of even more *toxic metabolites*. For example, it has been shown
by Vogel & McCarthy (1985) that under anaerobic conditions a sequential
reductive dehalogenation can occur from tetrachloroethene to trichloroethene, then to cis-1,2-dichloroethene and finally to vinyl chloride.

It is increasingly recognized that the ecotoxicological effect is determined by *"active" species of contaminants*, which may become partly
mobilized under varying chemical conditions. In landfills these are,
for example, for inorganic contaminants and partly also for ionic organic pollutants: pH, redox, inorganic salts, organic chelators. Mobility of metals seems predominantly be affected by these changes; however, examples will be given which indicate that significant effects on
the reactivity of critical organic substances cannot be excluded at
certain stages of landfill evolution.

While biological and hydrological processes are discussed in other
contributions, the present compilation deals with *geochemical aspects*
of landfills. Emphasis is given to *qualitative long-term perspectives*
of processes and mechanisms controlling the interactions of critical
pollutants with organic and inorganic substrates both in *"reactor landfills"* and in deposits, which already consist of *rock-like material*
("final storage quality").

2. CONCEPT OF POLLUTANT MOBILITY

From the experience with heavy metals the following questions have been
raised with respect to the *mobility* and bioavailability of potentially
toxic chemicals in landfills (Förstner, 1987a):

(1) How reactive are the pollutants introduced with *solid materials*
    from anthropogenic activities (hazardous waste, sewage sludge) in
    comparison to the natural compounds?

(2) Are the *interactions* of critical substances between solution and
    solid phases comparable for natural and contaminated systems?

(3) What are the factors and *processes of remobilization* to become
    particularly effective, when either the solid inputs or the
    solid/solution interactions induced weaker bonding of pollu-
    tants in the contaminated system?

2.1. Sorption Processes

Dissolved contaminants may *interact with the aquifer solids* encountered along the flow path through adsorption, partitioning, ion exchange and other processes. For some contaminants, such as ionic species of heavy metals and certain organic solutes, the degree of interaction can be predicted from factors such as the concentration and characteristics of the contaminant, the characteristics of the aquifer solids, the pH of the groundwater, and the presence of other dissolved constituents (Mackay et al., 1985). A rough estimate of the *retention behavior* of an organic compound in a groundwater aquifer may be obtained if the partition constant K_p (which for nonpolar compounds is directly related to the organic carbon content of the sorbent - for organic carbon concentrations exceeding approx. 0.1%) is known (e.g., Farrington & Westall, 1986). For porous media it has been shown that only the fine fraction of the aquifer material (fraction passing through a 125 m sieve) predominates for sorption (e.g., Karickhoff et al., 1979; Schwarzenbach & Westall, 1981; Hellmann, 1983)).

With regard to the sorption of *heavy metals* there is manifold evidence, that the colloidal fraction of the soil, consisting of clay particles, amorphous oxides of Fe and Al, and organic colloids, acts as a negative ion exchanger. Two effects relevant for the behaviour of metals both in natural and polluted systems are still not satisfactorily explained as yet: One relates to the competition between organic and oxidic adsorption processes, the other is concerned with the discrimination between adsorption/desorption and precipitation/dissolution processes:

It has been stressed by Salomons (1985) that from an *impact point of view* it is important to know whether the concentrations in the pore waters are determined by adsorption/desorption processes or by precipitation/dissolution processes. If the latter is the case the concentrations of pollutants in the pore and surface waters are independent of the concentrations in the solid phase. At present, it seems that calculations using equilibrium data (i.e. K_D-factors of solid/solution coefficients) are inadequate to model natural conditions, because of complexitiy of interactive mechanisms and, in particular, due to the lack of data on reaction kinetics of sorption/desorption processes.

The question to what extent *interactions between dissolved and solid metal species* are affected by either sorption or precipitation processes is still being discussed (Brümmer et al., 1986). There is a typical temporal evolution of the sorption processes, i.e. for those processes, which cannot be explained by a direct precipitation of metals from solution; four different types of evolution (rapid or slower adsorption to nearly 100%; rapid or slow adsorption at a lower level) have been distinguished from experiments using radioisotopes (Schoer & Förstner, 1984). These processes are influenced by the hydrological and chemical conditions; sorption of cesium, for example, is typically lowered in the presence of Ca- and Mg-ions. In the few cases, where kinetics were investigated, surface reactions were not found to be a single step reaction (Chen et al., 1973; Anderson, 1981). Experiments by Benjamin & Leckie (1981) showed a rapid and almost complete metal uptake process perhaps lasting no more than one hour, followed by a second, slower uptake process perhaps lasting days, or possible months; the first effect was thought to be true adsorption, and the second to be slow adsorbate diffusion into the solid substrate. For specific adsorption, binding strength typically depends on adsorbend concentration, since there exists a range of site-binding energies (Leckie et al., 1980). Particularly in systems containing organic substances, a reduced reversibility of metal sorption has been observed (Lion et al., 1982).

The data thus far obtained suggest that *organic-metal interactions* play an important role in the initial distribution of a trace element between the solid and the soluble phase, either by competition or enhanced adsorption of organic-metal species or through films or organic matter on inorganic (e.g. mineral) particles. Figure 1 from Benjamin & Leckie (1982) illustrates the various ways in which adsorption of the element

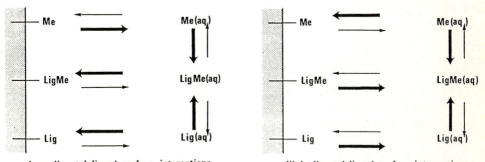

Figure 1. Schematic Representation of the Reactions Leading to Enhanced Adsorption of a Metal at Low and High pH in the Presence of a Complexing Ligand (from Benjamin & Leckie, 1982)

is affected when metal ions bind to dissolved ligands to form soluble complexes at different pH-conditions; an interesting finding for the present subject is, that at lower pH-values ("acetic phase" in the evolution of the municipal landfill) there should be an enhanced sorption of metal-organic complexes. Subsequent diagenetic effects (aging of metastable phases, decay of organic matter) cause a redistribution of trace elements. Experiments on sorption of radionuclides on reducing sediments by Maes & Cremers (1986) indicate that oxidation of samples leads to a significant increase in solid-liquid distribution coefficients. This effect is ascribed to the involvement of ferric oxides which are generated in the solid phase and which lead to a displacement of the metal from the humic acid sink.

Removal of *organic chemicals* from solution and their partitioning onto particulates may either be the result of chemical interactions between reactive solid constituents and the pollutant ("bound residues") - producing solid/water-equilibria being far on the side of the "fixed" fraction or of sorptive processes of the hydrogen-bond or physisorptive type. Some sorption phenomena such as positive sorption enthalpies or linear partition as in a Nernst-partioning in non-miscible liquids, lead to the discussion, wether it may not be so much of a sorption process but one of exclusion and solubilization of the organic chemical into an organic lipid-like surface layer on particulates (Chiou et al., 1979; DiToro and Horzempa, 1982). Several mechanisms are inolved, in particular so-called "hydrophobic bonding" where organic molecules are "squeezed" from their cooordination with water molecules onto hydrophobic solid surfaces (Walker & Crawford, 1968; Calvet, 1980; Khan, 1980). The sorption of organic chemicals on solid surfaces is dependent on their functional groups, the size and shape of the molecule and - if there is any - their charge. Based on these properties, the following categories can be set up:

(a) Cationic or basic compounds, such as herbicides paraquat and diquat, which interact with negatively charged particles and are strongly or irreversibly bound;

(b) acidic compounds, such as the herbicides 2,4-D- or 2,4,5-T- (phenoxy) acids, which were repelled by the negative charge of mineral or organic material;

(c) non-polar, volatile substances, such as toluene or tetrachloro-
ethen, which are weakly interacting with particles by hydrophobic bonding; and

(d) non-polar, non-volatile organic substances, such as HCH, HCB
and DDT, showing stronger hydrophobic bonding forces with particles than (c), increasing with decreasing water solubility.

Chlorinated phenols, which are hydrophobic weak acids (category "b")
exhibiting octanol/water partition coefficients between 10^2 and 10^5,
constitute a class of compounds which is of growing concern. At relative low overall sorption rates pH is a dominant factor, and non-linear
sorption isotherms were obtained which were interpreted to be the result of the superposition of several different sorption processes (Zierath et al., 1980; Farrington & Westall, 1986). For *non-polar organic
contaminants* (category "d") interactions can be described empirically
by simple relationship to the content of organic matter (Karickhoff,
1981). The partition coefficient normalized to organic carbon contents
of the solids (K_{OC} = K_D/fraction of organic carbon) should be highly
invariant over a wide range of substrate types (Karickhoff et al.,
1979). From experiments on the specific retardation of some organic
contaminants by slurry trench materials it has been argued by Wienberg
et al. (1986) that this relationship cannot be generalized. Ground slate, for example, having only about one tenth of the organic carbon of
fly ash (2.8 %), sorbed HCB in the same amount as fly ash. On the other
hand, for the more soluble parathion fly ash had the same sorbing properties as Ca-bentonite, which contained no organic carbon. These findings support suggestions by Farrington and Westall (1986) in their review on the behaviour of organic chemical pollutants in groundwater
that further work is needed to establish quantitative data for estimating partition coefficients of non-polar organic compounds with orga-
nic-poor sorbents. Another area which needs more research is the influence of colloidal and/or dissolved organic material (e.g. fulvic and
humic materials, solvents, detergents, etc.) on the transport of contaminants in landfills.

2.2. Mobilization of Pollutants from Solids

"Mobilization", in a wide sense, comprises changes in the chemical environment which usually are affecting *lower rates of precipitation or
adsorption* - compared to "natural" conditions - rather than active releases of contaminants from solid materials. With respect to particle-

bound metals, their solubility, mobility and bioavailability can be increased by five major factors in terrestrial and aquatic environments:

- *lowering of pH*, locally from mining effluents, regionally from
  acid precipitation;

- *changing redox conditions*, e.g. after land deposition of polluted
  anoxic dredged materials;

- *microbial solubilization* by accelerating the oxidation of metal
  sulfides; formation of organometallic compounds by *bio-methyla-
  tion*;

- *increasing salt concentrations*, by the effect of competition on
  sorption sites on solid surfaces and by the formation of soluble
  chloro-complexes with some trace metals;

- increasing occurrence of natural and synthetic *complexing agents*,
  which can form soluble metal complexes with trace metals that are
  otherwise adsorbed to solid matter.

Another type of "mobilization" is the release of trace metals from sediments via pore water by *mechanical resuspension* from bottom deposits,
e.g. by bioturbation, erosion or by dredging activities.

With respect to the *chemical conditions* in municipal "reactor" landfills, changes of *pH* and *redox* values as well as variations of the concentration of *organic chelators* should be of prime importance. Initial
conditions are characterized by the presence of oxygen and pH-values
between 7 and 8 (see contribution by Ehrig in this volume). During the
subsequent *"acetic phase"*, pH-values up to 5 were measured due to the
formation of organic acids in a more and more reducing milieu; concentrations of organic substances in the leachate are high. In a transition time of 1 to 2 years chemistry of the landfill changes from acetic
to *methanogenic conditions*; the methanogenic phase is characterized by
higher pH-values and a significant drop of BOD_5-values from more than
5.000-40.000 mg/l in the acetic phase to 20-500 mg/l. *Long-term* (i.e.
hundreds to thousands years) conditions in a municipal landfills are
not known; it can be expected that after transformation of all reactive
organic substances the intruding waters will gradually oxidize mineral
matter, including sulfides, which may in turn lead to a lowering of pH-
values.

Microbial processes in landfills are discussed by M. Aragno in this book. However, one mechanism should be mentioned here in particular, which influences mobility of trace metals by changing pH- und redox-conditions: *Acidic drainage* from coal and ore mines has been recognized as a serious environmental pollution problem since long, and there are now similar effects in less buffered low-carbonate dredged sludges (Salomons & Förstner, 1988). A simple chemical mechanism, however, could not explain the rapid production of acidic mine drainage by the oxidation of metal sulfides. A simple chemical mechanism could not explain the rapid production of acidic mine drainage by the oxidation of metal sulfides. According to Singer & Stumm (1970) ferric iron is the major oxidant of pyrite in the complex natural oxidation sequence and it is mainly Thiobacillus ferrooxidans, an iron-oxidizing acidophilic bacterium, which accelerates metal sulfide oxidations 10^6 times over the abiotic rate.

It has been stressed by Plant and Raiswell (1983) that the predominance of simple mineral solution equilibria explains the concentration of major elements in the surface environment, but the behaviour of many trace elements is more complex and is also determined by coprecipitation, surface effects and interactions with organic phases. Many relationships, however, can be estimated from the simple scheme in Figure 2:

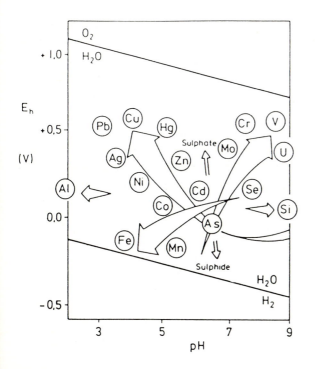

Figure 2

Schematic Presentation of Major Trends for Increasing Element Mobilities (Broadening Arrows) as a Function of Redox and pH Changes in Solid Waste Materials.

Lowering of pH, either by inputs of *acid precipitation* or by *oxidation of sulfidic minerals*, generally represents the most importants influencing factor on the mobilization of transition metals and related elements (Fe, Mn, Co, Ni, Zn, Cd, Pb, Hg; see Figure 2). *Changes from reducing to oxidizing conditions*, which involve transformations of sulfides and a shift to more acid conditions, will increase the mobility of elements such as Hg, Zn, Pb, Cu, and Cd. On the other hand, the mobility is characteristically lowered for Mn and Fe under oxidizing conditions. Elements exhibiting anionic species, such as S, As, Se, Cr, and Mo are appreciably solubilized, for example, from fly ash sluicing/ponding systems at neutral to alkaline pH-conditions (Dreesen et al. 1977; Turner et al., 1982). Special attention should be given to *arsenic*, which - according to calculations based on laboratory experiments by Blakey (1984) - shows least attenuation and greatest mobility under mildly reducing conditions in the pH range of 5 to 9. These conditions are typical for the transition stage of aerobic and acetic anaerobic phases in municipal solid wastes. Greatest attenuation and least mobility of arsenic occurred under strongly reducing conditions and at neutral pH. Precipitation of arsenic as sulfide (or co-precipitation with iron sulfide) provided sufficient attenuation for all the wastes studies in order to maintain a leachate concentration of less than 10 mg As/L (Blakey, 1984).

An example of the effect of variable pH- and redox-conditions on metals in *solid waste compost* has been studied by Herms and Brümmer (1978). At lower pH-values, one can observe a typical antagonistic behaviour for cadmium and iron under different redox conditions (Figure 3a). For example, at pH 5, iron was most mobile at redox values below zero, and concentrations in solution were up to 600 mg Fe/L; cadmium, on the other hand, had highest dissolved concentrations - up to 1 mg Cd/L - at redox conditions of + 500 mV. It seems, that the study of such coupled cycles will be particularly promising with respect to the interactions of large redox-controlled systems like iron and manganese with critical trace metals such as cadmium, mercury and arsenic.

Effects of organic substances on the pH-dependent solubility of cadmium in a soil sample are illustrated by Figure 3b (after Herms & Brümmer, 1980). By addition of 5% fermented organic substances solute concentrations of cadmium in the pH-range of 4-6 are significantly lower than for sterile soil material (original sample), whereas at higher pH the organic material provides higher Cd-concentrations. The latter effect, which is probably due to formation of complexing substances, seems to be particularly relevant for the behavior of copper (Herms & Brümmer, 1980). From literature data, Jonasson (1977) has established an order of binding strength for a number of metal ions onto *humic or fulvic acids*: $Hg^{2+} > Cu^{2+} > Pb^{2+} > Zn^{2+} > Ni^{2+} > Co^{2+}$.

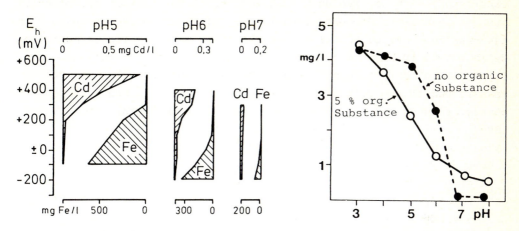

Figure 3. Solubility of Cd and Fe in Waste Compost at Different pH- and Redox Conditions (after Herms & Brümmer 1978) (Fig. 3a). Effect of Organic Material on Cadmium Remobilization from A Soil Sample (after Herms & Brümmer, 1980) (Fig. 3b).

Inorganic complexing ligands may also increase the mobility of metals in soils. This is particularly valid for cadmium and mercury (Hahne & Kroontje, 1973; Kinniburgh & Jackson, 1978), whose dissolved concentrations are significantly increased by the formation of chloro-complexes; cadmium and nickel may also form soluble sulfato-complexes (Garcia-Miragaya & Page, 1976; Bowman & O'Connor, 1982).

Compared to the wide experience with mobilizing processes on trace metals knowledge on desorption of organic contaminants from solid substrates in landfills is still relatively poor. Regarding at first before-mentioned parameters controlling sorption/desorption processes of inorganic pollutants, it can be expected that *pH and ionic strength* of landfill leachates should predominantly affect partitioning of ionic organic compounds. In fact, studies performed by Schellenberg et al. (1984) and Westall et al. (1985) indicate that for some environmentally important chlorinated phenols, i.e., for *tetra- and pentachlorophenol*, ionic strength and pH-values have to be considered in relation to octanol/water distribution. At low pH, the neutral form of the chlorinated phenol predominates in both the aqueous and nonaqueous phases, and the observed distribution ratio is approximately independent of pH. At pH values up to a few units above the acidity constant (pK_a = 4.75 for pentachlorophenol, 6.35 for 2,3,4,5-tetrachlorophenol) the phenolate ion is the predominate species in the aqueous phase and in this region the observed partition ratio is proportional to the hydrogen ion concentration; with increasing pH, aqueous species becomes more dominant. At high pH values, the phenolate ion is the predominate ion in both the aqueous and nonaqueous phases, and the distribution ratio is again independent of pH. In the high pH domain the distribution ratio is also dependent on ionic strength, since the cation of the salt must be transferred to the bulk of the non-aqueous phase to compensate the charge of the phenolate anion (see Figure 4, from Westall et al., 1985).

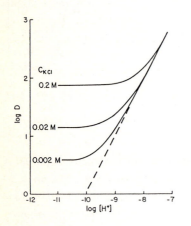

Figure 4.

Distribution Ratio of Pentachlorophenol (PCP) between Octanol and Water as Function of pH and Ionic Strength. If Phenolate Ions were not Present in the Octanol Phase (Corresponding to C_{KCl} = O), the Observed Log D vs. Log H^+ Relationship would Follow the Broken Line. (After Westall et al., 1985)

Field data on the migration of chlorophenols given by Johnson et al. (1985) from a chemical waste disposal site at Alkali Lake in South Central Oregon indicate retention factors which were about a half of those predicted from calculations of the retardation factor (Schwarzenbach & Westall, 1981). Possible causes for this apparent enhanced mobility of TeCP and PCP in the field are: (i) co-solvent effects within the plume itself, (ii) spatially or temporally non uniform release of the chemicals, (iii) fractures in the aquifer and (iv) non-uniformity of groundwater flow.

Since leachates of improper hazardous waste disposal sites usually comprise a great variety of organic contaminants it can be expected that desorption is effected by co-contaminants such as *dissolving agents* or *surface active substances*. Owing to the lack of data from solid/leachate interactions in landfills, experimental results are presented here

from a study of Wienberg & Heinze (1986) on interactions of hydrophobic
organic compounds such as *parathion* and *hexachlorobenzene* with remedial
construction materials (ground slate, fly ash, slurry mixtures) under
the influence of ethanol as well as anionic (TBS), cationic TCAB) and
nonionic ("Tween 20") surfactants. *Ethanol*, at concentrations up to 20
g/l, did not show any influence on the sorption of highly sorptive
hexachlorobenzene with all substrates used. *Tensids*, on the other hand,
indicate marked effects at concentrations exceeding 0.1 to 1 g/l (Fig.
5). These concentrations correspond with the "critical micelle concentrations" ("cmc"), which are 0.105 g/l for TBS, 0.14 g/l for "Tween 20"
and 0.34 g/l for TCAB, each at 25°C. Parathion in combination with TCAB
showed an increase in sorption with increasing concentration of ionic
compounds not exceeding the "cmc" (Figure 5). This can be interpreted
in that the hydrophobic "tails" of the tenside molecules produce new
hydrophobic sorption sites and thus enhance hydrophobic sorption until
the "cmc" is attained.

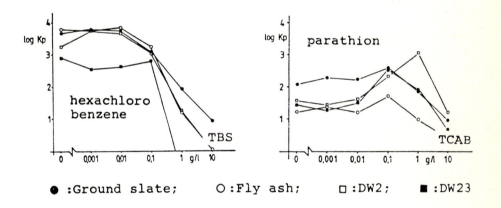

Figure 5. Influence of an Anionic Detergent (TBS = Tetrapropylene Ben-
zene Sulfonate; Left) and a Cationic Detergent (TCAB = N,N,N-
Trimethyl N-Cetyl Ammonium Bromide; Right) on the Sorption
Coefficients of Hexachlorobenzene and Parathion (Wienberg &
Heinze, 1986).

2.3. Mobility of Pollutants in Interstitial Water and Leachates

The composition of interstitial waters is perhaps the most sensitive
indicator of the types and the *extent of reactions* that take place between pollutants on waste particles and the aqeous phase which contacts
them. Particularly for fine-grained material the large surface area related to the small volume of its entrapped interstitial water ensures

that minor reactions with the solid phases will be indicated by major changes in the composition of the aqueous phase.

Significant enrichment of trace metals in pore waters has been found in (anoxic) sediment samples from Southern California basin, Saanich Inlet (Britsh Columbia) and Loch Fyne (Scotland), and has been explained by effects of *complexation* by organic substances (Brooks et al. 1968; Presley et al. 1972; Duchart et al. 1973) and by polysulfides (Jacobs & Emerson 1982). Nissenbaum & Swaine (1976) found that with the exception of iron, nickel and cobalt (which mostly occur as sulfides under anaerobic conditions) the elements concentrated in the interstitial solution are those which are also enriched in sedimentary humates. There are few data on the effects of higher concentrations of organic ligands in porewater.

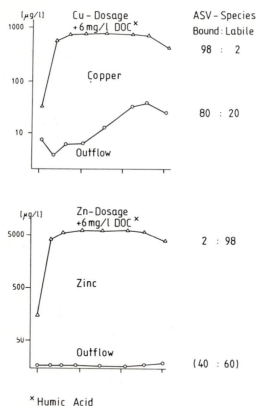

Figure 6. Sand Filter Column Experiments with Ruhr River Water, Humic Acid (6 mg/l DOC), Copper (1 mg/l; above) and Zinc (10 mg/l; below). Ratio of Labile to Bound Metal Species from Anodic Stripping Voltammetry Data (pH 4.7, Acetate Buffer). (After Förstner et al., 1982).

Studies on the interactions between natural substrates and metal-bearing solutions was performed with *laboratory filter columns* and semi-technical systems (Förstner et al., 1982). In an experimental series four identical sand filter columns were continually fed with water from the Ruhr River (Figure 6): Humic acid was added in concentrations of approx. 6 mg/l DOC, Cu with an 1 mg/l, Zn with an additional 10 mg/l. The data in Figure 6 indicate, that although the amount of zinc in the originial concentration is 10 times that of copper, no increase on the zinc concentrations were to be observed after passing through column. After a 10-hour flow period through the column however, the copper concentrations had increased significantly. Anodic stripping *voltammetry* determinations showed that zinc in the inflowing solution is basically labile-bound. A nearly total elimination of zinc during passage through the column can be explained as a result of an active interaction with the Fe/Mn-hydroxides of the filter particles (which were analyzed by sequential chemical leaching; see below). The copper component, on the other hand, is in a more stable association with the complexing (organic) ligands, and remains partially unaffected by the physico-chemical interactions with the filter material. From other experiments it is demonstrated that there are effects of competition between individual elements; for example, higher inputs of copper ions affect partial mobilisation of cadmium ions into deeper sections of the filter column; this effect, however, is reversed by addition of humic acids (as complexing agent).

Attenuation of potential pollutants in landfill leachate has been studied by Campbell et al. (1983) in an *in-situ lysimeter* constructed in calcareous clayey-sandstone for a period of nearly six years: Considerable reduction of heavy metals in the unsaturated zone was found to be attributable to a combination of processes including adsorption and precipitation. Leaching with a blank carboxylic acid solution (volatile fatty acid containing no heavy metals) resulted in limited remobilization; it is suggested that only the more labile soluble, exchangeable and, to a lesser extent carbonate, phase are remobilized, leaving a residue of metals in the sediment. The more mobile elements (in the order of mobility of the metals Ni > Cd > Zn > Cu > Cr > Pb), which are associated with these labile phases, are therefore also released to a greater extent. Relative high heavy metal concentration at the surface of the lysimeter were found to cause *inhibition of carboxylic acid degradation*, due to reduction of population density of microorganisms by a factor of 10^3. However, at depths in the lysimeter, below the limit of heavy metal penetration, anaerobic decomposition of acetate, propionate and butyrate were not affected.

Mobility of dissolved *organic pollutants* through porous media is gover- ned by water velocity, by diffusion, dispersion and sorption/desorption (e.g., Wienberg, 1988). Using a simple transport equation for the flux of organic contaminants through water saturated porous media (e.g., Van Genuchten et al., 1974), the order of magnitude of the contaminants mo- vement in relation to water velocity can be calculated. Several papers deal with the verification of calculated retention factors from diffe- rent soils by laboratory experiments and in the field, and they gener- ally show good agreement (Davidson & Chang, 1972; Wilson et al., 1981; Roberts et al., 1982; Schwarzenbach et al., 1983). Most of the studies were confined to contaminants of low to medium hydrophobicity, i.e., compounds exhibiting octanol/water partition coefficients between 10^2 and about 10^4. Interference of geochemical mechanisms with biochemical - i.e. microbial degradation - processes has to be studied on well- equipped sites (Reinhard et al., 1984). Another area which needs more research, is the influence of *colloidal* and dissolved organic material (detergents, solvents, fulvic materials) on the transport of non-polar organic contaminants, in particular, with regard to their remobiliza- tion from landfills.

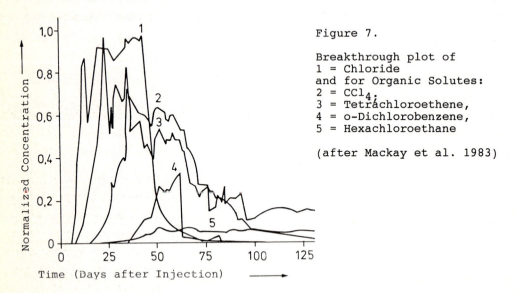

Figure 7.

Breakthrough plot of
1 = Chloride
and for Organic Solutes:
2 = CCl_4,
3 = Tetrachloroethene,
4 = o-Dichlorobenzene,
5 = Hexachloroethane

(after Mackay et al. 1983)

A field investigation of *halogenated organic solutes* was conducted by Mackay and his group (1983) in a relatively uncontaminated portion of an unconfined sand aquifer. The injection of 12 m^3 of an aqueous solution of several halogenated organics was so chosen as to allow detection of solute plumes for a period of more than 1 year by ion and gas chromatgraphy techniques. From the analyses on 9000 samples, the information is used to construct "breakthrough curves" for each solute. Data at a point 2,5 m downgradient are shown in Figure 7, which clearly illustrate the varied mobility of different solutes underground.

3. NEW METHODS FOR ESTIMATING LONG-TERM MOBILITY OF POLLUTANTS

Various laboratory techniques have been described for generating leachate from waste materials and are generally grouped into batch and column extraction methods. The batch extraction method offers advantages through its greater reproducibility and simplistic design, while the column method[1] is more realistic in simulating leaching processes which occur under field conditions (Jackson et al., 1984; Fuller & Warrick, 1985).

Initial estimation of potential release of metal from contaminated solids is mostly based on *elutriate tests*, which - apart from the characterization of the reactivity of specific metals - can provide information on the behaviour of metal pollutants under typical environmental conditions. Common single reagent leachate tests, e.g. U.S. EPA, ASTM, IAEA, ICES, and DIN use either distilled water or acetic acid (Theis & Padgett, 1983). A large number of test procedures have been designed particularly for soil studies; these partly used organic chelators such as EDTA and DTPA, both as single extractants or in sequential procedures (Sauerbeck and Styperek, 1985). Physiological solutions have been applied, e.g. for the estimation of organo-specific effects of atmospheric particulates (Harris and Silberman, 1983). Best results with respect to the estimation of short-term effects can be attained by *"cascade" test* procedures at variable solid/solution ratios: A procedure of the U.S. EPA (Ham et al., 1979) designed for studies on the leachability of waste products consists of a mixture of sodium acetate, acetic acid, glycine, pyrogallol, and iron sulfate.

---

[1] An elegant way for studying contaminant diffusion in sediments, soil and waste materials has been described recently by Van der Sloot et al. (1988) using a micro-column system (polyethylene tube: O = 8 mm, l = 50 mm) with suitable radiotracers and an interface marker.

Another standard leaching test has been developed by the Netherland Energy Research Centre (Van der Sloot et al., 1984) for studies on *combustion residues*. In the column test a column is filled with the material under investigation and percolated by acidulated demineralized water (pH = 4; for evaluating most relevant effects of acid precipitation) to assess short- and medium-term leaching (< 50 years). In the cascade test the same quantity of material is extracted several times with fresh demineralized water (pH = 4) to get an impression of long-term leaching behavior (50-500 years). As a time scale the liquid/solid ratio (L/S) is used; the relation between this relative time scale and the actual time scale depends on the time required for a L/S ratio to be reached in the actual situation. The maximum leachability is assessed by a shake experiment at L/S ratio of 100 under mild acid conditions (De Groot et al., 1987)[2].

3.1. Solid Speciation of Metals from Sequential Extraction

Since adsorption of pollutants onto particles is a primary factor in determining the transport, deposition, reactivity and potential toxicity of these materials, analytical methods should be related to the *chemistry of the particle's surface* and/or to the pollutant species highly enriched on the surface. Analytical tools are direct measurements on particles, i.e. on their surface (e.g. by X-ray fluorescence, electron microprobe, proton-induced X-ray emission analysis) or - indirectly - application of solvent leaching techniques (Keyser et al. 1978).

In connection with the problems arising from the disposal of solid wastes, particularly of dredged materials, *chemical extraction sequences* have been applied which are designed to differentiate between the exchangeable, carbonatic, reducible (hydrous Fe/Mn oxides), oxidizable (sulfides and organic phases) and residual fractions. The undisputed advantage of this approach with respect to the estimation of long-term effects on metal mobilities lies in the fact, that rearrangements of specific solid "phases" can be evaluated prior to the actual remobilisation of certain proportions of the element into the dissolved phase (Förstner, 1985).

---

[2] Recent improvements of this method have been achieved by comparing the "L/S-curves" for an individual element with its stability in a wider pH-spectrum; in some cases direct mineralogical evidence can be given for a distinct metal compound (Van der Sloot, private communication).

These changes can be measured, for example, *directly* in typical soils
and waste landfills. We are presently performing experiments where dialysis bags containing typical substrates with varying metal dosages
were inserted into boreholes to both anaerobic and aerobic groundwater
systems. Figure 8 shows results on the behaviour of low metal-dosage to
iron oxyhydrates and solid organic substrates under anaerobic conditions. In both examples lead is released during the 4-weeks experiment
particularly from easily reducible phases. For copper a slight increase
is observed during the experimental time period, again in the easily
reducible fraction (Förstner, 1988).

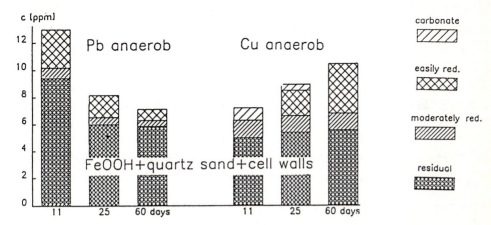

Figure 8. Changes of Total Concentrations and Chemical Forms of
Pb and Cu on Organic and Inorganic Substrates Inserted
into Anaerobic Groundwater (Förstner & Carstens, 1988)

With the *direct method* relationships between anthropogenic, xenobiotic
substances and natural soil substrates can be evaluated for:

(i) *site-specific properties* of different disposal systems
     which can be compared with respect to their compatibi-
     lity with the waste material, containing a wide range
     of pollutants;

(ii) influences of a *stabilized system* of hydrological and
     chemical factors, mechanisms and processes - including
     biochemical interactions - on the behaviour of indivi-
     dual critical pollutants.

3.2. Estimation of Long-Term Behaviour of Metals in Solid Wastes

Single-extractant procedures are restricted with regard to prediction of long-term effects in waste deposits, since these concepts neither involve mechanistic nor *kinetic considerations* and therefore do not allow calculations of release-periods. This lack can be avoided by controlled significative intensivation of the relevant parameters pH-value, redox potential and temperature combined with an extrapolation on the potentially mobilizable "pools", which are estimated from sequential chemical extraction before and after 10-weeks treatment of the waste material.

An experimental scheme, which has originally been used by Patrick et al. (1973) and Herms & Brümmer (1978) for the study of soil suspensions and municipal waste materials, was modified by inclusion of an *ion-exchanger system* for extracting the metals released within one week time each (Schoer and Förstner, 1987). Solutions adjusted - partly self-regulating - to combinations of pH 5/8 and redox 0/400 mV circulate with 2 litres per day through columns, containing 1:4 mixture of solid waste with quartz sand, the latter component to improve permeability (simulating flow-through conditions). The system can be modified for different intensities of contact between solid materials and solution, by using shakers (erosion of the depot by rivers) or dialysis bags (flow-by conditions). Before and after these treatments sequential chemical extraction is performed to differentiate between the exchangeable, reducible (hydrous Fe/Mn oxides), oxidizable (sulfides and organic phases) and residual fractions.

Extrapolations with respect to the total *mobilizable metal "pool"* can be performed by comparing data from sequential extraction experiments. Taking the example of zinc in Figure 9 the more labile "exchangeable" fractions will be released at first ("phase 1"), whereas during "release phase 2" - which is much slower than initial mobilization of acetate-extractable zinc - part of oxalate-reducible compounds will be dissolved.

Using this system *metal oxide residues* from heat processing, which were deposited into borrow pits for sand removal, has been studied at a 225 hectare site in the lower Rhine valley near Düsseldorf (Förstner and Schoer, 1987). Experimental conditions were pH 5 and 8, and E_h 0 and +400 mV, according to the borderline conditions at different phases in municipal waste deposits. For most elements *acidity* is more effective

than changes in redox conditions. Maximum element remobilization was
20% to 40% of total concentration for Zn, Co and Cd at pH 5 (both at 0
mV and 400 mV); higher release of Pb (29%) and Cu (8%) was found at
+400 mV rather than at 0 mV (both 2%). Maximum mobilization of thallium
occurred at pH 8 and 400 mV (25%). Solubility of Cr was less than 1%
for all studied conditions. Temporal release patterns are different for
the individual elements: While at pH 5/400 mV release of cadmium seems
to be completed within the experimental period mobilization of copper
is still going on and the end point cannot be estimated from data of
the "kinetic" experiments.

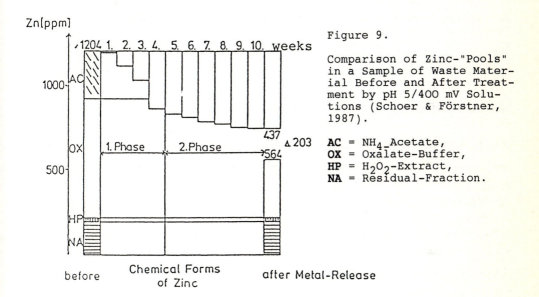

Figure 9.

Comparison of Zinc-"Pools"
in a Sample of Waste Material Before and After Treatment by pH 5/400 mV Solu-
tions (Schoer & Förstner,
1987).

AC = NH_4-Acetate,
OX = Oxalate-Buffer,
HP = H_2O_2-Extract,
NA = Residual-Fraction.

Taking into account both element contents released during the 10 week
experiments and those extrapolated from reducible pools concentrations
can be *calculated* for the interaction of 1 kg of solid waste with 1400
litre water of pH 5/ 400 mV (Tl: pH 8/ 400 mV) solution (Table 1). At
these extreme assumptions with respect to both solution/solid contact
and chemical conditions most metal concentrations would be expected in
the order of magnitude of the *guideline values* set by the authorities
of Nordrhein-Westfalen for aqueous elutriates of waste materials class
1 (partly derived from drinking water standards). From these data it is
indicated that further considerations with respect to more realistic
hydrological conditions should focus on the examples of zinc and lead.

Table 1   Calculation of Dissolved Metal Concentrations in Landfill Leachates from Direct Determinations (Treatment of 100 g Sample
with 140 Litre Extraction Solution) and from Estimations of
Mobilizable Metal Pools. For Comparison: Permissible Concentrations in Elutriate of Waste Materials Depot Class I (Nordrhein-Westfalen)

Mobilization at Flow-Through Conditions* per kg	pH/Redox	Mobilization-Relevant Metal-"Pool" in mg/kg Acetate (1)	Oxalate (2a)	(2b)	H₂O₂ (3)	Results in Dissolved Concentration mg/L	Depot Class I Criteria mg/L**
Cd 6 mg	pH 5/ 400	[1		2	1]	0.004	0.005
Co 100 mg	pH 5/ 400	[6	90]	40	5	0.07-0.1	0.05
Cr 3 mg	pH 5/ 400	0		250	15	0.002	0.01[a]
Cu 50 mg	pH 5/ 400	[15	35]	185	50	0.03-0.20	0.10
Pb 600 mg	pH 5/ 400	[20		350	200]	0.40	0.05
Tl 20 mg	pH 8/ 400	[10	10]	30	5	0.01-0.03	0.01
Zn 6000 mg	pH 5/ 400	[1000		5000]	300	4.0	1.0

*140 L solution/100g solid waste   **Nordrhein-Westfalen (in prep.)
[a] value for Cr(VI)

3.3. Assessing Specific Retardation of Organic Contaminants by Batch Equilibrium Experiments and Thin Layer Chromatography

Extrapolations to *long-term effects of organic pollutants* still are
very scarce. Laboratory experiments on the *specific retardation* have
been performed by Wienberg et al. (1986) in order to test the behavior
of remedial construction materials for landfills, such as cement, fly
ash, ground slate, bentonite, micaceous clays and slurry mixtures; some
of the latter mixtures contained special additives, such as Dynagrout
silicate gels (Hass, 1986). Batch equilibrium experiments were conducted using *C-14 labelled organic compounds* which were dissolved in distilled water to about half of their solubility and diluted in the order
of 1:2:4:8:20. 25 ml of these solutions were transferred into centrifu-
ge vials containing 0.1 g (strongly sorbing media) to 10 g (weakly sorbing media) of solids. The vials were shaken over head for 20 hours.
The supernatant was centrifugated and the sorbed fraction was extracted
using a non-gelating scintillation fluid. Both fractions were analyzed
by *liquid scintillation counting*. The experiments on the additional effects of dissolving agents (e.g. ethanol) and surfactants (e.g. tetrapropylene benzene sulfonate; see sect. 2.2) were undertaken in concentration steps of a power to ten in the range of 10^{-3} to 10 g/l (Wienberg
& Heinze, 1986).

Thin-layer chromatography was conducted according to the precriptions
of Helling & Turner (1968) and Poon et al. (1984) using slurry compo-

nents or mixtures as solid phase. The solids were spread as a 0.25 mm layer on 20 x 20 cm glass plates. Detection of the chemicals was performed by (i) several *spray reagents* and (ii) *autoradiography*, using radiolabelled compounds. The best-suited spraying method was application of a solution of 1 g $AgNO_3$ plus 5 ml of conc. ammonia, filled up to 100 ml with ethanole[3]. Examples in Figure 10 indicate that the organic acids (e.g. 2.4-D) are not retained by conventional slurry mixtures whereas Na-aluminate treated quartz shows high retardation efficiency.

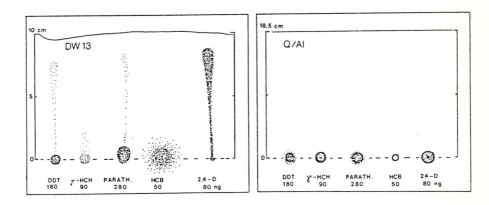

Figure 10. Thin Layer Chromatogram (Autoradiography) of Organic Sorbates on Slurry Components (Wienberg et al. 1986)

4. GEOCHEMICAL PROCESSES - REACTOR LANDFILL VS. FINAL STORAGE

In the previous sections different *types of waste materials* have been mentioned and the behavior of toxic chemicals was described in organic-rich solid waste from domestic and industrial sources, sewage sludge, dredged material, fly ash, and inorganic processing residues. Considerations on the interactions between dissolved and solid phases has been extended to the surrounding areas, i.e. soils and groundwater zones.

Within the conceptual perspectives of the present Workshop, which involves two major alternatives of future landfill operations, the long-term geochemical implications can be evaluated from a comparison of the inventory of either "reactor landfill" or the "final storage material"

[3] For the autoradiography, a Kodak XAR5 X-ray film was used. Time of exposure was 24 hours, activity of the compounds about 20 nCi.

with the earth crust composition and the processes taking place in the different sites (Table 2).

From a geochemical point of view the *"reactor landfill"* is characterized by labile conditions during the initial aerobic and acid anaerobic phases, the former mainly due to uncontrolled interactions with organic solutes. Subsequent to landfilling, the raw waste compounds undergo a variety of *early diagenetic processes* accompanying microbially mediated degradation of the organic compounds. The metabolic intermediates of organic matter decay (e.g. HCO_3^-, HPO_4^{2-}, carbohydrates and other low molecular organic acids) and those of the coupled inorganic reduction processes (e.g. Fe^{2+}, Mn^{2+}, S^{2-}, NH_4^+) accumulate in the interstitial water until concentrations are limited by physical convection/dispersion, by subsequent microbial utilization, or by diagenetic formation of secondary ("authigenic") minerals such as metal sulfides. This secondary inventory of a reactor landfill is critical both in buffering leachate water chemistry (Stumm & Morgan, 1981) and in affecting transport of pollutants to underlying groundwater aquifers. On the other hand, *final storage quality*, which is defined by the composition of

Table 2: Comparison of Inventories of Chemical Components in the Two Landfill Alternatives and in the Earth Crust

Reactor Landfill	Final Storage	Earth Crust
Major Solid Constituents		
Solid "Inert" Waste	Silicates, Oxides	Quarz, Fe-Oxide
Putrescible Waste	[Gypsum, NaCl][1]	Clay, Carbonates (Gypsum, NaCl)
Grease Trap Waste	(Char)[2]	Kerogenic Compounds
Minor Solid Constituents		
Organic Micropollutants	Organic Micropollutants	-
Metals in Reactive Chemical Forms	Metal-Bearing Minerals Mainly Oxides	Metals Mainly in Inert Forms
Dissolved Constituents		
Protons, Electrons	(Protons)	(pH: Acid Rain)
Organic Compounds	(Organic Residues)	(Humic Acids)
Dissolved Salts	[Dissolved Salts][1]	(Dissolved Salts)

[1] [Partial Extraction during Pretreatment]   [2] (Minor Constituent)

earth crust material, in most cases is not attained by simply incinerating municipal waste, i.e. by reduction of organic fractions only. There is, in particular, the problem of easily soluble minerals, such as
sodium chloride. Nonetheless the type of *inorganic residue deposits*
will increasingly receive prevalence as a method of final storage for
municipal wastes in the future.

The possible implications with *toxic chemicals* will be presented for
the *various types of landfills* such as deposits of sewage sludge (section 4.1.), municipal solid waste ("reactor") landfills (4.2.), sub-
sediment deposition of dredged materials (4.3.), borrow pit storage of
metal oxide residues from industrial processes (section 4.4.), deposits
of coal combustion ashes (4.5.) and finally, residues from incineration
of municipal waste materials (4.6.).

4.1. Storage of Fresh Sewage Sludge and Sludge Incineration Ashes

In their study of the two alternative approaches to sanitary landfilling - final storage or reactor deposit - it has been suggested by
Lichtensteiger & Brunner (1987) with respect to the latter approach
that a landfill with sludge only is a less complex and more homogenous
reactor which can be sampled, analyzed and interpreted with less difficulties than a municipal solid waste landfill. By comparison of their
long-term evolution to similar natural sediments (peat, organic soils)
and their diagenesis the following hypotheses, among others, have been
made by the authors

    (1) The transformation of organic material lasts for geological
        time scales (10^3 to 10^7 years);

    (2) Lime addition postpones the biochemical transformation for a
        few years only;

    (3) after 10^1 to 10^2 years, there are still xenobiotic organic
        substances which are only partly or not at all degraded in a
        sludge landfill.

Sludge-only landfills have lower permeabilities and interparticular porosities than municipal solid waste (MSW) landfills. Thus, landfill gas
and leachate control are more difficult and less effective than in MSW
landfills. Dissolved substances such as nitrogen and phosphorus reach
the supernatant water which consequently has to be treated; heavy metal

concentrations in the surface water of the studied sludge pond complied with the quality requirements for surface waters to be upgraded to drinking water (Lichtensteiger & Brunner, 1987).

With respect to the behavior of trace elements in *municipal sludge ashes* experimental data by Theis & Padgett (1983) on metal mobility during transition from raw sludge to incinerator ash indicate, that the addition of Fe(III) and/or Al(III) salts tends to partition metals from organic to inorganic forms in raw sludges. The inorganic fractions of Fe and Al become more crystalline on incineration, which acts to increase the leachability of metal cations (with the exception of Cd) if Fe and Al salts have been added during treatment. If these salts are not added the leachability of cations decreases as a result of incineration. Here too, arsenic was found to behave inversely to cationic elements, that is, it is less easily leached from both dewatered sludge and ash when Fe and Al have been added, whereas no addition of Fe and Al affect increased leachability in both form.

A form of organo-metallic compound particularly well-suited for disposal in an inorganic sanitary landfill is formed during pyrolysis of organic-rich wastes such as sewage sludge (Kistler et al., 1987). Metals in the *pyrolysis residues* ("char") are not likely to be mobilized by acidic leachates due to the highly buffered neutral to alkaline properties of the residue. Nethertheless, since these metals might be mobilized by acids combined with organic ligands, they should be deposited onto landfills where organic material is excluded ("monolandfill").

Recycling of valuable metals from waste materials (see "Outlook") could well start from sewage incineration ashes, due to the strong enrichment of gold and silver in these materials. Since 1986 Lasir Gold Inc. is recovering gold and silver from Toronto's sewage sludge ash; from ca. 300 tons of fly ash per day (80-100 tons of actual release from the plant plus materials piled up during the last three years) an annual recovery of 15,000 oz. of gold and 125,000-250,000 oz. of silver is expected (Anon., 1985). The ash is dumped on two 300-ton leaching pads in an old oil storage tank and is leached using dilute sodium cyanide solution. Total processing costs, including transportation, are expected to be in the order of $ 150 per oz. of gold produced, making it a very low cost operation.

4.2. Toxic Chemicals in Municipal Solid Waste (Reactor) Landfills

Implications with potentially toxic or adverse metal concentration in
reactor landfills have been described by Ehrig in this book. Even if
treatment of leachates from municipal waste landfills is mostly performed to reduce concentrations of dissolved organic carbon, and in
some cases for reduction of phosphorus and nitrogen compounds (Baccini
et al., 1987), there are manifold problems with both organic and inorganic toxic chemicals, in particular during the initial stage of these
landfills. Concentrations of dissolved iron and manganese are higher by
a factor of 50 in the acetic "phase" compared to methanogenic phase in
the landfill evolution (Ehrig, 1983), and this seems to favor plugging
of the pipe and liner system if the time period of the former phase is
extended. Anaerobic digestion of landfill leachate seems to be inhibi-
ted by elevated concentrations of metals such as zinc, copper and cadmium (Anderson et al., 1985), and similar effects on acetogenic bacteria may be expected within the landfill.

The long-term evolution of a reactor landfill, subsequent to the methanogenic phase, is still an open question. Oxidation of sulfidic minerals by intruding rainwater could mobilize trace metals, and the impact
on the underlying groundwater could be even higher if a chromatography-
like process, involving continuous dissolution and reprecipitation during passage of oxidized water through the deposit, would preconcentrate
critical elements prior to final release with the leachate.

4.3. Sub-Sediment Deposition of Anoxic Sludges

Incorporation in naturally formed minerals, which remain stable over
geological times, constitutes favourable conditions for the immobiliza-
tion of potentially toxic metals in large-volume waste materials both
under environmental safety and economic considerations. There is a particular *low solubility of metal sulfides*, compared to the respective
carbonate, phosphate, and oxide compounds. One major prerequisite is
the microbial reduction of sulfate. Thus, this process is particularly
important in deposits containing degradable organic matter. Such conditions are common in the marine environment, whereas in anoxic fresh-
water milieu there is a tendency for enhancing metal mobility due to
the formation of stable complexes with ligands from decomposing organic
matter.

Marine sulfidic conditions, in addition, seems to repressing the formation of *mono-methyl mercury*, one of the most toxic substances in the
aquatic environment (Craig and Moreton 1984). There are indications,
that degradation of highly toxic *chlorinated hydrocarbons* is enhanced
in the anoxic environment relative to oxic conditions (Sahm et al.
1986; Kersten 1988). A summary of the positive and negative effects of
anoxic conditions on the mobility of heavy metals, arsenic, methyl mercury and organochlorine compounds in dredged sludges is given in Table
3.

In a review of various marine disposal options Kester et al. (1983)
suggested that the best strategy for disposing contaminated sediments
is to isolate them in a permanently reducing environment. This type of
waste deposition under stable anoxic conditions, where large masses of
polluted materials are covered with inert sediment became known as

Table 3. Summary of positive and negative effects of anoxic (sulfidic
conditions on the mobility of heavy metals, metalloids, methyl
mercury, and organochlorine compounds in sludges (Kersten
1988).

Element or Compound	Advantageous Effects	Disadvantageous Effects
Heavy Metals (e.g.cadmium)	Sulfide precipitation	Formation of mobile polysulfide and organic complexes under certain conditions with low iron-oxide concentrations, strong increase of mobility under post-oxic and acidic conditions
Metalloids (e.g.arsenic)	Capture by sulfides	Highly mobile under post-oxic and neutral to slightly alkaline conditions
Methyl mercury	Degradation and inhibition of CH_3Hg^+ formation by HgS precipitate formation	Formation of mobile polysulfide complexes, especially at low Fe-Fe concentrations
Organochlorine compounds	Initiation of bio-degradation by reductive dechlorination (methanic environment is more favourable)	Formation of harmful terminal residues with certain compounds, especially in sulfidic environments; remobilisation through colloidal matter suspended in pore water

"subsediment-deposit"; the first example was planned for highly contaminated sludges from Stamford Harbour in the Central Long Island Sound
following intensive discussions in the U.S. Congress (Morton 1980).
From studies performed by Brannon et al. (1984) is seems that even a
50-cm sand layer is an effective barrier against the transfer of PCB-
compounds into the surface water.

For the disposal of approx. 10 million m^3 dredged sludge of from the
harbour area, the Port of Rotterdam and the Netherlands Waterways Administration - after several years of intensive and costly planning - has
now started to construct a *"sludge island"* in the form of a peninsula
as a containment for 150 million m^3 of sediment. The deposit will consist of a 20 m deep hole; the excavated material will form a 18 m high,
high-tide resistant wall. Model calculations suggest that the concentration of most contaminants will not affect significantly groundwater
composition; it is expected that pollutants discharged to the seafloor
will have only minor effects on the surrounding ecosystems (Municipality Rotterdam/Rijkswaterstaat 1984).

4.4. Storage of Purely Inorganic Industrial Wastes in Borrow Pits

For the long-term prognosis of the behaviour of *metal oxide residues
from heat processing* extreme assumptions with respect to both solu-
tion/solid contact and chemical conditions (pH 5/400 mV) initially have
been considered (section 3.2.). More "realistic assumptions" can be
introduced, when the *hydrological conditions* within and around the
deposit are included. It is suggested from data on permeability that
with the relative dense packing of these materials compared to the
surrounding sand aquifer one should expect *flow-by conditions* of the
surface water rather than a flow through the deposit. Simulation of
such conditions - which is experimentally performed by flow-by on dialysis bags - indicate significantly lower rates of remobilization (one
to two orders of magnitude for individual elements) than from the flow-
through experiments mentioned before. Even such mobilization cannot be
expected in reality, since there is no internal process - such as sulfide oxidation - which could affect pH 5 conditions in the deposit. Additional calculations on the basis of atmospheric *inputs of acidity* can
be interpreted in a way that with the present form of deposition below
groundwater level no significant change of metal mobility or release
will take place within several thousand years.

If long-term changes of pH and redox conditions cannot be excluded, *improvement of storage quality* of metal-containing waste materials may be achieved generally by maintainance of a neutral pH or slightly alkaline conditions - e.g. by application of lime or limestone - which favour adsorption or precipitation of soluble metals. For metal-finishing sludge disposal in limestone-lined segregated landfills may be an environmentally safe and economical alternative (Regan & Draper, 1987). The liner may act as a chemical barrier by precipitating the solubilized metals from the acidic drainage. Field studies have indicated that once the initial interstital water was removed from waste sludges only non-hazardous concentrations of metals were measured in the leachate, and that such a segregated landfill would not require an impermeable lining or an elaborate leachate treatment system[4].

4.5. Deposition of Coal Combustion Residues

Fly ash from coal burning power plants primarily represent a large-tonnage disposal problem. Typical changes of metal species occur during ash-pond disposal and in landfill leachates, which could affect ground-water quality (Förstner, 1986). In actual and laboratory-simulated disposal situations it has been shown that elements with anionic aqueous speciation such as S, As, Se, and Mo are appreciably solubilized in ash sluicing/ponding systems that are neutral to alkaline in pH (Turner et al., 1982). Obviously there is a correspondence between those elements exhibiting elevated concentrations in the effluent waters and those most extractable in water (Table 4).

Among the effects which coal fly ash has on the groundwater environment and particularly on the release of metals are *changes in pH*. In contrast with literature information recent investigations by De Groot et al. (1987) indicate limited solubility of anions at high pH (> 11). The metals Pb, Cu, Cd and Zn show a minimum solubility at high pH whereas major elements, like Al and Si, show two minima in the pH-range 7 to 9 and at pH values higher than 11. The latter was related to the formation of new mineral phases (ettringite).

---

[4] Several contributions in the book "Environmental Management of Solid Waste: Dredged Material and Mine Tailing" (Salomons and Förstner, 1988) describing methods of "geochemical engineering" emphasize the increasing efforts of using natural resources available at the disposal site for reducing negative environmental effects of all types of waste material, in particular of acid mine wastes.

Table 4.   Trace Element Concentrations in Ash-Pond Effluent Water
Relative to Cooling Lake Intake and Outlet (Dreesen et
al., 1977)

Concentration Ratio	Ash-Pond Effluent/ Cooling Lake Intake	Ash-Pond Effluent Cooling Lake Outlet
> 50	B, F, Mo, Se, V	Se
10-50	As	B
2-10	-	As, Cr, F, Mo, V
< 2	Cd, Cu, Zn	Cd, Cu, Zn

Trace-metal *migration in groundwater* from ash-pond seepage has been
studied by Theis et al. (1978) on a ponding system of the Northern
Indiana Service Company. Metals were released at low concentrations
into the groundwater, depending upon the ash-loading rates and procedures. Rapid attenuation of metal contents occurred close to the pond;
most metals were effectively scavenged by iron and manganese oxides or
were precipitated. During release typical *species variations* have been
determined by model calculations: For example, while lead hydroxide may
precipitate in the vicinity of the pond due to slightly elevated pH
levels, lead carbonate does not precipitate until the groundwater inorganic carbon levels have increased substantially.

As mentioned earlier, high *mobility of arsenic* is typical for the initial stage in many landfills. It has been stressed by Turner (1981)
from data on fly ash deposits, that since trivalent arsenic is likely
to be the predominant As species in ash pore water and groundwater and
is also the more toxic form in water, interactions of As(III) with
soils and landfill-liner material should be more closely examined to
ensure the protection of drinking-water aquifers in critical areas.

4.6. Deposition of Municipal Incineration Wastes

Waste incineration ashes usually exhibit relative high concentrations
of trace metals - Zn and Pb up to the percent range - and particularly
strong enrichment factors compared to natural contents have been observed for these elements and for cadmium and silver (Brunner & Zobrist,
1983). It is suggested that proportions of leachable elements, e.g. by
0.1 N HCl, are usually higher from municipal ash than from coal burning
fly ash (Austin & Newland, 1985). Sequential extractions performed by
Wadge & Hutton (1987) indicate (Figure 11), that about 20% of total Cd

and 1% of total Pb in coal fly ash was in exchangeable fraction; in contrast, the single largest fractions of Cd and Pb in refuse ash, at 72% and 41%, respectively, were present in the exchangeable form. Concentrations of trace metals extracted from refuse ash appear to be a function of the elemental boiling point and species that exist on combustion; the classification of elements for refuse ash - as studied by Cahill & Newland (1982) by various solvent extraction methods - strayed somewhat from the traditional geochemical classification scheme into which coal ash is placed (Klein et al., 1975).

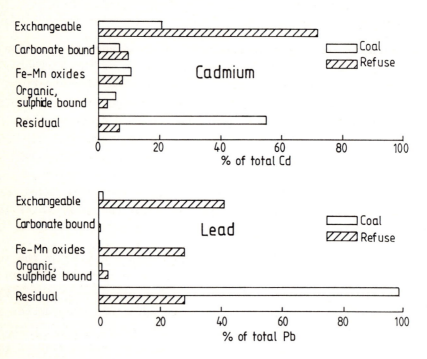

Figure 11. The Chemical Associations of Cadmium and Lead in Coal Fly Ash and Refuse Fly Ash (Wadge & Hutton, 1987)

A Soxhlet extraction of refuse fly ash with water was used by Karasek et al. (1987) to represent the leaching of *organic compounds* using water at acidic (pH 4), neutral (pH 7), and basic (pH 10) conditions. A number of chlorinated and other toxic compounds were identified in the extracts, and the pH of the water affected the type and amount of organic compounds removed. Although PCDD and PCDF were found in some of the extracts, their levels were very low.

Problems with *polychlorinated dibenzo-p-dioxins* (PCDD) and polychlorinated dibenzofurans (PCDF) in electrostatic precipitator ash from waste
incinerators may be solved prior to deposition of these materials. Experiments performed by Hagenmaier and his group (1987a) have shown,
that re-heating at $300^{\circ}C$ under oxygen deficient conditions affects a
dechlorination/hydrogenation of these compounds; with copper as best-
suited catalyzing compound decomposition of octaCDD and of all PCDD
formed as intermediates by the dechlorination reaction is 99.99% complete within 1 min. Polychlorinated biphenyls and chlorobenzenes also
were found to undergo decomposition in the presence of copper powder
(Hagenmaier et al., 1987b).

Baccini & Brunner (1985) have demonstrated, that even if *chlorine from
incineration plants* is only contributing to a minor degree to the contamination of this element in surface waters (approx. 4%), there are
several problems associated with this element in municipal waste incineration systems; as a (i) carrier of protons, which give rise to acidification of the atmosphere; (ii) carrier of metals, affecting higher
vapor pressure and solubility of certain metal chlorides; and (iii)
chlorination agent for aromatic hydrocarbons (see above). *Measures before incineration* include the separate collection of (organic) kitchen
and garden wastes (containing some chlorine and sulfur), which can be
transferred into compost; a major decrease of chlorine content, however, would require a significant reduction of PVC in municipal solid
waste. *After incineration* washing of the residues can be performed
either with neutral or acidified water. For example, material washed at
pH-4 in closed cycles can directly be deposited, since acid precipitation - as the major influencing factor on the geochemistry of inorganic
residue landfills - could not have any further effect on the mobility
of metals (Hämmerli-Wirth, 1987); however, it has to be considered that
part of the sorption sites, for example: iron oxides, are extracted by
this procedure as well. By the addition of alkaline binders such as cement, the initial leaching of some elements can be decreased also; care
must be taken not to apply too much alkaline material in order to prevent the dissolution of Pb, Zn and other metals at high pH-values, due
to the formation of soluble hydroxo complexes (Brunner & Baccini 1987).

5. OUTLOOK

Future efforts should not only be aimed for chemically stabilizing critical compounds in waste materials but - in particular - for recycling
valuable components such as strategic metals (e.g. Patterson, 1987). A
look to the data listed in Table 5 indicates that recovery of metals
such as lead, zinc, and silver from certain fractions of waste combustion products could well compete with natural resources of these elements.

Table 5. Trace Element Concentrations in Municipal Incinerator Suspended Particles (Nicosia/East Chicago, Ind.; Greenberg et al.
1978) Compared to Mean Element Composition in the Earth's
Crust (Bowen, 1979).

	Ag	Au	Cd	Pb	Sb	Zn
Refuse Inci-nerator Ash	110	0.43	1500	69.000	1600	114.000
Mean Content Earth's Crust	0.07	0.0011	0.11	14	0.2	75
EF	1.570	390	13.640	4.930	8.000	1.520

Economic feasibility of a state-wide hydrometallurgical recovery facility for metal bearing wastes - including metal sludge from electroplating, heat treating, inorganic pigment manufacture, lime treatment of
spent pickle liquor and emission control sludge - generated in Missouri
has been evaluated by Ball et al. (1987). A pilot test assembly, which
was originally developed by the Montana College of Mineral Science and
Technology, was capable of treating approximately 40 kg/day of mixed
metal sludge from metal finishing industries. Major unit processes were
(1) metal hydroxide dissolution (sulfiuric acid leach), (2) iron removal (jarosite precipitation), (3) solid/liquid separation by filter
pressing, (4) copper recovery by solvent extraction (chelating agent
dissolved into kerosene), acid stripping and copper sulfate crystallization, (5) zinc recovery (solvent extraction, acid stripping, zinc
sulfate crystallization), (6) chromium recovery (oxidation to dichroma-
te using chlorine gas, precipitation), and (7) nickel recovery by Ni-
sulfate crystallization. With an estimated charge of approx. 200 $/ton
for the studied sludges (metal finishing industries) the hydrometallurgical recovery facility would compete with 100-150 $/ton of disposal

costs. However, since the facility would have substantial environmental merits, it was recommended that the State of Missouri take initiatives to encourage development in 3 directions: (i) to determine the treatibility of wastes and "delisting" potential of the residuals generated by the process, (ii) monitor the regulatory environment, especially concerning the possible restrictions on land disposal of metal-bearing wastes, and (iii) review possible subsidies or incentive scenarios which would make the facility cost competitive with landfills.

In the process design for siumultaneous "detoxification" of wastes and recovery of precious metals biology-related technologies will play an increasingly important role (e.g., Patterson & Passino, 1987). The biotechnological potential of microbial metal transformations - including oxidation, reduction, alkylation, dealkylation, solubilization and precipitation - has not been used as widely as in other areas of biology; this may be because relatively few investigators have worked in the area and our knowledge is incomplete in terms of range and mechanisms of metal transformations (Olson & Kelly, 1986). Biohydrometallurgy is moving rapidly toward full-scale application for metals extraction from ores (Parkinson, 1985). The technology uses bacteria (or occassionally fungi) to implement or enhance the leaching step; in addition, inactivated or nonviable microorganisms can be used as metal-sorbing surfaces (Calmano & Ahlf, 1988).

REFERENCES

Anderson, G.K., C.B. Saw, and D.P. Purdue. "Effects of Heavy Metals on Anaerobic Digestion of Landfill Leachate". Proc. Int. Conf. Heavy Metals in the Environment, Athens, Vol. 1, pp. 592-594. CEP Consultants Edinburgh 1985.

Anderson, M.A. "Kinetic and Equilibrium Control of Interfacial Reactions involving Inorganic Ionic Solutes and Hydrous Oxide Solids". In: F.E. Brinckmann, and R.H. Fish (Eds.), Environmental Speciation and Monitoring Needs for Trace Metal-Containing Substances from Energy-Related Processes, pp. 146-162 (Washington D.C.: Dept. of Commerce, 1981).

Anonymus. "Groundwater Strategies". Environ. Sci. Technol. 14: 1030-1035 (1980).

Anonymus. "Laser to Revover Gold from Toronto's Sewage". Northern Miner, 2. 12. 1985.

Austin, D.E.. and L.W. Newland. "Time-Resolved Leaching of Cadmium and Manganese from Lignite and Incineration Fly Ash". Chemosphere 14: 41-51 (1985).

74

Baccini, P., and P.H. Brunner. "Behandlung und Endlagerung von Rest-stoffen aus Kehrichtverbrennungsanlagen". Gas Wasser Abwasser 65: 403-409 (1985).

Baccini, P., G. Henseler, R. Figi, and H. Belevi. "Water and Element Balances of Municipal Solid Waste Landfills". Waste Management & Research 5: 483-499 (1987).

Ball, R.O., G.P. Verret, P.L. Buckingham, and S. Mahfood. "Economic Feasibility of a State-Wide Hydrometallurgical Recovery Facili-ty". In: Patterson, J.W., and R. Passino (Eds.) Metals Specia-tion, Separation and Recovery, p. 690-709. (Chelsea, Michigan: Lewis Publ. 1987).

Benjamin, M.M., and J.O. Leckie. "Multiple-Site Adsorption of Cd, Cu, Zn, and Pb on Amorphous Iron Oxyhydroxide". J. Colloid Interface Sci. 79: 209-211 (1981).

Benjamin, M.M., and J.O. Leckie. "Effects of Complexation by Cl, SO_4, S_2O_3 on Adsorption Behaviour of Cd on Oxide Surfaces". Environ. Sci. Technol. 16: 162-169 (1982).

Blakey, N.C. "Behavior of Arsenical Wastes Co-Disposed with Domestic Solid Wastes". J. Water Pollut. Control Fed. 56: 69-75 (1984).

Bowen, H.J.M. "Environmental Chemistry of the Elements". (London: Academic Press 1979).

Bowman, R.S., and G.A. O'Connor. "Control of Nickel and Strontium by Free Metal Ion Activity". Soil Sci. Soc. Amer. J. 46: 933-936 (1982).

Brannon, J.M., R.E. Hoeppel, and D. Gunnison. "Efficiency of Capping Contaminated Dredged Material". In: Dredging and Dredged Material Disposal, Vol 2. Proc. Conf. Dredging '84, pp. 664-673 (Clear-water Beach, Florida, 1984).

Brooks, R.R., B.J. Presley, and I.R. Kaplan. "Trace Elements in the Interstitial Waters of Marine Sediments". Geochim. Cosmochim. Acta 32: 397-414 (1968).

Brümmer, G., J. Gerth, and U. Herms. "Heavy Metal Species, Mobility and Availability in Soils". Z. Pflanzenernährung Bodenkunde 149: 382-398 (1986).

Brunner, P.H., and P. Baccini. "The Generation of Hazardous Waste by MSW-Incineration Calls for New Concepts in Thermal Waste Treat-ment". In: Second International Conference on New Frontiers for Hazardous Waste Management, Pittsburgh, Pa., Sept. 27-30, 1987.

Brunner, P.H., and J. Zobrist. "Die Müllverbrennung als Quelle von Metallen in der Umwelt". Müll und Abfall 9: 221-227 (1983).

Cahill, C.A., and L.W. Newland. "Comparitive Efficiencies of Trace Metal Extraction from Municipal Incinerator Ashes". Int. J. Envi-ronmental Anal. Chem. 11: 227-239 (1982).

Calmano, W., and W. Ahlf. "Bakterielle Laugung von Schwermetallen aus Baggerschlamm - Optimierung des Verfahrens im Labormaßstab". Wasser + Boden 1/1988: 30-32 (1988).

Calvet, R. "Adsorption-Desorption Phenomena". In: R.J. Hance (Ed.)
    Interactions between Herbicides and the Soil, p. 1-29 (London:
    Academic Press 1980).

Campbell, D.J.V., A. Parker, J.F. Rees, and C.A.M. Ross. "Attenuation
    of Potential Pollutants in Landfill Leachate by Lower Greensand".
    Waste Management & Research 1: 31-52 (1983).

Capuzzo, J.M., J.M. Neal, and R.K. Bastian. "Ecological and Human
    Health Criteria for Cross Ecosystem Comparison of Waste Disposal
    Impacts". In: G. Kullenberg (Ed.) Role of the Ocean as a Waste
    Disposal Option, pp. 347-360. Dordrecht: D. Reidel Publ., 1986).

Chen, Y.R., J.N. Butler, and W. Stumm. "Kinetic Study of Phosphate
    Reaction with Aluminum Oxide and Kaolinite". Envir. Sci. Technol.
    7: 327-332 (1973).

Chiou, C.T., L.J. Peters, and V.H. Freed. "A Physical Concept of Soil
    Water Equilibria for Nonionic Organic Compounds." Science 206:
    831-832 (1979).

Craig P.J., and P.A. Moreton. "The Role of Sulphide in the Formation of
    Dimethylmercury in River and Estuary Sediments. Mar. Poll. Bull.
    15: 406-408 (1984).

Davidson, J.M., and R.K. Chang. "Transport of Picloran in Relation to
    Soil Physical Conditions and Pore-Water-Velocity." Soil Sci. Soc.
    Amer. Proc. 36: 257-261 (1972).

De Groot, G.A., J. Wijkstra, D. Hoede, and H.A. Van der Sloot. "Leach-
    ing Characteristics of Hazardous Elements from Coal Fly Ash as a
    Function of the Acidity of the Contact Solution and the Liquid/-
    Solid Ratio". Presentation at 4th Int. Hazardous Waste Symposium
    on Environmental Aspects of Stabilization/Solidification of Ha-
    zardous and Radioactive Wastes, May 3-6, 1987, Atlanta/Georgia.

Di Toro, D.M., and L.M. Horzempa. "Reversible and Resistant Components
    of PCB Adsorption-Desorption: Isotherms". Environ. Sci. Technol.
    16: 594-602 (1982).

Domsch, K.H., G. Jagnow, and T.H. Anderson. "An Ecological Concept for
    the Assessment of Side-Effects of Agrochemicals on Soil Microor-
    ganisms". Residue Review 86: 65-105 (1983).

Dreesen, D.R., E.S. Gladney, J.W. Owens, B.L. Perkins, C.L. Wienke, and
    L.E. Wangen. "Comparison of Levels of Trace Elements Extracted
    from Fly Ash and Levels Found in Effluent Waters from a Coal-Fi-
    red Power Plant". Environ. Sci. Technol. 11: 1017-1019 (1977).

Duchart, P., S.E. Calvert, and N.B. Price. "Distribution of Trace
    Metals in the Pore Waters of Shallow Water Marine Sediments".
    Limnol. Oceanogr. 18: 605-610 (1973).

Ehrig, H.-J. "Quality and Quantity of Sanitary Landfill Leachate".
    Waste Management and Research 1: 53-68 (1983).

Farrington, J.W., and J. Westall. "Organic Chemical Pollutants in the
    Oceans and Groundwater: A Review of Fundamental Chemical Proper-
    ties and Biogeochemistry. In: G. Kullenberg (Ed.) Role of the

Ocean as a Waste Disposal Option, pp. 361-425. (Dordrecht: D. Reidel Publ., 1986).

Förstner, U. "Chemical Forms and Reactivities of Metals in Sediments". In: R. Leschber, R.D. Davis, and P. L'Hermite (Eds.) Chemical Methods for Assessing Bio-Available Metals in Sludges and Soils, pp. 1-30. (London: Elsevier Applied Science Publ. 1985).

Förstner, U.: "Chemical Forms and Environmental Effects of Critical Elements in Solid-Waste Materials - Combustion Residues". In: M. Bernhard, F.E. Brinckman, and P.J. Sadler (Eds.) The Importance of Chemical "Speciation" in Environmental Processes, pp. 465-491. Dahlem Konferenzen. (Berlin: Springer-Verlag 1986).

Förstner, U.: "Metal Speciation in Solid Wastes - Factors Affecting Mobility". In: L. Landner (Ed.) Speciation of Metals in Water, Sediment and Soil Systems". Lecture Notes in Earth Sciences 11: 13-41. (Berlin: Springer-Verlag 1987).

Förstner, U. "Analysis and Prognosis of Metal Mobility in Soils and Wastes." In: K. Wolf, W.J. Van Den Brink, and F.J. Colon (Eds.) Contaminated Soil '88, pp. 1-10 (Dordrecht: Kluwer Academic Publ. 1988).

Förstner, U., and J. Schoer. "Langzeitverhalten von Spurenelementen in den Altablagerungen der Deponie Niederwallach". Expertise for the Ministry of Environment, Regional Planning and Agriculture, Nordrhein-Westfalen. September 1987.

Förstner, U., U. Schöttler, and C. Nähle. "Sorption of Heavy Metals in Sand Filters in the Presence of Humic Acids". In: K.H. Schmidt (Ed.) Artificial Recharge. Proc. Intern. Symp. Dortmund, 1979. DVWK-Bulletin 13: 95-125 (1982).

Fuller, W.H., and A.W. Warrick. "Soils in Waste Treatment and Utilization". (Boca Raton/Fla.: CRC Press, 1985).

Garcia-Miragaya, J., and A.L. Page. "Influence of Ionic Strength and Inorganic Complex Formation on the Sorption of Trace Amounts of Cd by Montmorillonite." Soil Sci. Soc. Amer. J. 40: 658-663 (1976).

Greeenberg, R.R., G.E. Gordon, W.H. Zoller, R.B. Jacko, D.W. Neuendorf, and K.J. Yost. "Composition of Particles Emitted from the Nicosia Municipal Incinerator". Environ. Sci. Technol. 12: 1329-1332 (1978).

Hagenmaier, H., M. Kraft, H. Brunner, and R. Haag. "Calaytic Effects of Fly Ash from Waste Incineration Facilities on the Formation and Decomposition of Polychlorinated Dibenzo-p-dioxins and Polychlorinated Dibenzofurans". Environ. Sci. Technol. 21: 1080-1084 (1987a).

Hagenmaier, H., H. Brunner, R. Haag, and M. Kraft. "Copper-Catalyzed Dechlorination/Hydrogenation of Polychlorinated Dibenzo-p-dioxins, Polychlorinated Dibenzofurans, and Other Chlorinated Aromatic Compounds". Environ. Sci. Technol. 21: 1085-1088 (1987b).

Hahne, H.C.H., and W. Kroontje. "Significance of pH and Chloride Concentration on Behaviour of Heavy Metal Pollutants: Mercury(II),

Cadmium(II), Zinc(II) and Lead(II)". J. Environ. Qual. 2:44-50 (1973).

Ham, R.K., M.A. Anderson, R. Stanforth, and R. Stegmann. "Background Study on the Development of a Standard Leaching Test". EPA-600/2-79-109. (Cincinnati: U.S. Environmental Protection Agency, 1979).

Hämmerli-Wirth, H. "Die Behandlung und Ablagerung von Kehrichtschlacken und Filteraschen". Phoenix International 6/87: 11-17 (1987).

Harms, H., and Sauerbeck, D. "Toxic Organic Compounds in Town Waste Materials: Their Origin, Concentration and Turnover in Waste Composts, Soils and Plants". In: R.D. Davis, and G.Hucker (Eds.) Environmental Effects of Organic and Inorganic Contaminants in Sewage Sludge, pp. 38-51. (Dordrecht: D. Reidel Comp. 1983).

Harris, W.R., and D. Silberman. "Time-Dependent Leaching of Coal Fly Ash by Chelating Agents". Environ. Sci. Technol. 15: 139-145 (1983).

Hass, H.J. "Allround Encapsulation of Hazardous Waste in the Soil by Means of Grouting Gels and Sealing Walls Resistant to Aggressive Chemicals". In: J.W. Assink, and W.J. van den Brink (Eds.) Contaminated Soil, Proc. Int. Conf. Utrecht, pp. 867-869. (Dordrecht: Martinus Nijhoff Publ. 1986).

Helling, C.S., and B.C. Turner. "Pesticide Mobility: Determination by Soil Thin Layer Chromatography". Science 162: 562-563 (1968).

Hellmann, H. "Korngrößenverteilung und organische Spurenstoffe in Gewässersedimenten und Böden". Fresenius Z. Anal. Chem. 316: 286-289 (1983).

Herms, U., and G. Brümmer. "Löslichkeit von Schwermetallen in Siedlungsabfällen und Böden in Abhängigkeit von pH-Wert, Redoxbedingungen und Stoffbestand". Mitt. Deutsche Bodenkundl. Ges. 27: 23-34 (1978).

Herms, U., and G. Brümmer. "Einfluß der Bodenreaktion auf Löslichkeit und toleriebare Gesamtgehalte an Nickel, Kupfer, Zink, Cadmium und Blei in Böden und kompostierten Siedlungsabfällen". Landwirtschaftl. Forschung 33: 408-423 (1980).

Jacobs, L., and S. Emerson. "Trace Metal Solubility in an Anoxic Fjord". Earth Planet. Sci. Lett. 60: 237-252 (1982).

Jonasson, I.R. "Geochemistry of Sediment/Water Interactions of Metals, Including Observations on Availability". In: Shear, H., and A.E.P. Watson (Eds.) The Fluvial Transport of Sediment-Associated Nutrients and Contaminants. pp. 255-271. (Windsor, Ontario: IJC/-PLUARG 1977).

Karasek, F.W., G.M. Charbonneau, G.J. Reuel, and H.Y. Tong. "Determination of Organic Compounds Leached from Municipal Fly Ash by Water at Different pH Levels". Anal. Chem. 59: 1027-1031 (1987).

Karickhoff, S.W. "Semi-Emirical Estimation of Sorption of Hydrophobic Pollutants on Natural Sediments and Soils". Chemosphere 10: 833-846 (1981).

Karickhoff, S.W., D.S. Brown, and T. Scott. "Sorption of Hydrophobic Pollutants on Natural Sediments". Water Res. 13: 241-248 (1979).

Kersten, M. "Geochemistry of Priority Pollutants in Anoxic Sludges: Cadmium, Arsenic, Methyl Mercury, and Chlorinated Organics". In: Salomons, W., and U. Förstner (Eds.) Environmental Management of Solid Waste: Dredged Material and Mine Tailings. (Berlin: Springer-Verlag 1988).

Kester, D.R., B.H. Ketchum, I.W. Duedall, and P.K. Park (Eds.). "Wastes in the Ocean". Vol. 2: Dredged-Material Disposal in the Ocean. 299 p. (New York: Wiley, 1983).

Keyser, T.R., D.S.F. Natusch, C.A. Evans, and R.W. Linton. "Characterizing the Surface of Environmental Particles". Environ. Sci. Technol. 12: 768-773 (1978).

Khan, S.U. "Physiochemical Processes Affecting Pesticides in Soil." In: S.U. Khan (Ed.) Pesticides in the Soil Environment, pp. 29-118 (Amsterdam: Elsevier Publ. Co. 1980).

Kinniburgh, D.G., and M.L. Jackson. "Adsorption of Mercury(II) by Iron Hydrous Oxide Gel." Soil Sci. Soc. Amer. J. 42: 45-47 (1978).

Kistler, R.C., F. Widmer, and P.H. Brunner. "Behavior of Chromium, Nikkel, Copper, Zink, Cadmium, Mercury, and Lead during the Pyrolysis of Sewage Sludge". Environ. Sci. Technol. 21: 704-708 (1987).

Klein, D.H., A.W. Andren, J.A. Carter, J.F. Emery, C. Feldman, W. Fulkerson, W.S. Lyon, J.C. Ogle, Y. Talmi, R.I. Van Hook, and N. Bolton. "Pathways of Thirty-Seven Trace Elements through Coal-Fired Power Plant". Environ. Sci. Technol. 10: 973-979 (1975).

Leckie, J.O., M. Benjamin, K. Hayes, G. Kaufman, and S. Altman. "Adsorption/Co-Precipitation of Trace Elements from Water with Iron Oxyhydroxide". Final Report EPRI RP-910 (Palo Alto/Calif.: Electric Power Research Institute 1980).

Lichtensteiger, Th., and P.H. Brunner. "Transformation of Sewage Sludge in Landfills" (In Prep., 1988).

Lion, L.W., R.S. Altman, and J.O. Leckie. "Trace-Metal Adsorption Characteristics of Estuarine Particulate Matter: Evaluation of Contribution of Fe/Mn Oxide and Organic Surface Coatings". Environ. Sci. Technol. 16: 660-666 (1982).

Mackay, D.M., D.L. Freyberg, M.N. Goltz, G.D. Hopkins, and P.V. Roberts. "A Field Experiment on Ground Water Transport of Halogenated Organic Solutes". Preprint Ext. Abstract. Div. Environ. Chemistry, American Chemical Society. 186th Nat. Meeting, Washington D.C. 1983. pp. 368-371.

Mackay, D.M., P.V. Roberts, and J.A. Cherry. "Transport of Organic Contaminants in Groundwater". Environ. Sci. Technol. 19: 384-392 (1985).

Maes, A., and A. Cremers. "Radionuclide Sorption in Soils and Sediments: Oxide-Organic Matter Competition". In: R.A. Bulman and J.R. Cooper (Eds.) Speciation of Fission and Activation Products in the Environment, pp. 93-100. (London: Elsevier Applied Science Publ. 1986).

Morton, R.W. "Capping" Procedures as an Alternative Technique to Iso-
      late Contaminated Dredged Material in the Marine Environment".
      In: Dredge Spoil Disposal and PCB Contamination: Hearings before
      the Committee on Merchant Marine and Fisheries. House of Repre-
      sentatives, Ninety-sixth Congress, 2nd Session, on Exploring the
      Various Aspects of Dumping of Dredged Spoil Material in the Ocean
      and the PCB Contamination Issue, March 14, May 21, 1980. USGPO
      Ser No 96-43, pp. 623-652, Washington DC.

Municipality of Rotterdam/Rijkswaterstaat. "Grootschalige locatie voor
      de berging van baggerspecie uit het benedenrivierengebied". Pro-
      jectreport/Environmental Compatibility Study, Oct. 1984, 334 pp.

Nissenbaum, A., and D.J. Swaine. "Organic Matter-Metal Interactions in
      Recent Sediments. The Role of Humic Substances". Geochim. Cosmo-
      chim. Acta 40: 809-816 (1976).

Olson, G.J., and R.M. Kelly. "Microbial Metal Transformations: Biotech-
      nological Applications and Potential". Biotechnology Progress 2:
      1-15 (1986).

Parkinson, G. "Metals Recovery Makes Big Splash in Canada". Chem. Engi-
      neering, Sept. 30:19-26 (1985).

Patrick, W.H., B.G. Williams, and J.T. Moraghan. "A Simple System for
      Controlling Redox Potential and pH in Soil Suspensions". Soil
      Sci. Soc. Amer. Proc. 37: 331-332 (1973).

Patterson, J.W. "Metals Separation and Recovery". In: Patterson, J.W.,
      and R. Passino (Eds.) Metals Speciation, Separation and Recovery,
      p. 63-93. (Chelsea, Michigan: Lewis Publ. 1987).

Patterson, J.W., and R. Passino (Eds.) Metals Speciation, Separation
      and Recovery. (Chelsea, Michigan: Lewis Publ. 1987).

Plant, J.A., and R. Raiswell. "Principles of Environmental Geochemi-
      stry. In: Thornton, I. (Ed.) Applied Environmental Geochemistry.
      pp. 1-39. (London: Academic Press 1983).

Poon, C.S. et al. "Assessing the Leaching Characteristics of Stabilized
      Toxic Waste by Use of Thin Layer Chromatography". Environ. Tech-
      nol. Letts. 5, 1-6 (1984).

Presley, B.J., Y. Kolodny, and I.R. Nissenbaum. "Early Diagenesis in a
      Reducing Fjord, Saanich Inlet, British Columbia. II. Trace Ele-
      ment Distribution in Interstitial Water and Sediment". Geochim.
      Cosmochim. Acta 36: 1073-1090 (1972).

Regan, R.W., and C.E. Draper. "Segregated Landfilling of Metal Finish-
      ing Sludge: Concept Evaluation Studies". Proc. Int. Conf. Heavy
      Metals in the Environment, New Orleans. pp. 245-247. (Edinburgh:
      CEP Consultants 1987).

Reinhard, M., N.L. Goodman, and J.F. Parker. "Occurrence and Distri-
      bution of Organic Chemicals in Two Landfill Leachate Plumes".
      Environ. Sci. Technol. 18: 953-961 (1984).

Roberts, P.V., J. Schreiner, and G.D. Hopkins. "Field Study of Organic
      Water Quality Changes during Groundwater Recharge in the Palo
      Alto Baylands". Water Res. 16: 1025-1037 (1982).

Sahm, H., M. Brunner, and S.M. Schobert."Anaerobic Degradation of Halo-
    genated Aromatic Compounds". Microbial. Ecol. 12: 147-153 (1986).

Salomons, W. "Sediments and Water Quality". Environ. Technol. Letts. 6:
    315-368 (1985).

Salomons, W., and U. Förstner (Eds.) "Environmental Management of Solid
    Waste: Dredged Material and Mine Tailings". (Berlin: Springer-
    Verlag 1988).

Sauerbeck, D., and P. Styperek "Evaluation of Chemical Methods for As-
    sessing the Cd and Zn Availability from Different Soils and Sour-
    ces." In: R. Leschber et al. (Eds.) Chemical Methods for Asses-
    sing Bio-Available Metals in Sludges and Soils, pp. 49-66. Else-
    vier Applied Science Publ. 1985.

Schellenberg, K., C. Leuenberger, and R.P. Schwarzenbach. "Sorption of
    Chlorinated Phenols by Natural Sediments and Aquifer Materials".
    Environ. Sci. Technol. 18, 652-657 (1984).

Schoer, J., and U. Förstner. "Chemical Forms of Artificial Radionucli-
    des and their Stable Counterparts in Sediments". Proc. First Int.
    Conf. Environmental Contamination, London, July 1984, pp. 738-745
    (1984).

Schoer, J., and U. Förstner. "Abschätzung der Langzeitbelastung von
    Grundwasser durch die Ablagerung metallhaltiger Feststoffe". Z.
    Vom Wasser 69: 23-32 (1987).

Schwarzenbach, R.P., and J. Westall. "Transport of Nonpolar Organic
    Compounds from Surface Water to Groundwater. Laboratory Sorption
    Studies". Environ. Sci. Technol. 15: 1360-1364 (1981).

Schwarzenbach, R.P., W. Giger, E. Höhn, and J.K. Schneider. "Behavior
    of Organic Compounds during Infiltration of River Water to
    Groundwater: Field Studies". Environ. Sci. Technol. 17: 472-479
    (1983).

Singer, P.C., and W. Stumm. "Acidic Mine Drainage: The Rate-Determining
    Step". Science 167: 1171-1173 (1970).

Stumm, W., and J.J. Morgan. "Aquatic Chemistry". 2nd Edition. (New
    York: Wiley 1981).

Tessier, A., P.G.C. Campbell, and M. Bisson. "Sequential Extraction
    Procedure for the Speciation of Particulate Trace Metals." Anal.
    Chem. 51: 844-851 (1979).

Theis, T.L., and L.E. Padgett. "Factors Affecting the Release of Tra-
    ce Metals from Municipal Sludge Ashes". J. Water. Pollut. Control
    Fed. 55, 1271-1279 (1983).

Theis, T.L., J.D. Westrick, C.L. Hsu, and J.J. Marley. "Field Investi-
    gation of Trace Metals in Groundwater from Fly Ash Disposal". J.
    Water Pollut. Control Fed. 50: 2457-2469 (1978).

Turner, R.R. "Oxidation State of Arsenic in Coal Ash Leachate". Envi-
    ron. Sci. Technol. 15: 1062-1066 (1981).

Turner, R.R., P. Lowry, M. Levin, S.E. Lindberg, and T. Tamura. "Leach-ability and Aqueous Speciation of Selected Trace Constituents of Coal Fly Ash". Final Report, Research Project 1061-1/EA-2588. (Palo Alto: Electric Power Research Inst. 1982).

Van der Sloot, H.A., O. Piepers, and A. Kok. "A Standard Leaching Test for Combustion Residues". Studiegroep Ontwikkeling Standaard Uit-loogtesten Verbrandingsresiduen (SOSUV). BEOP-31. June 1984.

Van der Sloot, H.A., J. Wijkstra, and G.J. De Groot. "Contaminant Dif-fusion in Sediments, Soil and Waste Materials". In: K. Wolf, W.J. Van Den Brink, and F.J. Colon (Eds.) Contaminated Soil '88, pp. 71-74 (Dordrecht: Kluwer Academic Publ. 1988).

Van Genuchten, M. Th., J.M. Davidson, and P.J. Wierenga. "An Evaluation of Kinetic and Equilibrium Equations for Predicting Pesticide Movement through Porous Media." Soil Sci. Soc. Amer. Proc. 38: 29-35 (1974).

Vogel, T.M., and P.L. McCarty. "Biotransformation of Tetrachloroethy-lene, Dichloroethylene, Vinyl Chloride, and Carbon Dioxide under Methanogenic Conditions. Appl. Environ. Microbiol. 49: 1080-1083 (1985).

Wadge, A., and M. Hutton. "The Leachability and Chemical Speciation of Selected Trace Elements in Fly Ash from Coal Combustion and Refu-se Incineration". Environ. Pollut. 48: 85-99 (1987).

Walker, A., and D.V. Crawford. "The Role of Organic Matter in Adsorp-tion of the Triazine Herbicides by Soils". In: Isotopes and Radi-ation in Soil Organic Matter Studies. Proc. 2nd Symp. Intern. Atomic Energy Agency, pp. 91-108 (1968).

Westall, J., C. Leuenberger, and R.P. Schwarzenbach. "Influence of pH and Ionic Strength on the Aqueous-Nonaqueous Distribution of Chlorinated Phenols". Environ. Sci. Technol. 19: 193-198 (1985).

Wienberg, R. "Assessment of Mobility and Transport of Organic Contami-nants through Mineralic Containment Materials. In: K. Wolf, W.J. Van Den Brink, and F.J. Colon (Eds.) Contaminated Soil '88, pp. 625-629 (Dordrecht: Kluwer Academic Publ. 1988).

Wienberg, R., and E. Heinze. "Spread of Organic Contaminants in Reme-dial Construction Materials as Effected by Dissolving Agents and Surfactants". In: J.N. Lester, R. Perry and R.M. Sterritt (Eds.) Chemicals in the Environment, Proc. Intern. Conf. Lisbon, pp. 733-744 (London: Selper Ltd. 1986).

Wienberg, R., E. Heinze, and U. Förstner. "Experiments on Specific Retardation of Some Organic Contaminants by Slurry Trench Mate-rials". In: J.W. Assink, and W.J. van den Brink (Eds.) Contamina-ted Soil, Proc. Inter. Conf. Utrecht, pp. 849-857. (Dordrecht: Martinus Nijhoff Publ. 1986).

Wilson, J.T., C.G. Enfield, W.J. Dunlap, R.L. Cosby, D.A. Foster, and L.B. Baskin. "Transport and Fate of Selected Organic Pollutants in a Sandy Soil." J. Environ. Qual. 10: 501-506 (1981).

Zierath, D., J.J. Hassett, W.L. Banwart, S.G. Wood, and J.C. Means. "Sorption of Benzidine by Sediments and Soils". Soil Sci. 129: 277-287 (1980).

WATER AND ELEMENT BALANCES
OF LANDFILLS

H.-J. Ehrig

ITW-Ingenieurberatung GmbH
Friedrich-Kaiser-Str. 23
D-5860 Iserlohn

Abstract

Municipal solid waste landfills represent an accumulation of material with less or more pollution potential. A part of this material is transfered to the environment by gas and leachate. To judge the environmental risk of landfills it is necessary to make statements about future emission streams. Today such statements are impossible but some less or more realistic estimations could be made. Results from different investigations are combined to extrapolate measured element transfer from an observed period into the future.

Conclusions:
- A decrease of leachate quantity could not be estimated. This is only possible with additional top sealing systems.
- The most important emission stream over a long time is leachte. For many elements the period until concentrations below todays limiting values is some hundred years and more. The combination of decreasing concentration slopes with time and decreasing pollution limits could multiply the period of environmental risks.
- The knowledge of processes in landfills in detail is too small. Therefore the presented results are only hypothesis without possibility of verification in an imaginable time.
- A final storage quality defined as near zero pollution potential of solids and pollution stream of liquids cannot reached during the next centuries. The possibility of such defination should be verified under the view of whole mass transport in the environment.

1. Introduction

Municipal solid waste landfills are a mixture of different organic and anorganic compounds which represents an important but local limited pollution potential. Experiences during the last decades have shown that parts of this potential transfered to the environmental by gas and leachate and thereby changed to a realistic pollution.

Leachate as liquid emission is the most important risk for the environment. Measurements of quality and quantity at different landfills have shown that leachate is a complex and high polluted wastewater.

Considering the limited mass potential of each compound in a landfill it could be estimated that the transfer rate must be decrease versus time. Today the most serious problem of estimating the time of transfer rates above a given limit is the knowledge gap of biological, chemical and physical processes in landfills. Most of these processes could be described in general. But transfered to the realitiy of a landfill and reaction times of at least some hundred years the complex combinations of all processes decreased to a black box.

This paper should give some indications for water and element balances estimated from a lot of laboratory, pilot and full scale experiments.

2. Basic investigations and methods

The presented data and results based on different research works which were performed at the Institut für Siedlungswasserwirtschaft of the Technical University of Braunschweig:

- Measurement of leachate quality and quantity at 15 different landfills over 1 to 5 years with weekly sampling. Comparing the effects of different operation conditions on leachate quality and quantity (1).
- Measurement of leachate quality and quantity at 8 test cells (diameter = 5 m; high = 6 m; 2 m layers) with different operation conditions. Leachate quality was analysed once a week below each layer. Refuse was analysed before dumping and also at 4 test cells after 5 years. The other 4 test cells are still operated (2).
- Operations of small test cells (volume 120 l) under optimised conditions as temperature of 30°C and water content of 65 %. Test of diffe-

rent operation conditions and domestic waste compositions (more   than 70 test runs over 100 to 400 days) (3), (4), (5).

This report is an effort to combine results from these investigations to estimate water and element balances of landfills. Basic datas are the analyses of leachate quality and measurement of flow rates at different landfills. The age of observed landfills was between 0 and 12 years. Today the oldest of these landfills have an age of up to 20 years and leachate are only analysed in large intervals of time. But it is extremly difficult to estimate future tendencies from 20 year long time series.

On the other side decreasing tendencies of most leachate components could be observed during laboratory scale experiments with optimised conditions over some hundred days. With a standardisation of both time intervals a combination and a better estimation should be possible.

3. Water balances

The water balances could be only observed at landfills which are sealed to surrounding strata. Fig. 1 presents leachate flow datas from different landfills as percentage of precipitation (operated and re-clamated). With exception of 2 columns all values are lower than 25 % of precipitation (average 772 mm). If also two special cases with flow rates lower than 10 % excluded (very young landfills) the average percentage will be 17.9 %. For following calculations this value is rounded to 20 %. The long term evaporation could be adopt from other measurements of groundwater production (Table 1).

Table 1: Evaporation with different vegetation covers
        as percentage of precipitation (6)

	% of precipitation average	range
without vegetation	30	18-55
crop land	40	30-55
gras land	60	42-87
wood land	65	45-95

Only a soil cover of wood could increase average evaporation up to 60 to 70 %. Consider that in most cases reclamated landfills are not an ideal subsoil for good growth of trees these values should be the

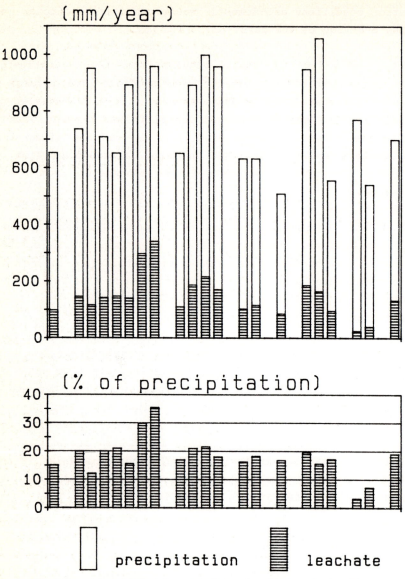

Fig. 1: precipitation (mm/year) and leachate flow
(mm/year and % of precipitation) at different
landfills and years

maximum. Another factor which could influence infiltration is surface
run off. But increasing run off increases also soil erosion and must
be prevented.

Fig 2 shows the water balance of a refuse columm in principle. The input to the column is only precipitation reduced by evaporation. In opposite to soils the capilarity of refuse is very low which prevented
an upflow of infiltrated water to evaporate.

Fresh dumped refuse in the FRG has an average water content of approximately 30 % wwt. The storage capacity is much higher (450-600 l/t
dry refuse = 39,1 - 46,1 % wwt) but decreases with age of landfill.
Today it is not possible to measure the dynamic process of fill up
storage capacity at such an unhomogenous system. But an end should be
sure. In Fig. 3 leachate flow rates of a reclamated landfill are presented. It could be seen that leachate flow increases sharply after
150 weeks of measurement and remains on this level with greater variation (values >25% in Fig. 1). The evaporation was not measured but
estimated to 40 to 60 % precipitation. As the time as the leachte
rates increases the additional storage could be in the range of 145 to
275 l/t dry refuse which could indicate the end of storage capacity.

Another process of water consumption is the anaerobic biological process and the water transport by warm gases (Fig. 4b). During times
with high gas production the water consumption could reach some ten
per cent of leachate flow. But the values decrease quickly to an unim-
portant level for the water balance. Fig. 4a shows the operation conditions of the refuse column for calculations of gas production, water
consumption etc. (This column is also the basis for calculations in
the following chapters).

Fill up of storage capacity and water consumption of biological process in essential manner decrease both after some years or decades. As
result the long termed leachate flow increase. Thereafter the decrease
of leachate emissions by flow reduction could not be estimated.

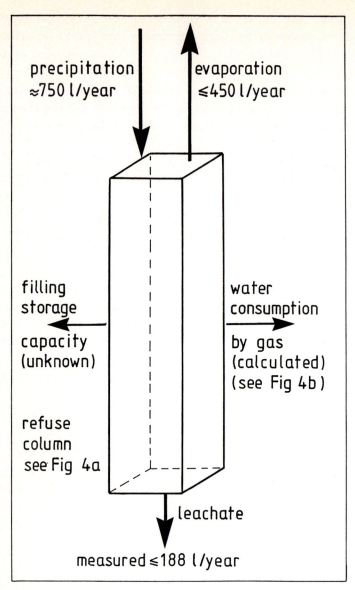

precipitation
≈750 l/year

evaporation
≤450 l/year

filling
storage

capacity
(unknown)

refuse
column
see Fig 4a

water
consumption

by gas
(calculated)
(see Fig 4b)

leachate

measured ≤188 l/year

Fig. 2:   water balance of a refuse column in principle

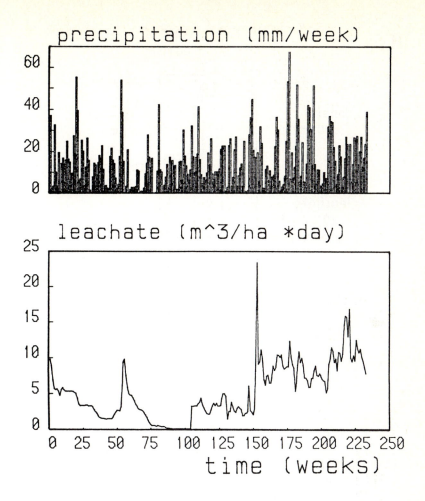

<u>Fig. 3:</u> measured precipitation and leachate flow at an
operated and thereafter reclamated landfill

a

year of
dumping

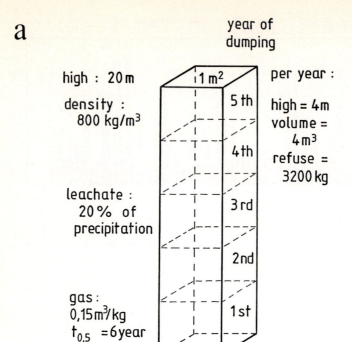

high : 20 m

density :
800 kg/m³

leachate :
20 % of
precipitation

gas:
0,15 m³/kg
$t_{0,5}$ = 6 year

1 m²

5 th

4 th

3 rd

2 nd

1 st

per year :

high = 4 m
volume =
4 m³
refuse =
3200 kg

b

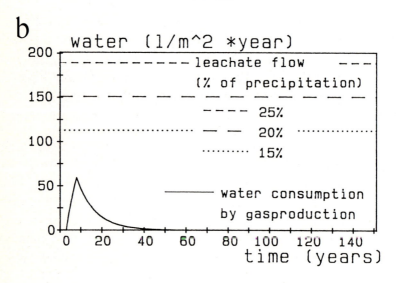

water (l/m^2 *year)

leachate flow
(% of precipitation)

---- 25%
— — 20%
········· 15%

——— water consumption
by gasproduction

time (years)

Fig. 4: a) used refuse column for all calculations in
this report
b) calculated water consumption by gas production (refuse column: see Fig. 4a)

4. Element balances

4.1 Basic data

To estimate element balances of landfills it is necessary to measure data of element potential and element transfer rates.

The element potential is the analyses of elemental refuse composition As consequence of the unhomogeneouse refuse structure it is extremly difficult to get comparable values. This report deals with different published datas (see Table 2). (10) reported that is was not possible to gather all metal loaded fractions for analysis. Real values can be much higher than published. The estimated percentage of the whole metal load is also given in table 2.

With exception of carbon the most important transfer is the element flow by leachate. Therefore it is necessary to give some data of leachate composition and of influences on the concentrations.

An important fact is the change of organics (BOD_5, COD) with the change from the acetic to the methanogenic phase in landfills (see Fig. 5). The average trend of BOD_5 is marked with "1" for landfills build up in 2-m layers with 2-4 m per year. A build up of more than 4 m per year (marked with "2") increases the BOD_5 and COD-concentrations and delays the change of microbiological conditions in landfills. On the other hand a slower build up and other enhancement technics decreases the organic contents (marked with "3") with an earlier change to the methanogenic phase. Besides, the absolute values of BOD_5 and COD the relationship BOD_5/COD is an important factor. Values greater than 0.4 during the acetic phase, indicates a good biodegradability. With the change to the methanogenic phase, the BOD_5 to COD-ratio decreases below 0,1, with low biodegradability of organic contents. High organic acid contents during acetic phase results in decreasing pH-values with increasing solubility of some anorganic substances (see table 3). The ammonium concentrations show only a low increase during the first years of operation and further relative constant values. An additional nitrogen component is organic nitrogen in the range of 30-150 % (average = 70 %) of ammonium.

In table 3 the analysis divided into two groups. The first group represents values during acetic phase with high organic loadings and low

Table 2: Analysis of solid waste

	(10)	(10)*	(3)	(12)	(14)	(16)
	g/kg		g/kg	g/kg	g/kg	g/kg
C	244		280/315	290	430	310-380
N			9,6	4	6	7- 10
P			3,3	1		
Cl				7,2	5	1,1-4,8
org.Cl	3,47					

			(15)		(13)	
	mg/kg		mg/kg	mg/kg	mg/kg	mg/kg
Ni					15	
Cr			2060-2810		28	5-100
Cu	238	305	411- 532	400	400	120-210
Pb	399	2660	210- 370	400	210	110-330
Zn	521	1628	588- 742	1200	1200	300-1000
Cd	3,48	8,7	40- 50	11	3,5	2- 14
Hg	0,64	1,6	0,3-0,4	4	1,1	1- 14

* from (10) calculated values for total domestic solid waste

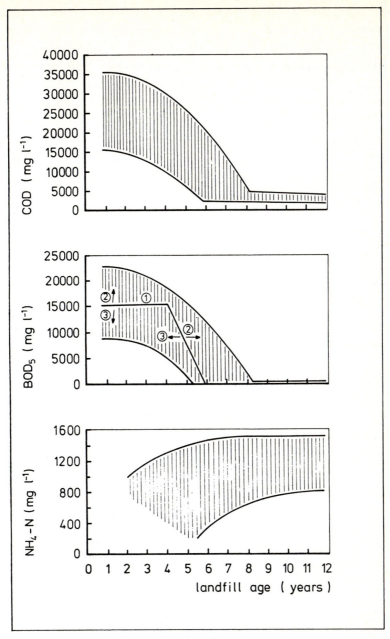

Fig. 5: general tendencies of COD, BOD_5 and NH_4 versus time

pH-values. The second group shows values from methanogenic phase with low biodegradable organics and higher pH-values. Parameters in table 4 don't differ between phases. At all landfills with an age between 0 and 12 years no time slope occurs. Additional the distribution of metals is presented in Fig. 6. The large variation of metal analysis is the result of differences between landfills and time depending randomised variation at each landfill. Until now differences between landfills cannot be explained by operation conditions etc.

Table 3: Leachate analysis
(parameters with differences between acetic and methanogenic phase)

	average	range
acetic phase		
pH (-)	6,1	4,5 - 7,5
BOD5 (mg/l)	13000	4000-40000
COD (mg/l)	22000	6000-60000
BOD5/COD (-)	0,58	
SO4 (mg/l)	500	70- 1750
Ca (mg/l)	1200	10- 2500
Mg (mg/l)	470	50- 1150
Fe (mg/l)	780	20- 2100
Mn (mg/l)	25	0,3- 65
Zn (mg/l)	5	0,1- 120
methanogenic phase		
pH (-)	8	7,5- 9
BOD5 (mg/l)	180	20- 550
COD (mg/l)	3000	500- 4500
BOD5/COD (-)	0,06	-
SO4 (mg/l)	80	10- 420
Ca (mg/l)	60	20- 600
Mg (mg/l)	180	40- 350
Fe (mg/l)	15	3- 280
Mn (mg/l)	0,7	0,03- 45
Zn (mg/l)	0,6	0,03- 4

From the presented data can be seen that some concentrations change between acetic and methanogenic phase. But this is only a short transition time of 1 to 2 years in the first decade of a landfill age. With exception of COD (including organic carbon) no long-term time slope could be observed from analyses of leachate from operated landfills. Differences in average concentrations are the result of the specific landfill conditions which often could not be explained in detail.

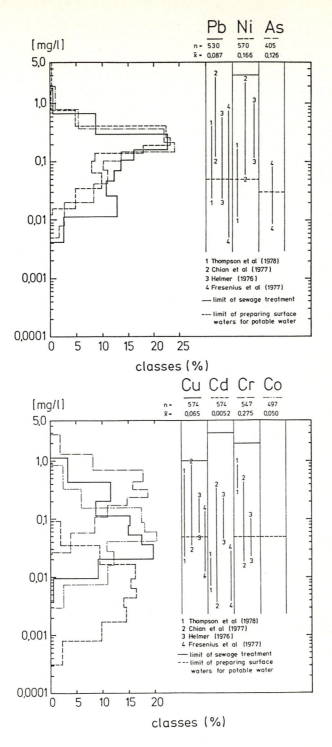

Fig. 6: distribution of metal analysis of different landfill leachates

Table 4: Leachate analysis
(no differences between phases could be
observed)

	average	range
Cl (mg/l)	2100	100- 5000
Na (mg/l)	1350	50- 4000
K (mg/l)	1100	10- 2500
alkalinity		
(mg $CaCO_3$/l)	6700	300-11500
NH_4 (mg N/l)	750	30- 3000
orgN (mg n/l)	600	10- 4250
total N (mgN/l)	1250	50- 5000
NO_3 (mg N/l)	3	0,1- 50
NO_2 (mg N/l)	0,5	0- 25
total P (mg P/l)	6	0,1- 30
AOX (μg Cl/l)*	2000	320- 3500
As (μg/l)	160	5- 1600
Cd (μg/l)	6	0,5- 140
Co (μg/l)	55	4- 950
Ni (μg/l)	200	20- 2050
Pb (μg/l)	90	8- 1020
Cr (μg/l)	300	30- 1600
Cu (μg/l)	80	4- 1400
Hg (μg/l)	10	0,2- 50

* adsorbable organic halogen

For full scale landfills it is not possible to enlarge observation period in a short time. It could be helpful to shorten the reaction time
in laboratory scale experiments under specific conditions to transfer
degradation or transport functions on full scale landfills to estimate
long term tendencies. Comparing leachate analysis from full scale
landfills (Fig. 7) with laboratory scale test cells (Fig. 8) shows for
both the change from acetic to methanogenic phase but after different
periods. As the concentrations of chloride and ammonium from full
scale are mostly constant the values from laboratory scale show a
significant decrease. It is necessary to find the connection between
both time scales eventually different for biological and chemical-phy-
sical processes. Exact values of gas production and the time depending
production function could be only gathered in lab. scale experiments.
Comparing specific leachate fix point from lab. to full scale could
help to transfer also gas production function on full scale landfills.

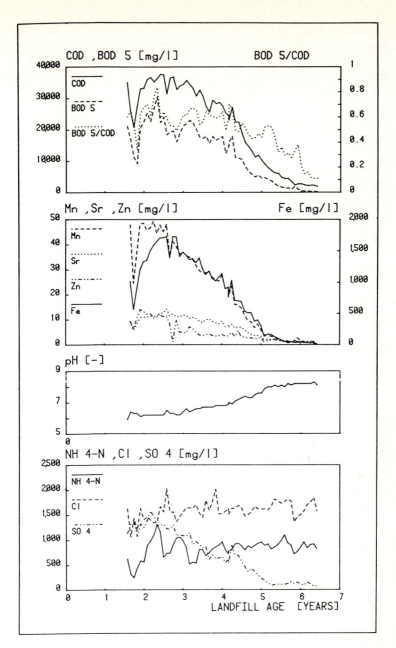

Fig. 7: Leachate analysis and time slope of a full
scale landfill

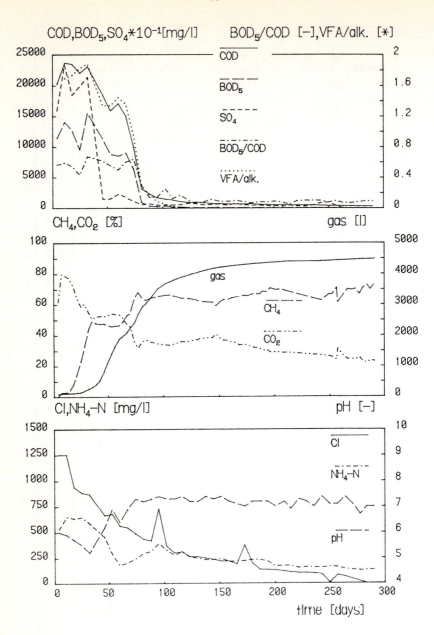

Fig. 8: Leachate and gas analysis and time slope of a laboratory test cell

4.2 Carbon

The carbon transfer from landfills to the environment is the most complex transfer system:

- Organic carbon in refuse is not a homogeneous fraction. There are extremly differences between carbon in kitchen waste and greens on one side and in plastics or lignin on the other side. Until now decomposition rates of plastic and lignin also under landfill conditions mostly unknown. Table 5 presents the carbon contents of different refuse fraction and the effort to estimate a transferable part. These values could be correct for a imaginable time. But what happens after this time? Observations at many organic rich geological formation cannot give a sure answer.
- Many of the organics in refuse are changed by biological, chemical or physical processes or by a combination of some or all to transfer the material in soluble conditions. These processes are known in principle but the complex acting together is well unknown.
- Organic carbon in leachates is the result of biological processes. That is right for the short observed landfill age. But sure there are also other processes as solubility etc. which effects are covered by the todays predominate biological processes.
- On the first view the carbon transfer by gas is a very simple way which could be easily calculated. But the whole gas production is only estimated today with differences between publishers of ± 50 %. All of these estimations are only used for a prediction of some decades. Statements for the long term biological processes are not possible.

Fig 9a presents COD data of operated landfills. Organic carbon was only measured in greater invervals. But it was possible to get a correlation between COD and TOC. It could be seen that after change to the methanogenic phase most values change to a more homogenous level of 3000 to 4000 mg COD/l. The level of methanogenic phase leachate shows also a slight time slope which could be extrapolated (Fig. 9b). Using a COD/TOC-relationship during acetic phase of 2.9 and during methanogenic phase of 2.15 the carbon transfer from landfills by leachate could be calculated.

For carbon transport dynamics it is necessary to define a typical landfill system (see Fig. 4a) presented with results of gas production in Fig. 10a. Fig. 10b shows the additional carbon transfer as TOC in

<u>Table 5:</u> Carbon contents and transfer

fraction	carbon potential (g/kg fraction)	(g/kg refuse wwt.)	carbon* transfer potential (%)	(g/kg refuse wwt.)
Paper/ cardboard	399,7	36,2	50	18,1
plastic	541,9	15,9	0	0
kitchen waste	311,9	62,9	100	62,9
fine fraction	192,3	7,8	50	3,9
residues	-	48,2	50	24,1
total refuse	246,9	171	63,7	109

calculated from (10) * only estimated

			**	
mixed solid waste	353	247	average 97	
			range 62-131	

data from (4)

** measured elimination rates with laboratory scale experiments

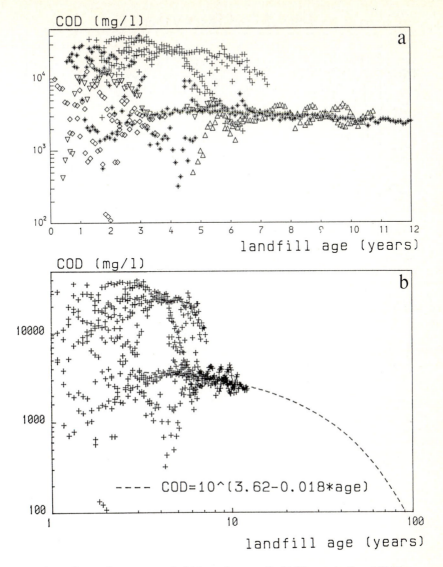

Fig. 9: a) measured COD-values of different landfill
leachates
b) COD-extrapolation of landfill leachates
(calculated from Fig. 9a)

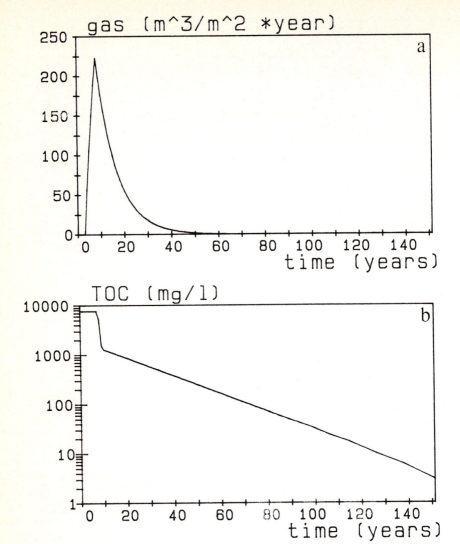

Fig. 10: a) gas transfer from refuse (column see Fig. 4a)
b) TOC-concentrations in leachates
(calculated from Fig. 9b)

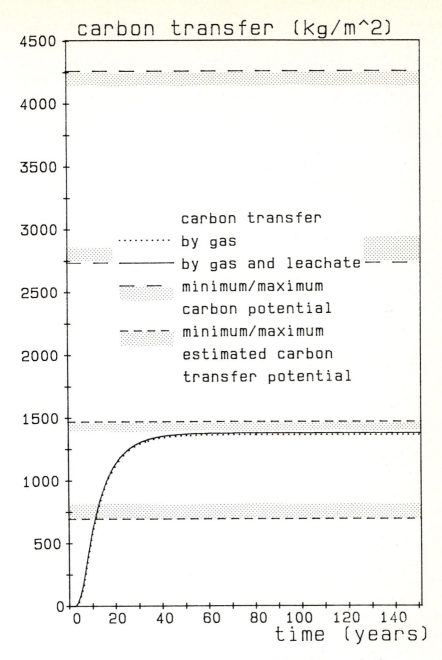

<u>Fig. 11:</u> carbon transfer by gas and leachate and the
potential or estimated transfer potential

leachates (calculated from Fig. 9b). The results of carbon transfer by gas and leachate over the first 150 years are shown in Fig. 11. The sum after 150 years is in the range of the transferable values in table 5. The following transfer rate by leachate as well as by gas could not be calculated but could be estimated as relatively low.

4.3 Chloride

In opposite to carbon transfer the transfer of chlorides based mostly on chemical processes but influenced by biological processes. Fig. 12 presents the comparison of chloride concentrations from aerobic and anaerobic operated test cells and also the low increase of chloride concentrations during the first years of operation at a full scale landfill. It could be seen that with increasing biological decomposition the solubility of chloride also increases. Therefore the offer of soluble chloride is limited by biological processes.

For the estimation of transfer rates two statements are necesarry:

- Observations at landfills with a maximum age of 20 years show no decrease of chloride concentrations.
- The chloride potential of landfills as batch reactors is limited and therfore a decrease of concentrations must occur.

To close these differences we used the results of laboratory scale experiments which are shown in Fig. 13a. But in opposite to real landfills the operation conditions are much more optimised. One connection point was the water regime of both with two possible dimensions:
- water contact per area unit
- water contact per volume unit.

The water contact per area unit could be excluded because the relationship was only 1 day lab. scale - 6,5 days full scale. With this relationship the differences between both cannot be explained. Using water contact per volume unit results in very much higher relationships: 1 day lab. scale = 1,16 years full scale basis for chloride and nitrogen calculations in Fig. 13b and 14b. The best fit of chloride

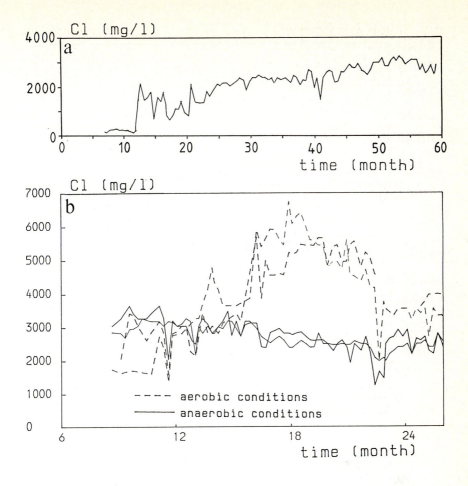

Fig. 12: a) increase of leachate chloride content
during the first years of operation at a
full scale landfill
b) comparison of aerobic or anaerobic effects
on chloride concentrations in leachates

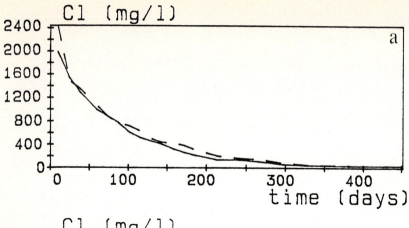

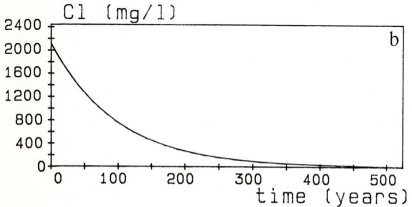

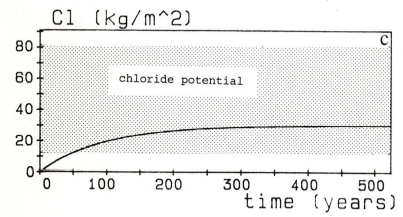

Fig. 13: a) measured chloride slope from lab. scale
test cells
b) estimated Cl-concentrations of a full scale
landfill (after Fig. 13a)
c) chloride transfer (after Fig. 13b and refuse
column Fig. 4b)

concentrations was an exponential function (Fig. 13a and b)

$$C = C_0 \cdot e^{(-k*t)}$$
t = time in years
k = coefficient = $\log_e 0.5/t_{0,5}$
$t_{0,5}$ = time of $C_0/2$
C_0 = initial concentration
= average chloride concentration of 2100 mg/l
C = concentration at time t.

The calculated values for $t_{0,5}$ were between 52 and 55 d (= 60 to 64 years). Fig. 13b and c presents the calculated concentration function and the sum of transfered chloride with leachate flow (20 % of precipitation; refuse column see Fig. 4a). The calculated transfer function only rises up in the lower part of the chloride potential area (calculated from Table 2). Not transferable chlorides (e.g. in plastics) or organic chlorides (up to 25%) could be an explanation for such differences. But also an increase of transferable potential as result of different biological, physical and chemical processes in future could not be excluded. The result would be longer $t_{0,5}$-values with an increase of chloride transfer. In opposite to leachate measurements chloride concentrations in Fig. 13b show a decrease also in the first years of operation. The constant values at operating land-fills could be the result of the semi-continuous refuse adding.

4.4 Nitrogen

The nitrogen transfer is a combination of carbon and chloride trans-port processes because most of nitrogen must be converted from organic nitrogen to ammonia by biological processes. As result the tendencies of ammonium concentration in Fig. 14a are much less exact as for chlo-ride. Using the same calculation procedure as for chloride we have got half-life values of 75 to 82 years. Similar to chloride the calculated concentration values are shown in Fig. 14b and the transfer function in Fig. 14c. Differences to measured nitrogen potential are much grea-ter than for chloride. That could be a result of the much less uniform composition (chemical forms of nitrogen) and thereafter different solubility conditions of nitrogen in refuse.

But similar to chloride discussion the knowledge gap is too large to explain such differences nearly exact.

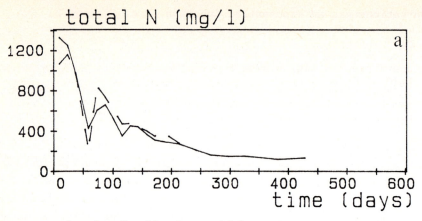

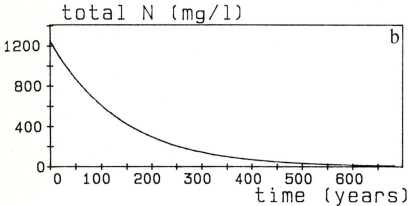

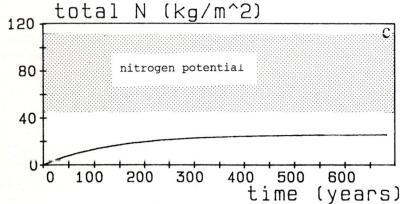

Fig. 14: a) measured nitrogen slope from lab. scale
         test cells
         b) estimated N-concentrations of a full scale
         landfill (after Fig. 14a)
         c) nitrogen transfer (after Fig. 14b and refuse
         column Fig. 4b)

4.5 Halogens and metals

The principle transfer functions must be nearly the same as for other compounds. But the relationship between measured initial concentrations and the potential in landfills is much higher. That means longer half-times must be estimated. Fig. 15 shows in principle organic chloride in refuse and transport values using the known concentrations of halogens in gas (average value 30 mg total organic halogen as chloride per m^3) and leachate (average value 2 mg adsorbable organic halogen as chloride per l).

Similar relationships are shown for some metals in table 6.

For halogens and for metals very long emission time with concentrations near initial concentrations must be expected.

Table 6: Metal contents of refuse and transfer rates

parameter	contents * mg/t dry refuse	transfer rate ** mg/(t dry refuse year)
Ni	≈ 15000	2,7
Cr	5000-100000	4,0
Cu	238000	1,1
Pb	399000	1,2
Zn	521000	8,0
Cd	3480	0,08
Hg	640	0,13

* Table 2 column 1, 5 and 6
** refuse column see Fig. 4a

5. Discussion

Chapter 4 presents some considerations of element transfer from landfills to the environment. Although the results base on measured datas they could only be hypothesis which could only be verified by better understanding of processes in landfills and much more by observation time. But overall it must be concluded that transfer by leachates occur over a long time (some hundred years) and concentrations only decrease very slowly.

Many technical influences on biological processes (enhancement etc.) are helpful for actual handling with landfills but the long term

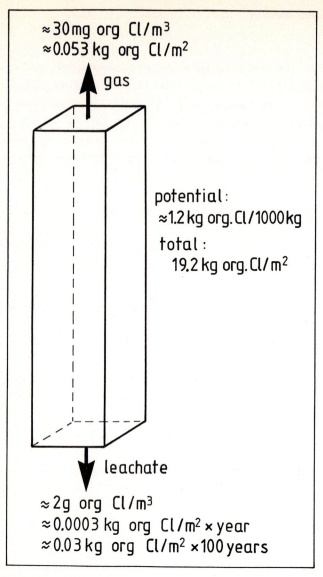

$\approx 30\,$mg org Cl/m^3
$\approx 0.053\,$kg org Cl/m^2

gas

potential:
$\approx 1.2\,$kg org. Cl/1000 kg

total:
$19.2\,$kg org. Cl/m^2

leachate

$\approx 2\,$g org Cl/m^3
$\approx 0.0003\,$kg org Cl/m$^2 \times$ year
$\approx 0.03\,$kg org Cl/m$^2 \times 100\,$years

Fig. 15: organic chloride potential of a refuse column
and measured transfer rates of gas and leachate

effects are very low. The aim of enhancement technic is to intensify gas production and therefore reduce leachate pollution during acetic phase. But this is only a change of flow medium from leachate to gas. Until now no effect on the long term emissions is known.

As an example intensifying biological processes - reducing half-time of gas production from 10 to 3 years - reduces the time were 99 % of gas was produced from 66.4 to 19.9 years (70 % reduction). This is a first-rate result for technical process but for the here discussed long period of time it is only of secondary importance.

On the other side biological processes have a great influence on all kinds of transfer processes, because it is the main process to destroy the structure of solid material and therefore increase the change of water and element contact. Overall as earlier the structure of solids destroyed and as more water passes the refuse in an uniform manner as higher initial concentrations and as shorter leaching times could be estimated. But this are hypothesis without any exact knowledge of time. It is questionable whether an important reduction of time could be reached. Under this point of view a stronger and longer acetic phase could be more helpful than every kind of enhancement technic.

With a stronger decrease of element concentration in leachate as of element potential the transfer time can increase in many cases. That seems to be so with halogens and metals. This fact could lead to the following statement: elements with a low solubility or transfer rate must be excluded from landfills as far as possible.

But on the other hand nobody could answer the question: what happened to concentrations and transfer time, if the potential of an element was halved? Is the result really an easier handling of landfills? Until now no element are known which existence prolong the biological decomposition in the measured concentration or which increase emissions in general. The knowledge of detailed processes in landfills - as black box - excluded any statement of new xenobiotic and/or hazardous compounds production. Our experiments with pilot scale test cells with addition of several hazaradous compounds to refuse have shown that negative effects on biological processes are possible. But in any kind of negative effects the added amount of hazaradous was extremely high and tenth power higher than in landfills excluding hazaradous waste landfills. If a compound shall be excluded from landfills a result should be an easier handling of landfill in future and/or an advantage for reduction of environmental pollution. It should be discussed very objective with all aspects of consequences.

If such long emission times from landfills are really the importance of an additional question increases dramatically: is it possible to transfer emissions out of landfills with technical systems (e.g. drainage system but also liner system)? The existent experiences with drainage systems have also a maximum age of 20 years with the additional problem of many unsuccessful ways. Some points of drainage destruction are known today but the lifetime of technical systems is limited in every case when repairs are not possible. That means control and transport must lean against more on natural and not on technical systems for the expected time of operating.

6. Final Storage

The main aim of all kinds of strength duration of emission streams from landfills is to define a situation where the environmental pollution decreases without any treatment process near zero. The presented data shows that such aim only could be reached in a very far future. The demand reaching "earth`s crust quality" for solids and "drinking water standards" for liquids in an imaginable time is not realistic. On the other hand the scope of landfills is to concentrate solid pollution material to unburden the environment. But such a nearly natural kind of concentrated pollution potential can not opperate without any spreading of pollution. Comparing all imperfect waste management technics an advantage of landfilling is that residuals were not diffused to the environment in some hours.

If the phase of final storage is not only accepted for a extremely far future the limits will have to be defined today and eventually adapt to changing conditions in future. Therefore general limits or specific limits are thinkable.

The phase before final storage could be devided into different phases:

1) biological processes	1a) acetic phase	1a und 1b : differences of biological processes; for long term considerations unimportant;
	1b) methanogenic phase	primary process: biological secondary process: geochemical

2) geochemical   primary process: geochemical
   processes   secondary process: biological

The transition between phases are not fixed or defined. Although the
used definations for the change from acetic to methanogenic phase is
not exact and don't correspond with the reality. The used practice to
define first time with leachate BOD/COD-ratio ≤ 0.1 and pH values at
the upper level is not the beginning of the methanogenic phase. At
this point most of the methane were already produced. On the other
hand until this time acetic acid are produced. That means over some
years both phases excist side by side.

The end of biological processes could be defined as end of gas production. But biological gas production is a process without time limit.
Therefore it is necessary to fix a technical end of gas production
(sewage sludge digester: 90 % of theoretical production). For landfills the percentage should be higher, e.g. 99 % or 99.9 %. But such a
definition is only of subordinate significance because it is only a
definition without a measureable change in the reality.

Fig 13 and 14 show that the slope of concentrations in the far future
could be very flat. That complicates the defining of final storage
quality, that means the change from a treatment to an untreatment
phase for leachate. Because the concentrations before final storage
quality point in many cases are only some percent higher (e.g. over
some hundred years). It would be difficult to maintain why the
leachate treatment is not necessary on one side or why the treatment
could not end several years before.

7. Conclusions

This paper should show that knowledge of landfill processes etc. for
technical purposes is sufficient but for all questions of the future
interactions between landfills and environment there is a large gap.
This deficiency cannot be closed because such long term processes cannot be shortened. All experiments with laboratory or pilot scale experiments could only give indication but not replace the reality. It is
necessary to know much more about true processes in full scale landfills. Experiences in the part often only engage values outside landfills - e.g. precipitation, leachate quality and quantity, gas composition etc. - but not interactions with processes in landfills. Until

now no landfills are so exact proved that they could used as basis for the future experiences.

With the todays knowledge from landfills as a less or more black box it is not possible to declaire landfills as a solid waste storage system with a definable environmental pollution capacity or as system which final storage quality could be predicted. A final storage quality defined as near zero pollution potential of solids and pollution stream of liquids cannot reached during the next centuries. It could be concluded that with such a demand the requirements for landfills are much higher than for all other waste management systems. Experiences of the past have often shown preferences of sophicated systems but also increasing failures of such systems. An advantage of landfills as nearly geological formation are the tendency of similar reaction times.

8. References

(1) Ehrig H.-J., Beitrag zum quantitativen und qualitativen Wasserhaushalt von Mülldeponien, Veröffentlichungen des Institutes für Stadtbauwesen, TU Braunschweig, 2nd edition (1980)

(2) Spillmann P., (publisher) Wasser- und Stoffhaushalt von Abfalldeponien und deren Wirkung auf Gewässer, VCH-Verlag (1986)

(3) Ehrig H.-J., Untersuchungen über den Einfluß der Hausmüllzusammensetzung auf Sickerwasser- und Gasemissionen (unpublished report), (1983)

(4) Ehrig H.-J., Anaerobic degradation of municipal solid waste-laboratory scale tests, in Wise D.L. (publisher), Global bioconversions Vol II, p. 121 (1987)

(5) Wolffson C., Untersuchungen über den Einfluß der Hausmüllzusammensetzung auf Sickerwasser und Gasemissionen (unpublished report), (1987)

(6) Brechtel H.-M., Beeinflussung des Wasserhaushaltes von Mülldeponien, Müllhandbuch Lfg 5/84, Kap. 4623

(7) Doedens H., Maßnahmen zur Verbesserung der Ausbeute des Gaspotentials an Deponien, Stuttgarter Berichte zur Abfallwirtschaft, Bd. 19 (1985), S. 231

(8) Ehrig H.-J., Untersuchungen zur Gasproduktion aus Hausmüll, Müll und Abfall, 1986, S. 173

(9) Ehrig H.-J., Was ist Deponiesickerwasser - Mengen und Inhaltsstoffe, ATV-Dokumentation Deponiesickerwasser - Ein Problem der Abwassertechnik, 1986, S. 19

(10) Greiner B., Chemisch-Physikalische Analyse von Hausmüll, Berichte 7/83, Erich Schmidt Verlag

(11) Henseler G., Figi R., Baccini P., Langfristige Entwicklung der Emissionen einer Siedlungsabfalldeponie, EAWAG-Jahresbericht 1985

(12) Belevi H., Baccini P., Water and element fluxes from sanitary landfills, in process technology and environmental impact of sanitary landfill, Cagliari, Sardinia, 19.-23.10.1987

(13) Nielsen R., Report on cadmium, Swedish Environment Board (1978)

(14) Tabasaran O., Separierung schwermetallhaltiger Hausmüllkomponenten durch Absieben, Müll und Abfall (1984), p. 15

(15) Fresenius, Schneider, Gorbauch, Poth, Ananlyse von Müll, Klärschlamm, Sickerwasser und Gas von Abfalldeponien, DFG 477/1-3

(16) Vogl J., Umwelteinflüsse von Abfalldeponien, Berichte aus Wassergütewirtschaft und Gesundheitsingenieurwesen, TU München, Nr. 19 (1978)

CONTROL OF REACTOR LANDFILLS BY BARRIERS

Walter H. Ryser
AG für Abfallverwertung AVAG

1. AN ACCOUNT OF THE DEVELOPMENTS OF THE "BARRIERS" IN TIME

There are four barriers which are important for landfill technique, which can be substantially influenced or rather determined by the engineer.

Barrier No. 4

The geological site provides the final hinderance to prevent the transportation of harmful substances into the ground water. Up until the end of the sixties more or less suitable sites were chosen for the dumping of waste, taking no further preventive measures into consideration. The majority of the landfills which now require sanitation date back to this time.

Fig. 1   Situation of the landfill technique up to the end of the sixties; more or less suitable sites - no technical precautions

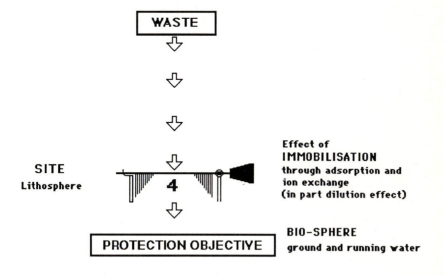

Barrier No. 3

At the beginning of the seventies the envelope was introduced, in view of the growing awareness that there was a lack of natural water tight sites in our land.

The leachate was in the main collected by means of an artifcially sealed base (liner) and subsequently conveyed to the sewage treatment plants. By sealing the surface, the land-ill could be dehydrated at a later date, thus reducing the leachate to a 'negligible' amount.

Fig. 2   Landfill technique in the seventies; envelope concept (base and cover)

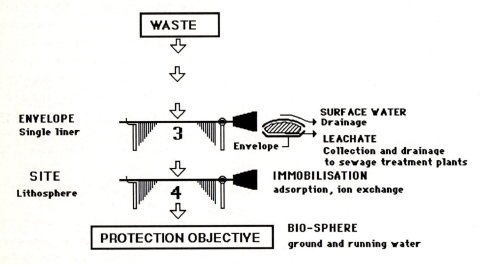

Barrier No. 2

Controlled reactions gave rise to great expectations. The lysimeter experiments could simulate extremely quickly aerobic, as well as anaerobic process, for example by the re-cycling of leachate.

In spite of intensive practical experiments in the seventies, the aerobic depositing in the so called decomposed landfills proved unsuccessfull (odour emmissions, gas, heavy metal, discharge, etc.), therefore in the eighties more emphasis was given to anaerobic depositing, the so called compacted landfills. However, as we will see, in the practice optimization of the controlled reaction was difficult and the knowledge of the long term behaviour of the landfill proved to be relatively poor.

Fig. 3   Situation of the landfill technique at the beginning of the eighties
("Controlled" reactor phase)

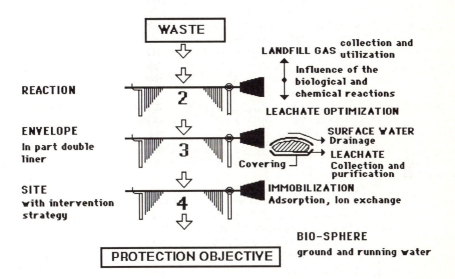

Barrier No. 1

With the continous increase of waste and harmful substances, always heavier demands were being made on the reactor landfill. The view that before tipping it is necessary to separate the waste for the purpose of utilization (wood, paper, old iron) or further conveyance (batteries, sewage sludge) was generally being accepted.

However, it will take some time before the essential measures are realised everywhere, but at the beginning of the nineties the barrier No. 1 should represent an essential factor for the control of the contents and thereby influence the long term behaviour of the landfill.

Fig. 4   Situation of the landfill technique at the beginning of the nineties . Preliminary separation of utilizable and harmful substances

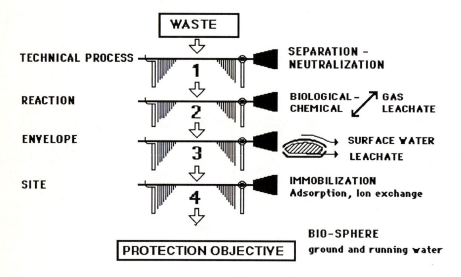

The following illustrates in short, the most important demands on the barriers.

2. DEMANDS ON THE BARRIERS

Fig. 5   Demands on barrier 4 - geological site

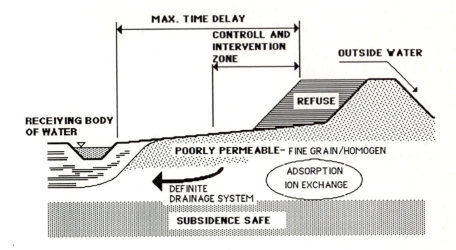

THE SITE DETERMINES THE QUALITY OF THE ENVELOPE

o   The base must be poorly permeable and safe against subsidence. Fine grained
    materials increase adsorption and ion exchange capacity.

o   No infiltration of outside water respective the flow of stratum water into the
    landfill.

o   Definite discharge towards main drainage channel

o   A good distance between the location of the landfill and rivers/ground
    water.

o   The possibility to intervene (e.g. if demage occurs through leachates)

Fig. 6   Demands on barrier 3 - envelope

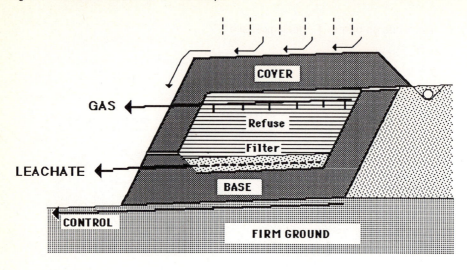

Principles for the bottom liner are, among others:

o   Firm ground

o   Sufficient slope for definite discharge towards the outside of the landfill (if possible shaftless base)

o   Low conductivity and high sorption and sorption of the liner with an underlying control system

o   Large dimensioned drainage system, which can be controlled from outside and flushed through (with verification of stability)

o   Drain and filter structure to prevent the dam up of leachate. The purpose of the final cover is to ensure dehydration of the landfill over a longer period of time. On top of which the landfill can be recultivated.

o   The landfill should have settled considerably before further construction can take place (possible pressure on the landfill).

o   Steeper slopes help to reduce density problems by further settling of the landfill

o   Waterproof sheeting (shingled liners), including underlying degassing system, plus drainage filter system for drainage of the surface water, must be resistant to settling of the landfill

o   Top cover and layer for recultivation should have a high water retention value - suitable planting increase rate of evaporation

Fig. 7   The effect of the "envelope" in respect of leachate quantity

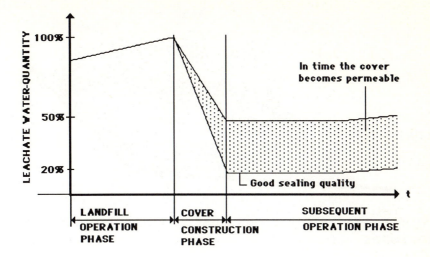

No long term experience is available as far as the efficiency of the final covering is concerned. However, it must be taken into account that depending on the construction, dehydration will only be partly successful and subsequent settling of the landfill, surface erosion, plus oxidation process will increase the porosity of the seal. In other words the final cover gives no absolute insurance against emission of harmful substances.

Fig. 8 Knowledge in respect of barrier 2 - Reactions in the landfill

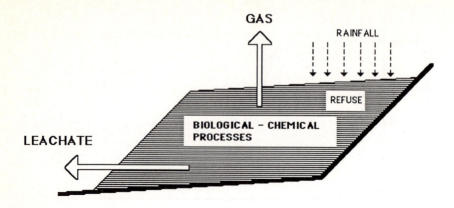

PRINCIPLE: LEACHATE AND GAS MUST BE COLLECTED AND PROCESSED

The reaction processes are influenced by:
WASTE: Type, chemical and mineralogical composition, age, density
HUMIDITY: Rainfall, leachate recycling, sludge deposits

On account of the above parameters the decomposition processes takes place.
Favourable fermentation conditions prevail when the pH value lies between 7 and 8 with temperatures between 30 to 40°C.

The control of the processes is limited however, through:

o   Operational possibilities (use of machines, driving over of vehicles)
o   Effect on neighbourhood (leachate water recycling can lead to unpleasent odour
     emissions)
o   Possibility of gas collection plant, rapid decomposition processes bring about
     short therm high gas production, witch in the practise is very difficult to bring under
     control (rate of gas collection)

Fig. 9   Example of an experiment to control the reaction processes

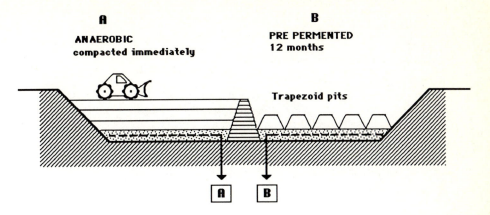

The following experiments were carried out on a large scale at the Steinigand landfill.

o   The base was divided into two equal areas A and B, each with separate
     dewatering systems
o   Area A :   The refuse was immediately dumped and compacted
     Area B :   The first layer was pre-fermented for 12 months and only then compacted
                (Phase 1)

Fig. 10   Preparation of preformented wastes and landfilling in two phases

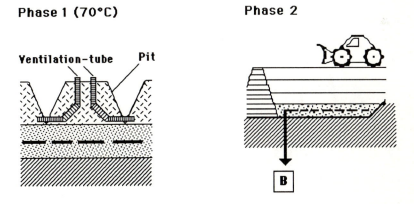

o   After 12 months the trapezoid pits were compacted and covered with further layers
     of compacted material (Phase 2)

Fig. 11 Leachate quality results - AVAG experiments Steinigand. Behaviour of the pH value as a function of time

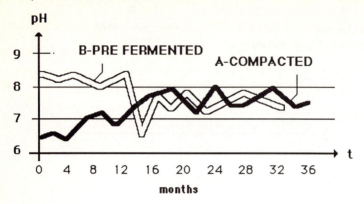

o   A   Compacted   - Acid phase took nearly one year
o   B   Pre-fermented - From the start a alkaline medium, short acid phase after 12
                         months, brought about subsequent compaction
o   After 20 months no significant difference could be detected

Fig. 12 Behaviour of light decomposable material (BOD5) in leachates as a function of time

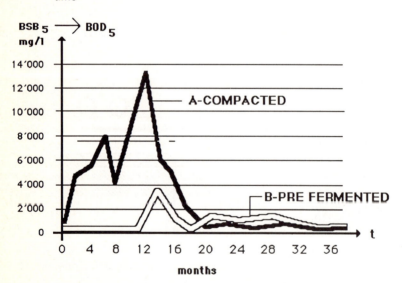

o   Direct compaction results in a high BOD5 (curve A)
o   With pre-fermentation, only a short peak after subsequent compaction
    (curve B)
o   After 20 months the behaviour of both landfills were similar

Fig. 13 The chloride concentration in leachates as a function of time

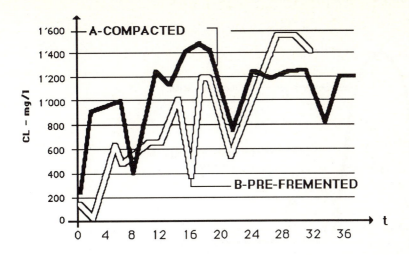

o   A lower value over a 20 month period with pre-fermentation (curve B)

o   Afterwards no more substantial difference

Fig. 14 Gas production results - comparison of gas production with/without leachate recycling

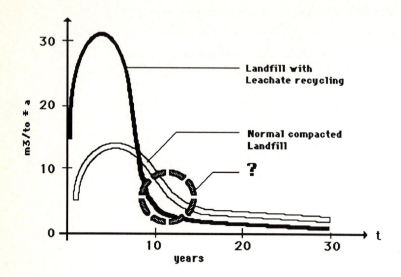

Observation:

o   In both cases, although only decomposition of approximately 50% of the biomass took place, the gas production has significantly decreased after approximately 10-15 years (Baccini, EAWAG).

In view of the foregoing the following conclusions are made:

o   It is impossible to model the reactor "landfill" on account of operational data and the "open" systems.

o   Controlling or altering the decomposition process has only a short term effect (at the most 2 - 3 decades).

o   Control of the reaction through sludge depositing, leachate recycling, pre-fermenting and decomposition is insignificant when related to long term behaviour. (Laboratory experiments can only to some small extent be compared to that which takes place in the field).

o   The waste determines the long term effect.

This observation does not mean that we cannot control the reaction, but that we cannot expect any great influence on the long term behaviour.

DEMANDS ON BARRIER 1 - TECHNICAL PROCESS

Landfill problems can to some extent be eliminated by:

o   Sorting and separation of re-utilizable and harmful substances
o   Pre-treatment and neutralization of refuse
o   Incineration, mineralization of the organic wastes
o   Concentration, fixation and separation of harmful substances

OBJECTIVE

o   Only admission of substances which have a low, relatively and harmless reaction potential in the landfill

FURTHERMORE

The landfill places demands on the technical process. Technical process stipulations

1   No environmental pollution (water protection, clean air, noise, traffic, etc.)
2   Large spectrum of input to deal with variable input (waste), defined output (end product)
3   Long time availability or enough reserve units (the reserve capacity poses financial problems which as yet cannot be solved).

"The greater the demands on the technical process, the longer the chain of procedure which in turn decreases the availability"

4   The technical process should not pollute the environment more than the reactor landfill, otherwise the investment is not oekologically justifiable.

3. BARRIER MODEL 2000 ?

According to Swiss guidelines the aim of the technical processes (barrier 1) could be so defined that no further (only minor) reactions without environmental impact occur after ultimate waste disposal (barrier 2) or that a technical envelope ist necessary (barrier3). The technical processes should transform wastes to products of final storage quality.

Fig. 15 Basic combinations for final storage landfills

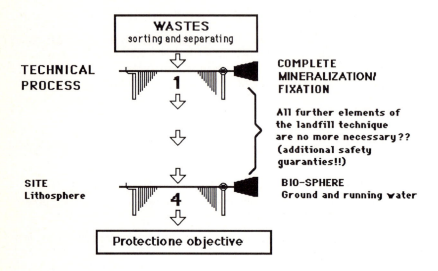

o   A large number of reactor landfills will be necessary within the next 2 to 3 decades because there is no guarantee of a efficient technical processes with long term availibility. Furthermore one must take into account the relatively long time required to reach decisions concerning the development of the necessary plants (see example hazardous waste Switzerland).

GROUP REPORT: BIOLOGICAL AND CHEMICAL PROCESSES

Preamble

The overall concept is for the landfill to convert the waste into an inert or stable mass so that additional emissions from the landfill are not of environmental concern. Group members came from different educational and philosophical backgrounds, plus from different countries, all of which combined to give a great diversity of opinion. Consequently, this report will attempt to describe not only the most popular opinions, but it will also give other opinions as appropriate.

Overall concepts

It was generally agreed that it is desirable to promote decomposition so most of the flow of gas and contaminants in leachate will occur as soon as possible, when gas and leachate control systems are most reliable, when monitoring is most active, and when organizations are most likely to be available to care for the landfill. The best landfill would have predictable and controllable decomposition and produce products of decomposition at an acceptable rate. Biological decomposition would maximize methane generation to promote methane use and take advantage of the lower leachate strength associated with methanogenesis.

The landfill design to promote decomposition and to minimize environment impact involves use of a permeable cover as sand to bring water into the landfill until methanogenesis is observed. Monitoring techniques are discussed under question 6 below. Then, a clay cover would be placed to reduce water infiltration, control the free venting of gas and aid gas collection. Finally, when methanogenesis is basically completed, a plastic sheet cover would be placed over the landfill to further reduce water flow in and gas flow out of the landfill. This would lower the flux rate of contaminants in leachate, provide more time for biological reactions to occur, and reduce oxygen access and hence the conversion of the landfill to oxic conditions.

There were two particularly troublesome aspects to the discussion which should be noted at the beginning. First, landfills are not perfect reactors dealing with a perfect (homogenous and specified) waste, so perfectly predictable and acceptable products of decomposition cannot be expected. The best we can hope to achieve is general control major constituents. Trace of minor constituents both in the waste and in emissions from the landfill can not be predicted or controlled at this time.

Second, modern landfill practice with modern wastes is by definition a recent development; consequently, there is no experience or data that can be used to describe what will happen beyond the first few years of a landfill's decompositional life.

Answers to specific questions

Six questions were given to guide the discussion. These questions were found to be adequate to describe the landfill as a biological and chemical reactor as discussed, so the questions and answers comprise the remainder of this report.

Question 1:

Which measures can accelerate the decomposition processes of organic material and by this shorten the time of the "reactor period" of a landfill?

1. Major methods of acceleration are:
    a) shredding or homogenizing the waste,
    b) controlled moisture addition (see "e" below),
    c) controlled sewage sludge addition (see "e" below),
    d) use of aerobic or other prior decomposition of the waste to reduce the acid phase (note that this will reduce methane generation),
    e) "controlled" in points b and c above refers to monitoring pH and not adding water or sewage sludge in amounts such that the pH becomes acidic (<6.5 or so).

2. Other less certain methods of acceleration were mentioned, but there
is no conclusive evidence to show that they will work in a landfill environment:

    a) sorting out metals and big pieces
    b) adjusting C/N ratio by adding N

3. In some specific situations, adding seed organisms, increasing pH to
lower any sulfide toxicity, or decreasing pH to lower any ammonia toxicity may promote decomposition. Research is needed in these areas.

4. Leachate recirculation is a good method to promote initial decomposition and, especially when the leachate is neutralized, it can
hasten the onset of methanogenesis; however, it often results in water
management problems and has little effect on methane production once
methanogenesis is underway. It is not recommended here except in special situations (for example, dry areas).

5. A landfill does not decompose in a homogenous sense. There are
many "micro environments" within a landfill at any given time which are
quite different. Homogenization of a landfill by removing large objects
or shredding the waste promotes more rapid and uniform decomposition
and a more uniform state at that point in the future when the waste
is "stabilized".

Question 2:

How can one define the transfer from biologically mediated processes to
geochemically mediated processes (i.e. practically only abiotic processes) in landfills?

1.Biological and geochemical processes form in fact an intricate net-
work of interactions, which evolves to a state of dynamic equilibrium
such as exists in the soil. There is, consequently, no transfer to be
expected from biologically to geochemically mediated processes.

2. It can be arbitrarily defined that when some important parameters,
such as CO_2 or methane production, have decreased below a given percentage of their maximum rates, the landfill can be considered as having reached this state of dynamic equilibrium.

3. It can be expected that changes from reducing to oxidizing condi-

tions, which involve transformations of sulfides and a shift to more acidic conditions, will increase the mobility of elements such as Hg, Zn, Pb, Cu and Cd. On the other hand, the mobility is characteristically lowered for Mn and Fe under oxidizing conditions. Elements exhibiting anionic species, such as S, As, Se, Cr and Mo are appreciably solubilized, for example, from fly ash sluicing/ponding systems at neutral to alkaline pH-conditions. Literature data has established an order of binding strength for a number of metal ions onto humic or fulvic acids as follows: $Hg^{2+} > Cu^{2+} > Pb^{2+} > Zn^{2+} > Ni^{2+} > Co^{2+}$.

Question 3:

Which compounds should be excluded from landfills due to their potential to
- prolong the reactor period
- increase emissions
- produce new xenobiotic and/or hazardous compounds?

1. There are no full-scale landfills known in which waste components have been observed to inhibit decomposition.

2. Emissions of concern occur both in the gas and leachate and include compounds or elements known or likely to be hazardous or toxic.

3. Due to their persistence and their leachabilty properties, some waste components will (independent of the type and age of a landfill) always create problems with respect meeting emission standards. Examples of such compounds include any forms of the elements Hg, Sn, Cd and As, as well as xenobiotic organic compounds such as halogenated hydrocarbons and other halogenated compounds, nitroaromatics and anilins. Such hazardous components may come from industry or small businesses, may be present in consumer wastes, or may even arise from pretreatment of wastes (as by incineration). It should also be noted that some organic compounds that are not of major concern may be transformed in the landfill to compounds that are more hazardous or toxic.

4. Items 1 - 3 above do not consider the overall decision of whether a given waste should be placed in a landfill or treated elsewhere as by incineration or composting.

Question 4:

What is the lifetime of landfill internal installations (e.g. drainage systems) to secure a controlled water household?

1. Plugging of leachate collection pipes is common in Germany, especially during the acidic decomposition period. There is also crushing of leachate collection pipes. Little such plugging or crushing of pipes has been experienced in the US using heavy-walled pipes placed in trenches for support, when oxygen is clearly excluded, and when pipe clean-out systems are installed and utilized. The US system which appears to be most successful uses a drainage layer over the entire liner surface.

2. Additional long term experience is needed in this area.

Question 5:

Which landfills (types, examples) are well documented so that they could serve as objects for multidisciplinary studies?

1. There are no full scale landfills operated to promote stabilization according to the concepts presented under question one. The only experience on this is from test landfills and laboratory experiments, plus theory.

2. There is some experience that aerobic pretreatment of waste, and constructing the landfill in layers so that the bottom layers are practically stabilized before upper layers are placed, shortens the acidic phase of decomposition. Once the methanogenic phase is underway, however, there appears to be no change resulting from these techniques.

3. There is no experience regarding emissions from landfills late in the decomposition cycle when geochemical processes become dominant.

Question 6:

Which "indicator parameters" should be selected to describe the state of evolution of a landfill?

1. Parameters which should be monitored routinely are

> - temperature
> - gas composition
> - leachate analysis for: pH, BOD/COD ratio, VFA (Volatile
>   Fatty Acids)/BOD ratio, chloride, sulfide-sulfate, nitrate-
>   nitrite-ammonia-organic N, manganese and cadmium.

2. Parameters of interest for more complete monitoring, as for research
purposes, include for leachate:

> - cellulolytic activity
> - ATP (measure of living biomass)
> - F420 (measure of fluorescent pigment from methanogens)
> - acidity/alkalinity
> - iron

plus volatile organics, such as solvents for example, in both the
leachate and the gas.

3. Other parameters may be measured for other reasons, such as for
leachate treatment.

Concluding Comment

The group agreed that it is unrealistic to expect that emissions
from a landfill, even if the waste is carefully selected or pretreated
by known methods, will be eliminated after 30 years. It is, therefore,
important to consider the result of this group along with the results
of the other groups in reducing emissions from a landfill to acceptable
levels according to each landfill's geological and geographical setting. It is also important to consider the total impacts of landfill
disposal with the total impacts of other technologies in a fair and objective manner to achieve better solid waste management in the future.
Such an analysis would include energy usage as well as environmental
and economic impacts.

MATERIAL TRANSPORT AND PROPERTIES OF REACTOR ENVELOPES

Chairman:	Rainer Stegmann (FRG)
Lecturers:	Grahame J. Farquhar (CAN)
	Donald H. Gray (USA)*
	Fritz T. Madsen (CH)
Troublemaker:	André Zingg (CH)
Rapporteur:	Eduard Höhn (CH)
Discussion participants:	Eckehard Bütow (FRG)
	Werner Kanz (CH)
	Peter Oggier (CH)
	Gabriel Piepke (CH)
	Walter Ryser (CH)
	Christian Schlüchter (CH)

SPECIFIC QUESTIONS:

(Definition of the term "Envelopes": Artificial or natural materials around deposits to function as barriers in transport processes)

1) Which "indicator parameters" should be used to describe transport processes across envelopes?

2) With artificial and/or natural envelopes, is tailor made hydraulic control possible?

3) What is the lifetime of a liner with respect to a certain barrier function (e.g. for non polar organic molecules)?

4) What is the influence of variations in design and configuration on liquid and contaminant transport through liners?

5) Which landfills (types, examples) are well documented with respect to barrier functions so that they could serve as study objects?

MATERIAL TRANSPORT AND PROPERTIES OF A REACTOR ENVELOPE

CONSEQUENCES FROM THE CURRENT KNOWLEDGE FOR THE DESIGN OF ENVELOPES

Rainer Stegman

Technische Universität Hamburg-Harburg

Theoretically emissions from landfills can be avoided by

- treatment of waste so that no emissions in the gas- and liquid phase can be produced
- total collection of the leachate at the bottom of the landfill by means of liners and drain systems; collection of the gas by extraction
- avoiding water penetration into the landfill that no materials can be dissolved in the liquid phase due to the lack of transport mechanism
- avoiding the deposit of organic waste, so no gas emission can occur.

Since none of the above mentioned measures is effective on its own today, a combination of several measures dependent upon the waste will be necessary for emission control.

The evaluation of the literature and own experiences in the following three papers show that lots of information are available today concerning the behavior of liners depending upon the physical, chemical and hydraulic impact, so that based on the validation the consequences for envelope design can be drawn. In addition requirements for waste quality (resp. leachate quality) should be made. This is necessary when no adequate liners for specific components can be found. Based on the literature review also requirements with regard to landfill operation may be necessary.

Specific questions and problems

The possibilities and limitations of an envelope (bottom liner and cap) are still very controversially discussed. To express this in a simplified way, two groups can be formed:

- group number one thinks that if an envelope is installed, there will be no more emission from the landfill; the problem is solved
- group number two sees the problems of the influence of chemicals in the leachate that increase the hydraulic conductivity of natural materials; and that plastic and natural liners cannot be put in place properly; plastic material does not last longer than 20-30 years. Using Darcy's law and pessimistic assumptions (high hydraulic gradients etc.) this group sees not much improvement in emission control by installing an envelope.

As in most cases the truth is in-between those two extreme positions. There is a long tradition in lining as for dikes, ponds, canals and dams. Based on these experiences landfill envelopes have been designed. Apart from the above mentioned aims of a liner, leachate from a landfill should not enter the subsoil and groundwater. We know today that this cannot be guaranteed with time so that it is our scientific and practical aim to improve envelopes. It is possible to minimize these emissions over hundreds or thousands of years.

But we can build an envelope today also in very different ways. The working group A2 saw its task in presenting the state of knowledge and experiences with envelopes and in drawing conclusions for optimum design and operation. The three detailed papers represent the state of knowledge in this field today. Each paper approaches the subject "envelope" from a different perspective. While Gray focuses on the geotechnical engineering aspects, Madsen and Mitchell present detailed information on the geochemistry of clay. Farquhar and Parker discuss in detail the influence of leachate quality on natural and synthetic liners. On the other hand there is an overlapping of subjects dealt in all the papers. The group chairman has not eliminated these overlappings - also in those cases where the authors come to the same conclusions - in order to show the reader that there is or is not controversy in a certain point of area.

These three excellent papers represent the basis and the knowledge that we used for our discussions and conclusions in our working group. The other members of this group added own experience and approaches during the discussions so that we were able to conclude our work in presenting general statements and answers to the questions that were formulated by the scientific organizing committee.

There are still many open questions and there are some gaps between what could be done today and what is done. Some of those questions and gaps are stressed below from the author's point of view:

- Natural materials that are used for envelope construction should be reproducible and homogeneous; that means that natural clays have to be treated (shredded, mixed, adjustment of water content) and - if necessary - mixed with other materials - f.e. bentonite - to guarantee material quality.

- Only little information is available with regard to the gaseous conductivity of mineral liners. Further research is needed.

- Long-term behavior of liner materials is still unknown. For this reason envelopes must be observed and - if necessary and possible - be repaired. The effectiveness of these envelopes has to be monitored and evaluated on a long-term basis. Actual leakage rates and influences on the liner systems have to be quantified.

- In addition to liner materials the total liner system is influencing the emission control (slopes of surface caps and bottom layer, design of gas and water drain systems, top soil quality and height of the layer, surface water capturing and transport, kind of vegetation avoiding of severe settling etc.).

- During construction the influence of weather has to be minimized on the liner quality (construction under a tent, immediate covering of clay with sand, storage of materials in silos etc.).

- The kind of single and/or multiple liner systems should not be regulated today by any authorities since there is too much development in this field. The effectiveness of a liner system should be quantified in regulations.

If all the basic knowledge presented in the papers and the recommendations from the working group are respected, design and operation of envelopes can be improved significantly.

Main subjects of the three fundamental papers

The following three papers are of a high scientific level. For this reason basic knowledge in geochemistry, geology, soil mechanics and civil engineering is helpful. The conclusions and the answers by the working group are subjected to a broader audience.

The paper of Gray deals with the influence of leachate and chemicals on the permeability of clay. In this context the influence of the kind of laboratory test procedure on the results is discussed. Gray also points out the influence of compaction on the hydraulic conductivity of clay. Of great importance - and not often published - is the approach to calculate the influence of voids and cracks on the water movement through clay layers. Is the transport of pollutants through mineral liners more influenced by the hydraulic or the concentration gradient (diffusion)? Answers and ways of calculation are presented. The possibilities of attenuation of leachate components in envelopes are presented based on the evaluation of the literature. The paper ends with the chapter that stresses the geotechnical factors that determine final storage quality.

Farquhar and Parker summarize field data experiences with regard to effectiveness of envelopes, pointing out that methods to extrapolate laboratory data to field conditions are poorly understood. The authors present fundamental information concerning the interactions of leachate, organic solvents and mineral as well as synthetic materials. Based on this information conclusions are drawn and further research is recommended. In this paper the possibilities and limitations of natural barriers as well as the lack of knowledge in this field are presented.

The paper of Madsen and Mitchell describes in the first chapter the factors that affect hydraulic conductivity by explaining the geo-chemistry of clay. The influence of different laboratory test methods on the prediction of the hydraulic conductivity of clay is shown by means of an example. In addition to the above described two papers the effect of inorganic and organic chemicals on the hydraulic conductivity of clay is presented in very great detail and, based on this informa-tion, conclusions are drawn.

The overlapping in the three papers results from the subject "influence of leachate and chemicals on liner material" that is dealt with in each

paper. But there is very little repetition of the same facts and studies. In contrast each author gives additional information so that this
great point of interest, which is discussed very controversially at
each seminar dealing with "envelopes", is handled in an adequate way
and depth.

The effectiveness of the envelopes is the key for the environmental impact of landfills. The time period this envelope will function cannot
be predicted. Since we cannot avoid landfills today - although we
should minimize the amount of waste that has to be deposited - we have
to focus our efforts on the improvement of the liner systems and their
long-term behavior together with landfill design and -operation as well
as waste separation and -pretreatment techniques that result in less
emissions from the deposited waste.

GEOTECHNICAL ENGINEERING OF
LAND DISPOSAL SYSTEMS

Donald H. Gray
The University of Michigan

ABSTRACT

Land disposal of solid waste entails the placement of
waste in above ground stockpiles or dike impoundments or
conversely, burial/encapsulation in below ground vaults.
Containment failure can result from either mass-stability
failure of above ground structures or leakage from both types
of disposal systems. Geotechnical considerations play an
important role in preventing either type of failure.

Interactions between the waste (or its leachate) and the
containment envelope significantly affect containment
integrity. Leachates can increase the hydraulic conductivity
of clay liners; however, many of the large (several orders-
of-magnitude) increases reported in the technical literature
were caused by experimental procedure e.g., the use of
extremely high gradients and employment of fixed-wall permeameters. Pure organic liquids (solvents) can interact
adversely with clays causing some shrinking and cracking with
large attendant conductivity increases. On the other hand,
dilute solutions of organics (in aqueous carriers) have
little or no effect.

Leakage through landfill liners - either compacted clays
or synthetic membranes - is dominated by flow through void
volume defects. These defects include desiccations and/or
syneresis cracks, fissures, root holes, pin holes, lift
joints, tears, faulty seams, and porous inclusions (clods).
The likelihood and the gravity of the consequences of void
volume defects are greatest in very thin membranes.

Careful compaction/composition specifications are
essential to insure both low permeability and good volume
stability in clay liners. The latter objective will minimize
the danger of desiccation/syneresis cracking. Generally both
these objectives can be achieved by compacting at or near
optimum water content to high relative densities, by using

compaction plants that thoroughly mix and "homogenize" the clay, and by excluding the use of high plasticity clays.

Movement of leachate across clay liners occurs by both advection and diffusion. In cases of very low seepage velocity (characteristic of dense, natural or compacted clays) diffusional transport can be quite important. Solute breakthrough or transit times under diffusion vary as the square of the barrier thickness. In the absence of solute partitioning or adsorption these breakthrough times may be very low (less than 10 years) for thin barriers.

INTRODUCTION

Land disposal can take many forms ranging from above ground, diked impoundments and waste stockpiles to below ground shallow burial vaults (landfills) and deep well injection. In virtually every case geotechnical considerations play an important role in the design and construction of these land disposal systems.

A critical factor in land disposal is the nature and consequence of interactions between a waste and the containment envelope. The choice of materials for the envelope, how these materials are emplaced, and the geometric design of the envelope are all factors that affect these interactions and the integrity of a disposal system.

The following geotechnical considerations are explored in this paper with the goal of establishing guidelines for improved disposal practice:

1. Modes of land disposal and containment failure.
2. The effects of chemicals or waste leachates on clay liner permeability.
3. The role and influence of compaction conditions on the permeability and integrity of clay liners.
4. The relative importance of "matrix" vs. "volume defect" permeability in clay liners and strategies to minimize the latter.
5. The role and importance of diffusional transport of leachate solutes across barriers.
6. Leachate attenuation by different liner systems.

MODES OF LAND DISPOSAL AND CONTAINMENT FAILURE

Utilization of inert solid wastes (e.g., flyash and mine tailings or detoxification and treatment of hazardous wastes (for recycling or reclamation) are desirable goals. There will always remain, however, a residual volume of wastes which must be contained an permanently isolated. Even wastes selected for recycling must still be stored in temporary facilities. The earth or ground is a logical repository

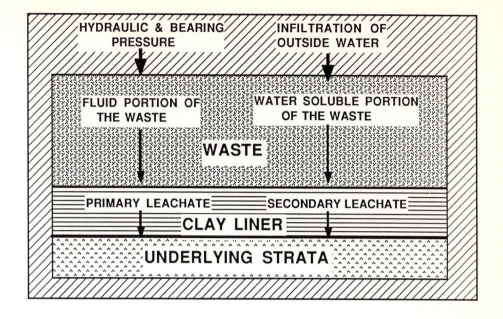

Figure 1. Sources of leachate that may come in contact with clay liners.

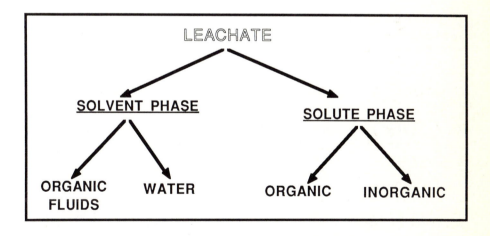

Figure 2. Composition of the leachate of a waste

for such storage or disposal.

Land disposal can take one of several forms, viz.:

1. Storage and containment in natural depressions or pits dug in the ground, e.g., flyash ponds.
2. Storage and containment is above ground, diked facilities, e.g., slurried tailings impoundments.
3. Waste stockpiles or embankments.
4. Shallow burial vaults, e.g., sanitary landfills.
5. Deep burial vaults, e.g., radioactive waste disposal in underground chambers in granite and rock salt.
6. Deepwell injection.

Each of these disposal methods poses unique problems and requires its own engineering analysis and design. Geotechnical considerations play an important role in nearly every case.

Improper disposal of wastes in stockpiles, impoundments and vaults can result in containment failures that degrade environmental quality and threaten human safety. Mass stability failures of stockpiles and impoundments can release huge quantities of potentially toxic wastes into the environment. Some 20 phosphate waste impoundment dam failures have occurred in the USA since 1947 releasing large amounts of phosphate slimes, and seriously, although temporarily polluting surrounding waters.

The consequences of catastrophic failures of waste impoundments and disposal fills have been brought dramatically to the attention of the public and the engineering profession in the last few decades. The failure of the coal mine refuse bank in 1966 at Aberfan in Wales killed 127 people, more than 100 of whom were school children. The coal refuse embankment which failed on Buffalo Creek near Saunders West Virginia killed 125 people and left nearly 4000 homeless in addition to causing serious erosion and sedimentation damage.

Leakage is a less spectacular but more insidious form of containment failure. Leakage can occur in one of several ways, viz.,

1. Downward
   Migration or seepage of contaminants by advection/diffusion through underlying containment, whether geologic strata or artificial liner (e.g., compacted clay and/or polymeric membrane) - into groundwater below site.

2. Sideways
   Migration/seepage through or under lateral containment dykes or cut-off walls or along fault

zones into both groundwater and surface
drainageways.

3. <u>Over the Top</u>
Flooding/seepage over the top of perimeter
dams and dykes or sides of waste pits. Occurs
when impoundments become too full or when excessive
infiltration into landfills overwhelms internal
drainage or leachate collection capacity.

CHARACTERIZATION OF LANDFILL LEACHATES

In order to design/construct secure landfills it is first
necessary to identify and characterize various wastes and the
leachates they produce. These leachates can affect permeability and degrade the integrity of landfill liners - be they
compacted clay or synthetic membranes. Leachates are produced from two sources, namely, the flowable constituents of
the waste itself and the flowables generated by outside water
infiltrating and percolating through a waste. These
"flowables" are referred to as primary and secondary leachate
respectively as shown in Figure 1. Chereminosoff <u>et al</u>.
(1979) have estimated that 90%, by weight, of industrial
hazardous wastes are produced as liquids and thus have a high
primary leachate potential.

Leachates can be divided into a solvent and solute phase
respectively. The predominant fluid or solvent phase of a
leachate may be water or an organic fluid (Figure 2).
Organic solvents can cover a spectrum of types such as
acidic, basic, neutral-polar, and neutral-nonpolar compounds.
Corresponding solutes in a leachate are any chemicals, salts,
or compounds that dissolve in the solvent.

Both the solvent and solutes in a leachate can affect the
permeability or containment integrity of a liner. The
mechanisms or manner in which leachates affect permeability
or apparent permeability of liner materials are discussed in
the next section.

EFFECTS OF LEACHATE ON CLAY LINER PERMEABILITY

The results of scores of different tests on chemical or
leachate influence on clay hydraulic conductivity have been
reported in the technical literature during the past decade.
Mitchell and Madsen (1987) provide a comprehensive review and
summary of all these findings.

Based on a review of sometimes contradictory findings
reported in the literature on this subject it appears that
the measured hydraulic conductivity of compacted clay during
leachate permeability tests can be affected strongly by the
following factors:

1.   The composition, porosity, and fabric of the clay
2.   The type of permeant
     - aqueous vs. organic solvent
     - pure solvent vs. mixture
     - solute type and concentration
3.   The type of permeameter
     - fixed-wall (compaction mold)
     - flexible-wall (triaxial type)
     - consolidation-cell
4.   The gradient

The composition and fabric of the clay are controlled by the type of clay selected and either its geologic origin and history in the case of natural clay liners or method of compaction (including the type of compaction, compactive effort, and moulding water content) in the case of man made or emplaced liners. Compaction considerations are discussed in the next section.

The type of permeant used in the permeability tests clearly plays an important role. The preponderance of evidence now indicates that large increases in measured permeability are only observed when pure organic solvents of low dielectric strength are employed. These increases are caused primarily by shrinkage cracking as a result of syneresis. Syneresis is the separation of a homogeneous colloidal system into two phases--a coherent gel and a liquid. Syneresis is caused by a change in the inter-particle force balance within the colloid. Syneresis can also be triggered by exposure or leaching of a clay mass to a highly saline aqueous solutions. This phenomenon and its consequences are discussed in a subsequent section of this paper.

The type of permeameter used during laboratory tests can critically affect test results. This one factor alone may account for a good part of the controversy about the effects of leachate on clay liner permeability. Large increases in permeability have been reported (Anderson, 1982) mainly in those tests employing fixed-wall (compaction mold) permeameters in conjunction with low-dielectric organic solvents that have caused shrinkage cracking of the clay bed. This type of permeameter does not simulate a confined liner in the field that is subjected to both a vertical and lateral confining stress. A fixed-wall permeameter also permits side wall leakage which is purely an artifact of the test procedure and not representative of field conditions.

The influence of gradient should also be considered carefully as well. Field gradients in a landfill are likely to be less than 10, in contrast to reported gradients in laboratory permeability tests that have ranged as high as 300. Gradients such as these can only be justified for experimental convenience; they bear no relation to actual field gradients in landfills. High gradients are partly responsible for some of the high relative permeabilities reported in the technical literature during replacement with organic liquids. High gradients facilitate desolvation or displacement of capillary water and artificially elevate the relative permeability of the non-wetting organic phase.

On the basis of their comprehensive review Mitchell and Madsen (1987) reached the following general conclusions:

1.   The influence of chemicals on high water content clays such as slurry walls is likely to be much greater than on low-water content compacted clay.
2.   The effects of organic chemicals are influenced primarily by their water solubility, dielectric constant, polarity, and whether the clay is exposed to a pure chemical or a dilute solution.
3.   The type of test used may have a very large effect on the values of hydraulic conductivity that are measured.
4.   Pure organic liquids can interact adversely with clays by causing some shrinking and cracking with attendant large conductivity increases; however, dilute solutions of organics have little or no effect.

Haxo (1980, 1982) conducted a series of long-term liner exposure tests to different types of wastes under conditions that more nearly simulate the actual situation in a landfill or lined impoundment. His test consisted of exposing a square foot of surface of selected liner materials under one foot of a variety of hazardous liquid wastes. Haxo tested the durability of a compacted clay, various soil admixes, and polymeric membranes to six different types of hazardous waste. The amount of seepage below the liner, penetration of waste components into the liner, and permeability changes in the liner were monitored for periods up to three years.

Results from Haxo's work indicate that no single available lining materials appears suitable for long-term impoundment of all wastes. There was considerable variation in the compatibility between wastes and liners. Generally, results of his study showed that wastes which are highly ionic, contain salts, strong acids, or strong bases can be aggressive to soils, soil cement, hydraulic asphalt concrete, and to some of the commercial polymeric membranes, particularly those containing plasticizers.

Wastes which had organic compounds were often aggressive toward polymeric membrane liners and towards asphaltic materials. Oily wastes, in particular, tended to cause swelling and degrade desirable physical properties (e.g., elongation and resiliency) in the membrane liners. The findings of Haxo's studies underscore the need to perform exposure tests prior to the selection of liners for containing specified wastes and to insure that only those specified wastes are added to a landfill after selection of a suitable liner.

INFLUENCE OF COMPACTION ON CLAY LINERS

The permeability of compacted clay liners is very sensitive to compaction variables. These variables include the type of compaction (kneading, static, and impact), compaction effort (roller weight and number of passes), and molding water content. Compaction conditions also affect other soil properties of importance to clay liner integrity such as volume stability (shrink-swell potential) as shown by the work of Seed and Chan (1959).

Mitchell et al. (1965, 1966) and Lambe (1954) have conducted extensive and definitive studies on the permeability of compacted clay. The following findings have emerged from these studies: (1) Permeability is governed more by clay structure or fabric than any other single variable, (2) Structure is greatly influenced by shear strains associated with compaction wet of the line of optimum, and (3) Different methods and amounts of compaction induce different amounts of strain. These findings mean that clay microstructure, compaction conditions, and permeability are all interrelated.

Compaction at a lower effort can result in greatly increased permeability even for samples compacted to identical dry densities and at identical molding water contents. Daniel (1984) has pointed out the problems that can arise from failure to recognize this fact, particularly if compaction efforts in the field are less than those in the laboratory. Differences in compactive effort between the lab and field can thus lead to quite different permeabilities because of resulting differences in microstructure notwithstanding apparent similarities in other respects. Arguments about the effect of leachates on the properties of compacted clay liners are somewhat irrelevant if the properties of samples compacted in the laboratory, which are used for leachate permeability tests, are not at all representative of the properties of the compacted clay in the field.

Compaction conditions also influence volume stability or shrink-swell potential of clay liners. Very little attention has been paid to this question in promulgating design criteria for clay liners. In some cases volume stability may prove to be equally as important a consideration in the selection and emplacement of a clay liner as permeability. Volume stability is particularly important in landfills where a substantial danger exists from shrinkage cracking which can be caused by desiccation or syneresis.

Whereas compaction wet of optimum tends to reduce permeability sharply it also increases shrinkage greatly. This effect is even more pronounced for clays compacted by kneading compaction. As noted earlier compaction on the wet side tends to disperse the clay structure making it less permeable but also less resistant to volumetric contraction. This response can pose a serious dilemma because it means that low permeability which can be secured by compaction wet of optimum will also be achieved at the expense of high shrinkage. It may be necessary in this case to resort to other strategies such as compacting slightly dry of optimum, but to much higher density, or to specify that the clay be compacted no lower than its shrinkage limit.

CAUSES AND CONSEQUENCES OF VOID VOLUME DEFECTS (CRACKING) IN CLAY LINERS AND BARRIERS

A crucial issue with regard to containment effectiveness of any liner system is the relative importance of "matrix' vs. "volume defect" permeability in controlling flow through a liner. This issue is germane as well to concerns about the influence of leachates on liner permeability. Matrix permeability refers to flow through intra- and inter-particle pores whereas volume defect permeability refers to flow through gross void volume defects such as cracks, fissures, root holes, pinholes, tears, faulty seams, et. The former involves advective or diffusion controlled flow through microscopic pores that is described by Darcy's and Fick's laws respectively. The latter involves flow through gross void volume defects or macropores that is not well described by either of these flow laws.

Daniel (1984) has presented data from four projects in which rates of leakage from ponds lined with clay significantly exceeded the rates that would have been predicted on the basis of laboratory permeability tests. Actual hydraulic conductivities of the clay liner (back calculated from field leakage rates) were generally 10 to 10,000 times larger than those obtained from laboratory tests on either undisturbed or recompacted samples of the clay liner (see Table 1).

The main source of difficulty with laboratory permeability tests is the problem of obtaining a representative sample of the liner material for testing. Neither recompacted samples nor small, undisturbed samples are likely to contain a representative distribution of dessication cracks, fissures, slickensides, and other hydraulic volume defects (e.g., clods) that are present in the liner or that may develop later after its construction and emplacement in the field. The dilemma here is similar to that of trying to determine the permeability of a brick wall by running permeability tests on the bricks.

TABLE 1.   Comparison of Hydraulic Conductivities measured
in the Laboratory vs. Field (from Daniels, 1984)

HYDRAULIC CONDUCTIVITY, IN CENTIMETERS PER SECOND				
Project location	Measured in laboratory	Measured in field	Back-calculated from field leakage rates	k_{field}/k_{lab}
Central Texas	5×10^{-10} to 8×10^{-7}	4×10^{-5}	2×10^{-5} to 5×10^{-5}	25 to 100,000
Northern Texas	2×10^{-9} to 2×10^{-7}	2×10^{-6}	2×10^{-6} to 4×10^{-6}	10 to 2,000
Southern Texas	1×10^{-7} to 4×10^{-7}	---	1×10^{-5} to 2×10^{-5}	25 to 200
Northern Mex	1×10^{-8} to 2×10^{-7}	2×10^{-7}	1×10^{-6}	5 to 100

Note: k indicates hydraulic conductivity. The ratio k_{field}/k_{lab} is the value of k back-calculated from field leakage rate divided by the value of k measured in the laboratory.

The importance of hydraulic volume defects on permeability and the magnitude of the problem of predicting (and controlling) field leakage rates from compacted clay liners can be seen by noting the many pathways and mechanisms for cracking of clay liners as shown in Table 2.

The phenomenon of syneresis is of particular interest as a mechanism or cause of induced shrinkage cracking in clay liners. Syneresis can occur following exposure of the clay to certain types of fluids or leachates. It is highly likely from the description of test procedures and results reported by many investigators (Michaels and Lin, 1954; Anderson et al., 1981, 1982; Green et al, 1984) that the large increases in permeability that were observed following desolvation of water by low-dielectric organic solvents were caused by syneresis.

Syneresis and desiccation cracking both manifest themselves by development of shrinkage cracks; however, one occurs as the result of subaerial exposure or evaporation whereas the other occurs subaqueously. Burst (1965) has shown that cracks can develop in clastic deposits containing montmorillonite during periods of uninterrupted water cover by increasing the salinity of the aqueous environment. Plummer and Gostin (1981) conducted an extensive investigation to determine the most likely cause of shrinkage cracks which have been observed in sedimentary deposits. They showed that shrinkage cracks can form not only at the sediment-air interface by dessication processes but also at the sediment-water interface (and also substratally) by syneresis.

TABLE 2.   POTENTIAL CRACKING MECHANISMS IN COMPACTED CLAY LINERS

MECHANICALLY INDUCED CRACKING:

1.   Construction Cracking

> Tensile separation caused by compacting on weak subgrade or in laterally unconfined areas.

2.   Settlement Cracking

> Cracks induced by excessive settlement or differential movement within a clay mass after placement.

3.   Compaction Cracking (lift planes)

> Transverse cracks induced by poor compaction at boundaries between successive vertical lifts of compacted clay.

PHYSICO-CHEMICALLY INDUCED CRACKING:

1.   Syneresis

> Subaerial or substratal gel separation and shinkage cracking in a colloidal system induced by a change in the interparticle force balance resulting from:

> a) replacement of interstitial water with a low dielectric organic solvent.

> b) exposure or leaching with a highly saline aqueous solution.

2.   Dessication Cracking

> Subaerial shrinkage cracking induced by thermal removal (evaporation) of adsorbed water.

Daniel (1984) attributed a large part of the difference
in measured field vs. laboratory permeabilities of clay
liners to the presence of hydraulic volume defects. He felt
that these defects were not adequately sampled nor modelled
in the laboratory tests. He also noted that in all cases
involving relatively high rates of leakage, the liners were
relatively thin (less that 24 inches) and with only one
exception had been subjected to dessication cracking.

The persistence of cracks once they are initiated is a
matter of serious concern in the case of clay liners and
covers. The following case study is cited to underscore this
problem (Anderson & Kneale, 1984). A clay embankment
(LL=73%) was instrumented in order to record soil moisture
suction and surface crack development. A detailed crack
survey was made in a test plot and soil water suction
recorded at depths to 1 meter. Figure 3 shows the smoothed
response throughout the year at 60 cm depth.

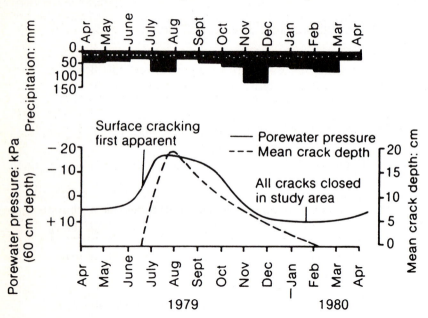

Figure 3. Soil suction change and crack development in a
test plot in a clay embankment (from Anderson
and Kneale, 1984).

Crack initiation as a result of dessication was observed
to correspond to an average soil suction of 6 kPa (approximate air entry value for clay); subsequent closure was a
function of rainfall pattern as shown in Figure 6. All
cracks appeared to close with the advent of winter rains.
The permeability of the intact (uncracked) zones relative to
those locations that had exhibited cracking during the
previous summer were of particular interest. Following
surface closure of cracks in February, field permeability

tests revealed permeability values of 5×10^5 m/sec for former cracked locations versus 4×10^7 in/sec for intact sites. In addition, a number of core samples were taken and subjected to x-ray radiography (Anderson, 1979). The samples were taken from sites that had cracked, but had subsequently closed. Although closed, the cracks showed up in the x-ray radiograms as zones of lower density even at depths up to 18cm. The implication of these findings is that cracks once formed continue to persist. In spite of apparent closure at the surface, vestiges of the crack remain, and are manifested as zones of lower density and significantly higher permeability below the surface.

In view of these findings it would be extremely useful to have some method of estimating the equivalent or composite permeability of a liner system that contains cracks--whether they are caused by syneresis or dessication. A tentative procedure for estimating this equivalent permeability is presented herein. This procedure is based on determining the equivalent permeability of a jointed mass in which the joints (or cracks) are either (1) clear and smooth sides, or (2) infilled with a soil of known permeability.

A schematic model of a liner that contains cracks (or joints) which partially penetrate the mass is shown in Figure 4. The cracks have a half-width "a" and are uniformly spaced a distance "b" apart. The liner thickness is "L" and the cracks penetrate a distance "L_c". The flow is laminar and parallel to the plane of the crack or joint. The equivalent permeability is calculated stepwise; first in the zone containing the crack or joint and then for the entire mass as follows:

Equivalent Permeability of Topmost Cracked Zone:

Case i) Cracks infilled with soil of know permeability, k_c

$$k_I = \frac{k_c A_c + k(A_c - A)}{A} \tag{1}$$

where: k_I = equivalent permeability of cracked zone
k_c = permeability of crack infill
k = permeability of intact liner
A = total unit flow cross sectional area
A_c = flow cross-sectional area of crack

if

$$A_c \ll A \quad \text{then} \quad A - A_c = A$$

and

$$k_I = k_c (A_c/A) + k \tag{2}$$

Case ii) <u>Clear, smooth sides cracks</u>

$$k_I = \frac{2a}{b} \frac{a^2 g}{3y} + k \qquad (3)$$

where: a = half width of the crack (m)
         b = crack spacing (m)
         g = gravitational constant (9.81 m/sec)
         y = kinematic viscosity of fluid

<u>Equivalent Permeability of Entire Mass (Zone I + II)</u>:

$$\overline{k} = k_{I,II} = \frac{L}{(L_c / k_I) + (L - L_c)/k} \qquad (4)$$

$$\overline{k} / k \ (\%) = \frac{100 \ L}{\dfrac{L_c \ A}{[A + (k_c/k)A_c]} + (L - L_c)} \qquad (5)$$

Let x = crack penetration ratio (L_c/L)
        r = crack area ratio (A_c/A)

Then

$$\overline{k} / k \ (\%) = \frac{100 \ [1 + (k_c/k) \ r]}{x + (1 - x) \ [1 + (k_c/k) \ r]} \qquad (6)$$

The ratio (k/k) represents the ratio of leakage rate through a "cracked" liner relative to that through an intact liner under a unit hydraulic gradient. This ratio is plotted in Figures 5 and 6 as a function of the crack penetration ratio (L_c/L) and crack permeability ratio (k_c/k). Plots are shown for two different crack area ratios $(A_c/A = 10^{-4}$ and $10^{-3})$ which are equivalent to crack (joint) widths of 0.1 and 1.0 mm spaced uniformly 1 m apart.

Several interesting conclusions can be drawn from the plots shown in Figures 5 and 6, namely:

1.   Cracks do not significantly affect or increase the leakage rate until they penetrate at least 80% of the liner thickness.

2.   Cracks which do penetrate all the way through a liner can greatly increase leakage rates. This increase can reach 3 orders-of-magnitude when the crack permeability ratio exceeds 10^5 and the crack area ratio equals or exceeds 10^{-3}.

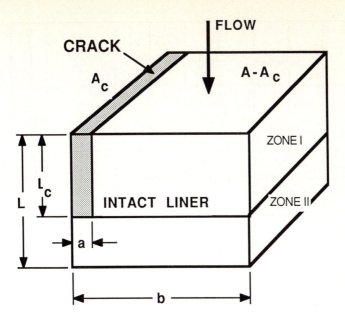

k   =   permeability of intact liner
k   =   permeability of crack
A   =   flow cross sec. area of crack
A-A =   flow cross sec. area of intact liner

Figure 4.   Idealized model of a bottom liner with uniformly spaced,
partially penetrating cracks (or joints). The cracks are
treated as either (1) infilled slots with a soil of known
permeability or as (2) clear, smooth sided joints.

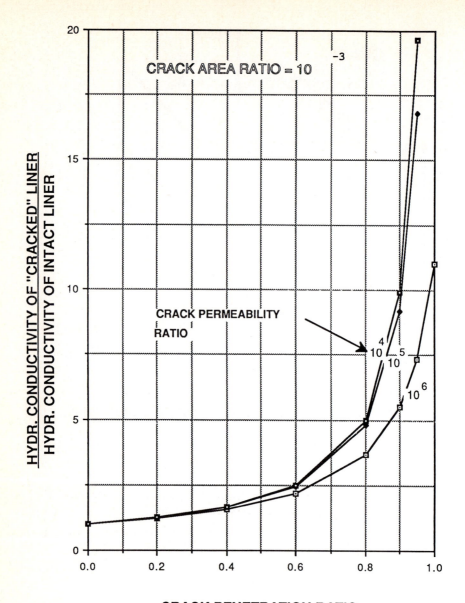

Figure 5. Relative permeability of "cracked" liner vs. crack penetration ratio and permeability ratio. Crack area ratio (width/spacing) equals 10^{-3}

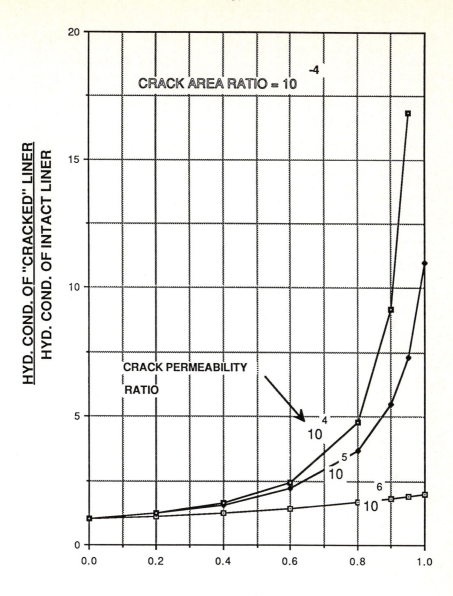

Figure 6.   Relative permeability of "cracked" liner
            vs. crack penetration ratio and permeabi-
            ratio.   Crack area ratio (width/spacing)
            equals 10^{-4}.

Both the likelihood and the gravity of the consequences
of fully penetrating cracks, fissures, etc. are greatest in
thin liners or membranes. There is little redundancy or
backup in a liner that is only a few millimeters or less in
thickness at the outset. This fact needs to be kept firmly
in mind when evaluating and selecting a liner system.

DIFFUSION OF LEACHATES ACROSS LANDFILL ENVELOPES

Leachate migration studies have tended to focus almost
exclusively on convective fluid movement and hydraulic
conductivity. The movement of leachate solutes, on the other
hand, is governed by both convection and diffusion. In cases
of very low seepage velocity (characteristic of dense natural
or compacted clays) diffusional transport can be quite
important (Goodall and Quigley, 1977; Folkes, 1982; Tallard,
1984; and Desaulniers et al., 1984).

Diffusion flow is also the main cause of solute movement
across polymeric membranes (Haxo, 1984). In contrast to clay
soils and other porous materials, polymeric membranes or
liners are non porous. Haxo (1984) as identified the
following steps that control the movement of liquids or
fluids through non porous membranes:

1.   Dissolution of the liquid in the membrane

2.   Diffusion of the liquid through the membrane

3.   Scavenging or removal of the liquid on the
      downstream side of the membrane.

Thus, the principal driving force for the transmission of
liquid solutes across membranes is a concentration gradient,
not a hydraulic gradient. Results of pouch tests conducted
by Haxo showed that commercial polymeric membranes with
negligible hydraulic conductivities were, nevertheless,
permeable (in a diffusion sense) to water, and some organic
liquids, e.g., oils, acetone, xylene, organic dyes, and
hydrogen and hydroxyl ions.

Recent work (Desaulniers et al., 1981) has shown that at
very low pore water velocities characteristic of water
move-through clayey deposits, solute transport is largely
diffusion controlled. Desaulniers and his co-workers used a
one-dimensional solution to the advection-dispersion equation
to show that chlorides in a Paleozoic bedrock underlying a
clay till were able to move upward by diffusion along a
concentration gradient against a downward flow of groundwater
through the till. The bedrock was a source of high chloride
concentration.

An analogous situation exists in a landfill containing
leachate which has solute concentrations exceeding those in

the surrounding groundwater. A schematic illustration of a hazardous waste leachate/clay barrier/groundwater system is shown in Figure 7. A clay barrier (slurry cut-off) wall is shown ringing the waste to prevent lateral migration of the leachate and another clay barrier (compacted clay liner) underlies the waste to prevent downward migration. An additional margin of safety against diffusion efflux of solutes is presumed to exist by maintaining a counter hydraulic gradient into the landfill. The latter is accomplished by keeping the level of leachate in the landfill lower than the GWT outside by means of an internal leachate collection system.

Breakthrough curves are shown plotted in Figures 8 and 9 for parameters and conditions typical of clay slurry cutoff walls in use at present. A breakthrough curve is a plot of the exit/source concentration ratio (C/Co) vs. time (t) on the exit side of a clay barrier of thickness X. The simulation shown in Figures 8 and 9 is based on a mathematical model of advection-diffusion which describes the one-dimensional flux of a single solute through a saturated porous medium. Details of this mathematical model, including assumptions and boundary conditions used for obtaining a solution, are described by Gray and Weber (1984).

The following parameter values were used in the mathematical simulation:

$$X = 3 \text{ ft}$$
$$D_e = 0.102 \text{ ft}^2/\text{yr} \ (3.7 \text{x } 10^{-6} \text{ cm}^2/\text{sec})$$
$$\bar{K} = 0.1 \text{ ft/yr} \ (10^{-7} \text{ cm/sec})$$
$$n = 25\%$$
$$= 1.65 \text{ gm/cc}$$
$$i = 0, -0.02, -0.05, -0.2, -0.5$$
$$K_p = 0, 5, 10, 20$$

These data are based on actual data and information furnished to The Michigan Dept. of Natural Resources as part of a permit application to construct clay slurry cut-off walls around the Granger Land Development Company Landfill #2 in Watertown Township, Clinton County. Because of extremely low flow velocities in a clay barrier, the dispersion coefficient (D) was assumed equal to the effective diffusion coefficient (D_e). Values for the latter were obtained from published data (Gilham and Cherry, 1982) for sand-bentonite mixes. The effect of a counter hydraulic gradient on solute transit time or breakthrough was modelled by varying the gradient from 0 to -0.5. The effect of solute retardation caused by solute sorption (partitioning) was studied by assuming local equilibrium and linear sorption, and by letting the corresponding partition coefficient (K_p) vary from 0 to 30. This variation covers the range of values reported in the technical literature (Kay and Elrick, 1967) for partitioning of non-ionic, organic solutes on sand-silt-clay mixtures with low organic contents (less than 4%). In contrast,

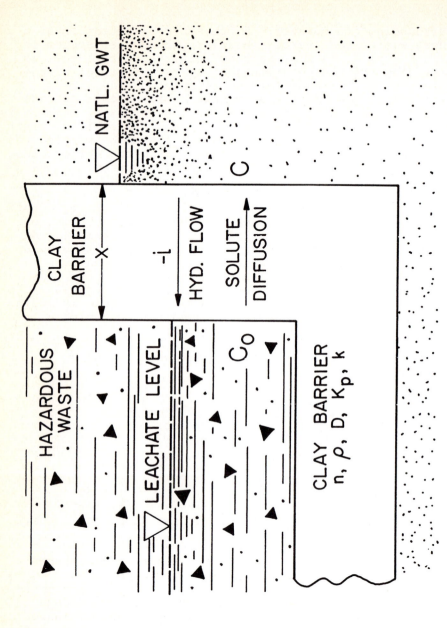

Figure 7. Schematic illustration of diffusive and advective mass transport across a clay barrier separating landfill leachate from groundwater.

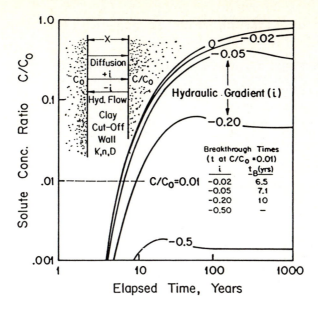

Figure 8.   Solute transport across clay cut-off wall showing
effect of counter hydraulic gradient (i) for case
of non-reactive solute (K_p = 0).

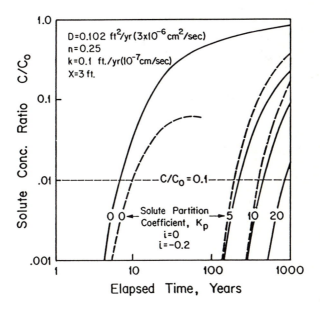

Figure 9.   Solute transport across cut-off wall showing
effect of solute partitioning (K_p > 0).

an ionic solute like chloride is non-reactive and has a partition coefficient equal to zero.

Breakthrough curves are plotted in Figure 8 for <u>nonreactive</u> solute transport ($K_p=0$) across a 3-foot thick clay slurry wall with different counter hydraulic gradients. Chloride is an example of a non-reactive solute. Breakthrough times, i.e., the time required for the exit concentration to reach 1% of the source concentration, are less than ten years...inspite of counter hydraulic gradients as high as -0.5.

Breakthrough curves are plotted in Figure 9 for <u>reactive</u> solutes ($K_p>0$) for both diffusion alone (i=0) and for a counter hydraulic gradient (i= -0.2). Lindane, an organochlorine insecticide (hexachlorocyclohexane), is an example of a reactive solute. Solute partitioning (i.e., sorption) increases breakthrough times, but not dramatically. Solute partitioning coefficients of 5 and 10 correspond to breakthrough time (t @ Co/C = 0.01) of 200 and 400 years respectively. A counter hydraulic gradient is ineffective in impeding solute breakthrough in this case.

The results of these preliminary analyses are significant for the following reasons:

1. Breakthrough times for non-reactive solutes (e.g. Cl) are very low--less than ten years--inspite of counter hydraulic gradients as high as -0.5.

2. Solute retardation can increase breakthrough times substantially ($t_B>1000$ yrs) only if the partition coefficient (K_p exceeds 20. The ration term (/n) varies between very narrow limits for earthen media; thus the retardation factor (R) is mainly affected by K_p alone.

3. Whereas clays are good exchangers/adsorbers for heavy metal cations (e.g., Cd, Zn, Hg), they exhibit poor sorption for non-ionic, organic solutes from aqueous solution. Elrick and Kay (1967) measured a partition coefficient value of only 3.0 ml/gm for Lindane sorption on Ca-bentonite. Hence, conventional slurry cut-off walls may provide little retardation or attenuation against diffusion of organic solutes.

4. In contrast to the unfavorable sorptive properties of pure clays, slightly organic clays or soils have much higher linear partition coefficients. Weber <u>et al</u>. (1983) have shown that organic carbon content of the solids appears to be the major factor controlling sorptive capacity.

5. For diffusion controlled transport, breakthrough time (t_b) varies as the square of the barrier thickness X. Hence, inadvertent "thinspots" in a slurry cut-off wall will greatly decrease breakthrough times. A 50% decrease in wall thickness, for example, would decrease the breakthrough time to one quarter of the original time.

These preliminary findings are of particular concern when viewed in the context of present and intended uses of clay slurry cut-off walls or thin, compacted clay liners for containment of hazardous waste leachates. Slurry wall systems are increasingly regarded as permanent, long-term remedies for containment and isolation of leachate from groundwater (Tallard, 1984).

ATTENUATION OF LEACHATES BY LANDFILL ENVELOPES

Attenuation has been defined by Folkes (1982) as the permanent or temporary decrease in the maximum concentration of a solute in transport due to various physical, chemical, and biological processes. Possible attenuation mechanisms or reactions between a liner and solutes include:

1. Precipitation and filtration

2. Sorption and exchange of polar and ionic solutes on clay minerals

3. Degradation of organic compounds to inert forms

4. Neutralization of excess acidity

Some of these attenuation mechanisms are reversible upon an increase in solute concentration or changes in chemistry, pH, and temperature (Griffin and Shimp, 1978).

Attenuation capacities of sand and calcium-saturated clay mixtures have been reported by Griffin and Shimp (1978). Attenuation increased with clay content, and was greatest for montmorillonite, followed by illite, then kaolinite... primarily due to the relative cation exchange potentials of the clay minerals. In the case of landfill leachates these same clay sand mixtures tended to completely attenuate (remove) heavy metals (Hg, Pb, Cd, Zn) from leachate passed through the mixture. Cartwright et al. (1981) reported a "hardness halo" in the leachate after it passed through these clayey media. The hardness halo was caused by preferential adsorption of heavy metal ions which displaced the original calcium ions on the exchange sites.

Cartwright et al. (1981) recommended that clay liners be designed to optimize attenuation rather than containment in

order to avoid the bathtub (overflow) problem in a landfill.
Their data suggested that over-all pollution would be
decreased if landfill liners were designed to achieve higher
permeability and to selectively attenuate the most toxic
pollutants from the leachate. They presented a procedure for
determining the point of "optimal" attenuation for each
chemical constituent in a leachate in terms of liner
hydraulic conductivity, exchange capacity, and liner
thickness.

Several investigators (Karickhoff et al. 1979; Weber et
al 1983) have noted the overwhelming importance of carbon
content of a soil or sediment in the adsorption of neutral,
organic compounds. This class of organics includes many of
the EPA priority pollutants, e.g., PBB, HBB, MCB, TCB. Karickhoff et al. (1979) showed that the magnitude of the
partition coefficient for a specific organic pollutant in a
particular sorbent system was directly related to octanol;
water partition coefficient of the pollutant and the organic
content of the solid phase onto which the sorption occurs.
Weber et al. (1983) verified this finding by showing that
pure clay minerals or natural clays stripped of their organic
content (by peroxide oxidation) had significantly lower partition coefficients. Kay and Elrick (1967) had observed the
same effect for Lindane sorption on clay-silt-sand mixtures
with and without organic matter.

These finding suggest that the sorption (attenuation)
properties of clay barriers might be improved by
incorporating organic additions or carbonaceous matter in
clay liners or clay slurry wall backfill mixes.

Unlike clay liners, polymeric membranes are non porous
and offer little prospect of attenuation. They are designed
for containment primarily and have relatively little or no
capacity for sorption and ion exchange of polar and ionic
solutes. Thus, the advantage gained in containment is lost
by poor attenuation. Moreover, if one also accepts the
likelihood that most leakage through a liner--be it compacted
clay or polymeric membrane--is through hydraulic volume
defects, then even the containment advantage of a membrane is
comprised. Thin, polymeric membranes are just, if not more,
vulnerable to this problem than are compacted clay liners.

GEOTECHNICAL FACTORS DETERMINING FINAL STORAGE QUALITY

From the preceding discussion a number of important
factors can be identified that determine the geotechnical
quality of final storage. Safety against mass stability
failure of impoundment dykes and waste stockpiles depends
largely on traditional geotechnical engineering design
considerations. Safety against leakage problems, on the
other hand, depends to a large extent on waste/leachate/

envelope interactions. Factors which significantly affect
this interaction are briefly cited and discussed below:

1. <u>Selection of liner/barrier materials</u>

It appears that no single lining material is superior
for the long term impoundment and/or storage of all wastes.
Instead, liners should be selected for specified wastes
based on compatibility/permeability tests with the waste in
question.

Much of the criticism that has been directed at the use
of compacted clay liners appears to be based on permeability
tests that are not at all representative of field conditions.
Reasons for high leakage rates that have been reported with
compacted clays include a) use of pure organic solvents as
opposed to aqueous leachates, b) use of fixed-wall permeameters, and c) use of abnormally high gradients.

Although there is no single lining material that is
superior for the containment of all wastes, it does appear
that compacted clays offer several advantages over polymeric
membranes. These advantages include the following:

a) <u>Lower cost</u>. Earthen materials can be obtained at
considerably less cost than polymeric materials. A
true cost comparison would require, however,
comparison of emplaced or installed costs including
costs of acceptance testing and inspection.

b) <u>Greater attenuation</u>. Clay can adsorb and attenuate
a variety of leachate constituents in ways that
synthetic, polymeric membranes cannot.

c) <u>Less vulnerability to defects</u>. The main advantage
claimed for polymeric membranes is positive
containment. Yet, membranes are just, if not more,
vulnerable to hydraulic volume defects than clays.
The likelihood and gravity of the consequences of
cracks or tears penetrating all the way through are
greater in thin membranes than in thick, well
compacted clays.

Realization of the advantages offered by compacted clays
will require, however, that closer attention be paid to the
type of clays that are used, how they are laid down and
compacted, what type of permeability testing is required,
what compaction/composition requirements are imposed to
mitigate shrinkage problems, and what restraints are necessary (e.g., minimal thickness) to minimize the effects of
diffusion transport of leachate. The organic content of a
clay soil strongly controls the adsorption or partitioning of
neutral, non-polar organic compounds. These findings suggest
that sorption (attenuation) properties of clay barriers for

organic leachates might be improved by incorporating organic additions or carbonaceous material (e.g., high carbon flyash) in clay liners or clay slurry wall backfill mixes. The consequences of tears or holes in membranes can also be mitigated significantly by using a composite liner system in which a synthetic membrane is placed on top of a compacted clay layer.

2. Emplacement/compaction

The permeability of compacted clay is very sensitive to compaction conditions, viz., compactive effort, molding water content, and type of compaction. Compaction conditions also influence volume stability or shrink-swell potential of clay liners. Volume stability may be equally important as permeability in the selection, design, and construction of a clay liner because shrinkage cracking can radically and irreversibly increase leakage rates.

Compaction slightly wet of optimum tends to reduce permeability sharply but it also increases shrinkage. Compaction procedures should be specified carefully to improve volume stability without sacrificing low permeability. Suggested guidelines in this regard include the following:

a) Compact to high relative density just slightly wet of optimum.

b) Avoid prolonged exposure of compacted lifts to sun and cover or moisten clay surface to prevent dessication.

c) Avoid use of high plasticity CH clays.

d) Avoid clay contact with pure solvents with low dielectric constants or highly concentrated brines.

3. Geometric design of envelope

Geometric design refers to such considerations as the thickness of a liner or cut-off wall, the size and shape of the envelope, degree of slope of liner and amount of crown on the cover, side slopes of dikes or pits, and how different elements of the envelope (cover, liner, dikes, cut-off walls, etc.) are connected or keyed into one another.

Considerations such as thickness are important with respect to the margin of safety or redundancy in event of puncture, tearing or erosion. Barrier thickness is also very important with regard to solute breakthrough times under diffusion which vary with the square of the barrier thickness.

The slope and height of retaining dikes affect their mass stability and the degree of slope on the sides of shallow pits or burial vaults affect membrane support and resistance to sliding.

Other factors also determine the geotechnical quality of final storage. Foremost among these are the issue of drainage provisions, both internal, such as leachate collection systems, and external, such as measures to intercept and divert surface water away from a landfill. Proper siting of a landfill in a geologic setting is also a critical consideration. A discussion of these factors, although important, lies outside the scope of this paper.

REFERENCES CITED

Anderson, M.G. (1979). On the potential of radiography to aid studies of hillside hydrology. Earth Surface Processes, 4: 77-83.

Anderson, M.G. and Kneale, P.N. (1984). The influence of vegetation on the swelling and shrinking of clay. Discussion in Geotechnique, 24(@): 139-171.

Anderson, D. (1982). Does landfill leachate make clay liners more permeable? Civil Engineering-ASCE, 52|(9):66-69.

Anderson, D. and Brown, K.W. (1981). Organic leachate effects on the permeability of clay liners, In Land Disposal: Hazardous Waste. Proceedings, 7th Annual Research Symp., U.S. Envl. Protection Agency, Philadelphia, pp. 119-130.

Anderson, D. Brown, K.W. and Green, J. (1982). Effect of organic fluids on the permeability of clay soil liners. In Land Disposal of Hazardous Waste, Proceedings, 8th Annual Research Symposium, U.S. Envl. Protection Agency.

Brown, K.W., Thomas, J.C., and Green, J.W. (1984). Permeability of compacted soils to solvent mixtures and petroleum products. In Land Disposal: Hazardous Waste, Proceedings, 10th Annual Research Symposium, U.S. Envl. Protection Agency, Cincinnati, OH, pp. 1-14.

Burst, J.F. (1965). Subaqueously formed shrinkage cracks in clay. Journ. of Sedimentary Petrology, 35(2):348-353.

Cartwright, K., Griffin, R.A., and Gilkeson, R.H. (1981). Migration of landfill leachate through glacial tills. Groundwater, 15(4):294-305.

Cheremisonoff, N.P. et al. (1979). Industrial and Hazardous Waste Impoundments. Ann Arbor Science Publications, Inc., Ann Arbor, MI.

Daniel, D.E. (1984). Selected case histories of field performance of compacted clay liners. Journ. Geotech. Engineering Division, 110(2):285-300.

Desaulniers, R.E., Cherry, J.A., and Fritz, P. (1981). Origin age, and movement of pore water in argillaceous quarternary deposits of southwestern Ontario. Journ. Hydrology, 50:231-257.

Folkes, D.J. (1982). Control of contaminant migration by use of clay liners. Can. Geotech Journ., 29:320-344.

Garcia-Bengochea, I., Lovell, C.W., and Altschaeffl, A.G. (1979). Pore distribution and permeability of silty clays. Journ. Geotechnical Engineering, 105(7):839-856.

Gilham, R.W. and Cherry, J.A. (1983). Predictability of solute transport in diffusion-controlled hydrogeologic regimes, Proceedings, Symposium on Low-Level Waste Disposal, U.S. NRC, NUREG/CP-0028, Conf-820911, Vol. 3, pp. 379-410.

Goodall, D.C. and Quigley, R.M. (1977). Pollutant migration from two sanitary landfills near Sarnia, Ontario. Canadian Geotechnical Journal, 14:223-236.

Gray, D.H. and Weber, W.J. (1984). Diffusional transport of hazardous waste leachates across clay barriers. Proceedings, Seventh Annual Madison Waste Conf., University of Wisconsin-Extension, Madison, pp. 373-389.

Green, W.M., Lee, F.G., and Jones, R.A. (1981). Clay-soils permeability and hazardous waste storage. Journal of WPCF, 53(8):1347-1354.

Griffin, R.A. and Shimp, N.F. (1978). Attenuation of pollutants in municipal landfill leachate by clay minerals, Illinois State Geological Survey, report prepared for U.S. Envl. Protection Agency, Report No. EPA-600/2-78-157.

Griffin, R.A. and Chian, E.S.K. (1979). Attenuation of water-soluble polychlorinated biphenyls by earth materials, Illinois State Geological Survey, Environmental Geology Notes, No. 86, Urbana, IL., 98 pp.

Haxo, H.E. (1980). Interaction of selected lining materials with various hazardous wastes. In Disposal of Hazardous Wastes. Proceedings, 6th Annual Research Symposium, U.S. Environmental Protection Agency, EPA-600/9-80-010, pp. 160-180.

Haxo, H.E. (1981). Durability of clay liners for hazardous waste disposal facilities. In Landfill Disposal: Hazardous Waste, Proceedings, 7th Annual Research Symposium, U.S. Envl. Protection Agency, Philadelphia, pp. 140-156.

Haxo, H.E. (1984). Permeability characteristics of flexible membrane liners measured in pouch tests. Proceedings, 10th Annual Research Symposium, U.W. Environmental Agency. Cincinnati, OH, pp. 230-237.

Karickhoff, S.W., Brown, D.S., and Scott, D.A. (1979). Sorption of hydrophobic pollutants on natural sediments. Water Resources, 13:241-248.

Kay, B.D. and Elrick, D.E. (1967). Adsorption and movement of Lindane in soils. Soil Science, 104(5):314-322.

Lambe, T.W. (1954). The permeability of compacted fine-grained soils. In ASTM Special Technical Publication No. 163, pp. 56-67.

Mitchell, J.K.,Hooper, D.R. and Campanella, R.G. (1965). Permeability of compacted clay. Journal of Soil Mechanics and Foundations Division, ASCE, 91(SM4): 41-65.

Mitchell, J.K. and Madsen, F.T. (1987). Chemical effects on clay hydraulic conductivity. Proceedings, ASCE Geotechnical Engineering Specialty Conference, Ann Arbor, Michigan, Geotechnical Special Publication No. 13, pp. 87-116.

Seed, H.B., Mitchell, J.K. and Chan, C.K. (1960). The strength of compacted cohesive soils. Proceedings, ASCE Geotechnical Engineering Division Specialty Conference, Boulder, Colorado, pp. 877-963.

Tallard, G. (1984). Slurry trenches for containing hazardous wastes, Civil Engineering, ASCE, 54(2):42-45.

Weber, W.J. et al. (1983). Sorption of hydrophobic compounds by sediments, soils, and suspended solids - II. Sorbent evaluation studies. Water Resources, 17(10):

INTERACTIONS OF LEACHATES WITH NATURAL AND SYNTHETIC ENVELOPES

Grahame J. Farquhar and Wayne Parker
Department of Civil Engineering
University of Waterloo (Canada)

ABSTRACT

This review paper deals with interactions between barriers including clay soils, either natural or amended, and synthetic materials in contact with leachates and other liquids. It appears that MSW leachate does not adversely affect the permeability of clay soils or many of the synthetic materials although the latter do experience certain physical property changes including swelling and organic uptake. The impact of these changes on permeability in the longer term is not known. It was also found that high concentrations of many inorganic contaminants and dilute concentrations (< 0.1%) of several organic compounds do not appreciably increase the permeability of either material type. Concentrated, generally pure organic liquids, have been shown to increase the permeability of clay soils. Chlorinated solvents have had a similar effect on most synthetic barrier materials. Both situations should be avoided.

The paper examines contaminant attenuation in natural barriers and concludes that, while some contaminants are well attenuated, others are not. Poorly attenuated compounds can diffuse through the barrier at velocities exceeding the theoretical fluid transport rate. Accurately predicting contaminant migration through barriers requires input information which is not presently available. It is acknowledged that, on one hand the performance requirements of barriers are stringent but on the other, the ability to quantify their behavior and the factors which affect it is limited. The paper concludes that additional and more lengthy field testing is essential for improved barrier understanding and design.

1. INTRODUCTION

The intention of this paper is to review information on the effects of leachate exposure on natural and synthetic envelope materials. It is intended to be of use in assessing the suitability of these materials as barriers against long term contaminant transport at waste disposal sites. Ideally, one would want to consult only those sources of information dealing with long term, field investigations. However, very few field studies of landfill/barrier interactions have been reported upon, and of these, none extend long enough to represent full term behavior at most sites. Consequently, the review was expanded to include experimentation done in laboratories with liquids in addition to landfill leachate in order to broaden the data base. The work was directed only toward liquid/barrier interactions although it is acknowledged that restriction of vapour movement is also important at some sites. The barriers considered in this review included mostly landfill liners, cut-off walls and trenches.

2. BARRIER PERFORMANCE EXPECTATION

Barriers at waste disposal sites, are designed with impedance to leachate transport as the main performance criterion. Since most barrier materials are permeable to some degree, design involves maintaining a permeability standard throughout the design life of the barrier. For aqueous leachates, hydraulic conductivities less than 10^{-7} cm. s^{-1} are usually required. Consequently, barrier failure occurs when overall permeability increases above the design value (Peterson and Gee, 1985) or some specified upper limit.

Barriers have also been designed to retard leachate contaminants from a more mobile liquid phase (Griffin, 1983; Gordon et al., 1984; Weinberg, et al., 1985; Bowders and Daniel, 1986). The main processes which provide retardation include adsorption, precipitation, biodegradation and filtration. Failure in this regard occurs when the barrier is unable to achieve the necessary contaminant retardation to meet environmental standards.

In the following sections, attention is focussed on the impact of leachates and other liquids on the permeability and attenuation capacity of barrier materials.

3. INTERACTIONS WITH NATURAL BARRIERS

In this paper, natural barrier materials have been taken to include soils mainly clays and soils amended with such additives as cement, polylectrolytes, fly ash and others.

3.1 FIELD INVESTIGATIONS

A Sarnia, Ontario Landfill

Quigley and Rowe (1986) have presented one of the few reports written on field investigations of landfill leachate impacts on barrier systems. The landfill was near Sarnia, Ontario, and had received municipal solid waste (MSW) to a depth of 7.5 m for 15 years prior to the investigation. The barrier consisted of in situ, massive grey clay to a depth of approximately 30 m. The top 3 m of clay were characterized on average as:

Minerology:	34% carbonate, 25% illite, 24% chlorite, 15% quartz and feldspar and 2% smectite
Particle Size:	40-45% < 0.002 mm
CEC:	15 meg. 100 g^{-1}
Moisture:	Actual water content 28%, PL = 11% LL = 32%
Fissures:	None visible
Permeability:	Not reported

The zone of saturation was mounded in the landfill 4.5 m above the clay producing a hydraulic gradient of 0.25 $m.m^{-1}$ and an estimated downward fluid velocity of 0.24 $cm.yr^{-1}$.

Core samples of the clay below the fill were taken and analyzed to establish contaminant transport distances. The results of this showed that Cl and Na had migrated to distances of 1.5 m below the fill, further than all other contaminants. Concentrations as high as 1200 $mg.l^{-1}$ and 2890 $mg.l^{-1}$ were measured for Cl and Na respectively although they were often equivalent on a $moles.l^{-1}$ basis. Since the advective transport distance had been estimated at 3 cm, molecular diffusion appeared to be the major transport mechanism. Unfortunately, the chemical composition of the leachate was not reported. Concentrations of Cl in excess of 1000 $mg.l^{-1}$ are common in MSW based leachate but concentrations of Na greater than 2500 $mg.l^{-1}$ are not. It may have been that Na was released from the clay in exchange for other cationic species in the leachate.

All heavy metals were retained within the upper 15 cm of the clay with Fe, Zn, Cu, and Pb being above background levels in this zone. Concentrations of Fe as high as 40,000 $g.g^{-1}$ were measured on the clay.

Organic contaminants up to 40 moles.m^{-3} (as acetate?) were found in pore water at the interface. However this reduced to background concentration in a depth of 1.1 m. The upper 20 cm of the clay were black and oily, suggesting anaerobic microbial activity as the mechanism responsible for the reduction of the leachate organic matter. The redox potential (Eh) of - 300 mV measured at the interface is consistent with this activity.

The effect of the leachate on the permeability of the clay at the interface was not assessed. One might guess that the organic residue created by the microbial activity would in fact reduce the permeability of the clay. It was clear that the clay deposit worked well as a barrier in this case with one condition. Diffusional transport of Cl and Na did reach 1.5 m into the clay. This would continue further with time, especially if further additions of MSW were to be made. The rate of advance would be minimized however if leachate would be pumped off the clay barrier thus reducing both driving forces, the concentration and the hydraulic gradients.

Boone County, Kentucky Test Site

A second field investigation, this at the Boone County, Kentucky site, was reported upon by Emcon Associates (1983). Several barrier materials including a clay liner were exposed to MSW leachate for 9 years in a full scale landfill. The clay liner was 61 cm thick with PL = 20%, LL = 42% and a moisture content at placement of 25%. The hydraulic conductivity in place was measured at 4×10^{-7} cm.s^{-1} (2×10^{-8} cm.s^{-1} laboratory). Chemical analysis at depth within the liner showed very little change in metal composition. No deterioration of permeability appeared to have occurred since no increase in leachate flux through the clay was observed during the 9 years of experimentation. The effects of longer term exposure to the leachate and diffusion of more mobile contaminants were not commented upon.

Three Landfills in Wisconsin

Gordon et al. (1984) evaluated the effectiveness of clay liners at three Wisconsin MSW landfills installed in the mid 1970's. The liners ranged in thickness from 1.2 to 1.5 m with specifications for maximum in place hydraulic conductivities (k) of 10^{-7} cm.s^{-1} and compaction to 90% modified Proctor. Leachate collection systems including drainage plane slopes and materials as well as pipe slopes, spacings and diameters were designed to minimize leachate head build-up on the liners. The zone of saturation was a minimum of 7.6 m below the base of the liner.

The Eau Claire County landfill field tests using a standpipe apparatus produced some k's in excess of the 10^{-7} cm.s^{-1} requirement. Difficulty in achieving saturated flow with this method may have also been a problem such that the actual liner k's may be even higher. Undisturbed Shelby tube samples were not used to measure permeability at this site as was the case at the other Wisconsin sites. Leachate collection (suction) lysimeters were installed within the liner (depth not given) and 1.5 m below it at 2 locations. Groundwater monitoring wells were also installed approximately 30 m downflow from the perimeter of the fill area. Refuse placement was begun in 1979 and lysimeter monitoring was also initiated then. The quality of the liquid collected in the lysimeters at the oldest region of the site is summarized below (sample values from data set):

Parameter Sample Dates:	79/11/26	80/11/19	83/9/16
pH - liner lysimeter	7.9	"6.6"	6.6
- 1.5 m depth lysimeter	7.6	6.2	7.0
Conductivity (mhos. cm^{-1})	600	1080	2000
	200	350	520
Hardness (mg.l^{-1} as Ca CO$_3$)	190	528	870
	79	262	290
Cl mg.l^{-1})	10	12	130
	4	10	25
COD (mg.l^{-1})	28	114	140
	16	15	---

The data show that the quality of liquid collected in both lysimeters deteriorated significantly during the 4 year study period. Although no leachate contaminant concentrations were reported, it would appear the biodegredation of leachate organic matter at or near the liner surface was active with traces of poorly biodegradable material moving downward. Mobile Cl also moved downward as did divalent metallic ions either from the leachate or as a result of exchange mechanisms within the clay. The anion(s) to balance these cations were not reported but were likely to have been mainly HCO$_3$ and Cl. No significant changes in the groundwater quality were detected at the closest downflow monitoring well.

Based on somewhat pessimistic values of k, porosity (n) and leachate head on the liner, it was estimated that the seepage rate through the liner was on the order of 0.3 m.yr^{-1}. Velocities of poorly attenuated contaminants would be even larger because of molecular diffusion. These factors help to explain the appearance of leachate at the base of the 1.2 m thick clay liner within the 4 year period. This poor liner performance would have been improved by increasing liner thickness and compaction (to reduce k). There is no evidence that leachate/liner interactions were responsible for the high k values.

The Marathon County Landfill was begun in 1980 with the installation of a 1.2 to 1.5 m clay liner with remoulded laboratory k's < 10^{-7} cm.s^{-1}. Three 20 mil PVC basin lysimeters were installed immediately below the liner to measure percolate volume and quality. Some liquid was collected initially in one lysimeter but this declined steadily with time. The other two were dry throughout the 4 year study period. No measurable change in groundwater quality was detected in monitoring wells immediately down flow from the site. Thus, the liner appears to have been functioning effectively at the time of investigation but one would not expect leachate or its contaminants to have penetrated the liner within the study period. Future results will be of greater interest.

The third site, the Brown County Landfill, begun with a liner in 1976, exhibited slight increases in conductivity, hardness, and COD in the groundwater approximately 3 m below the liner. However, the monitoring wells were installed down through the liner and, in spite of attempts to prevent it, leachate is likely to have moved downward through the liner along the bore holes. The authors acknowledge that this is an unsatisfactory method of monitoring groundwater quality beneath a landfill.

3.2   LABORATORY SCALE EXPERIMENTATION

Laboratory scale experiments to study the impact of leachates on liners and other barriers have been preferred to field experimentation by most scientists and engineers because of comparative economies in time and money. However, it must be borne in mind that:

- most experimentation does not extend long enough to represent long term field conditions;

- many experiments are accelerated through the application of hydraulic and concentration gradients which are often much higher than field conditions;

- most experiments are not large enough to account for soil
  heterogeneities and installation inconsistencies that can occur in
  the field;

- methods to extrapolate laboratory data to field conditions are
  poorly understood.

In spite of these limitations, however, accelerated laboratory scale
experiments do provide insights into field performance and data which can
be correlated to real systems in the future.

3.2.1 LANDFILL LEACHATE

There have been several laboratory scale studies in which landfill
leachate has been contacted with clays and clay soils. These have
generally been done with some form of column, often operated as a
permeameter in which the soil is "undisturbed" taken from core samples or
remoulded and compacted in place. Leachate (hydraulic) gradients have
ranged from less than 1 to greater than 100 using liquid, air or some
other gas to pressurize. Permeameters have been of both the rigid and
flexible side wall varieties. Leachates have of course been highly
variable and study objectives have been to examine impacts on k and/or
attenuation of contaminants. Experimental duration has been usually
based on the numbers of pore volumes (PV) collected with 3 to 7 PV being
common. Because of this multiplicity of experimentation, it is
difficult to normalize the information presented. However, an extended
summary of experimental results is presented here.

Effects on Permeability

Most investigators have found that exposure of clay materials to
landfill leachate tends not to increase permeability (Anderson and Jones,
1985). In fact, permeabilities have decreased in many cases because of
biomass and precipitate accumulation at the clay surface (Griffin et al.
1976; Farquhar and Constable 1978; Biene and Geil, 1985), Wuellner et
al., 1985; These results are encouraging but are based on laboratory
studies and their attendant shortcomings as noted above.

The work of Wuellner et al. (1985) is worthy of some detailed
examination. The average properties of their clay samples included: LL
= 31.5%; PL = 17.5%; optimum moisture content = 15%; max dry density =
1,880 kg.m^{-3}. The leachate used was typified by the Omega Hills,
Wisconsin Sample NR7:

$$pH = 6.4 \qquad Fe = 165 \text{ mg.l}^{-1}$$
$$COD = 10,060 \text{ mg.l}^{-1} \qquad K = 1,080 \text{ mg.l}^{-1}$$
$$TKN = 1,280 \text{ mg.l}^{-1} \qquad Na = 1,950 \text{ mg.l}^{-1}$$
$$NH_4 = 990 \text{ mg.l}^{-1} \qquad Cl = 3,750 \text{ mg.l}^{-1}$$
$$TSS = 18,600 \text{ mg.l}^{-1}$$

This is comparatively strong being typical of a young MSW leachate. It also contained limited amounts of priority organics:

1,1 dichlorobenzene	$640 \text{ }\mu\text{g.l}^{-1}$
ethylbenzene	$3,600 \text{ }\mu\text{g.l}^{-1}$
methylene chloride	$2,700 \text{ }\mu\text{g.l}^{-1}$
toluene	$11,000 \text{ }\mu\text{g.l}^{-1}$
trichloroethylene	$2,800 \text{ }\mu\text{g.l}^{-1}$

Samples were tested in falling head, rigid wall permeameters with hydraulic gradients up to 80 created with N_2 pressure. Tests typically required 3 to 4 months to produce up to 10 PV of effluent. Initially, k's ranged from 1.5 to 8×10^{-8} cm. s^{-1}. As testing proceeded, k's remained constant or decreased slightly. Methane gas was produced during the test and created problems with maintaining the falling head liquid tubes. The upper 4 cm of the clay exhibited blackening along fine joints which developed during exposure. This was attributed to anaerobic biomass development. The result is somewhat disconcerting however since it showed the existence of fracture as opposed to bulk flow through the soil. The fractures or joints may have been due to the remoulding process or to Na substitution if swelling clays were present in the sample. In any case, longer term exposure may have produced higher k's due to flow induced along the joints. This may also have field implications in the longer term but little information is available to support this idea.

Contaminant Attenuation Capacity

Work on quantifying the capacity of clay materials to attenuate leachate contaminants was begun in the early 1970's by Griffin (Griffin et al., 1976) and Fuller (Fuller and Korte, 1976) among others. The general trends in the results produced appear to be reasonably well established (Griffin, 1983). Mechanisms include dispersion, adsorption, precipitation and co-precipitation and biodegradation. These are in turn influenced by several factors such that contaminant attenuation tends to increase when:

- pH is in the neutral or slightly alkaline range
- Eh is positive
- contact time is increased
- temperature is increased.
- CEC is increased

Heavy metals tend to be removed quickly and extensively in clay soils through precipitation as hydroxides and carbonates or by coprecipitation with Fe and Mn hydrous oxides (Griffin et al. 1976; Fuller and Korte,. 1977). Removal efficiency tends to improve with increased pH. Exchange of resident ions is also active in removing cations with alkaline earth metals being released in most cases and possibly being discharged from the liner. Elevated hardness downflow from landfills is common. However, pH, Eh and relative ion concentrations must be known before a cation replacement series can be established.

Attenuation of anions in clay soils is not as well documented as it is for cations (Griffin, 1983). Given "typical" MSW leachate, one can expect extensive retardation of PO_4, some retardation of SO_4 and no retardation of Cl beyond that provided by diffusion.

Conversion of biodegradable organics can be expected at the clay liner surface once biomass has been established. Nonbiodegradable acidic and nonpolar organic residuals are likely to pass through the clay barrier.

Some of these trends are evident in the recent laboratory work of Yong and his co-workers (1986). Clay taken from a landfill near Montreal was remoulded in permeameters and characterized by the following data:

Minerology   : feldspar > illite > quartz > kaolinite
                     > hornblende > chlorite
CEC   : 60 meg. $100g^{-1}$
Moisture   : PL = 26%, LL = 65%
Permeability : < 10^{-7} cm.s^{-1}

Liquid extracted from the clay was analysed with the following composition being determined:

HCO_3 = 180 mg. l^{-1} Na = 130 mg. l^{-1}
Cl   = 170 mg. l^{-1} Mg = 29 mg. l^{-1}
TOC = 50 mg. l^{-1} Ca = 22 mg. l^{-1}
pH   = 7.5   K = 14 mg. l^{-1}
                             Fe < 0.1 mg. l^{-1}

Leachate for the investigation was taken from a collection basin at the landfill and characterized by the following concentrations:

BOD	=	450 mg. l^{-1}		Mg	=	350 mg.l^{-1}
COD	=	860 mg. l^{-1}		Ca	=	180 mg.l^{-1}
TOC	=	190 mg. l^{-1}		Fe	=	5 mg.l^{-1}
NH_3	=	21 mg. l^{-1}		Zn	=	2.5 mg.l^{-1}
Phenol	=	0.04 mg.l^{-1}		Cu	=	1.7 mg.l^{-1}
Cl	=	190 mg. l^{-1}		Pb	=	1.0 mg.l^{-1}
HCO_3	=	303 mg. l^{-1}		Hi,Cr,As < DL		
pH	=	6.9 mg. l^{-1}				

This is particularly weak leachate especially with respect to NH_3, Fe, and BOD and atypical of MSW leachate to which a clay barrier would be exposed in the early years of active leachate production. An explanation for the state of the leachate which could have included long term exposure to air, dilution by surface water or the age of the site was not provided by the authors.

Leachate was applied to 10 cm dia x 10 cm deep clay plugs at a pressure of approximately 30 kPa (gauge) in several replicates. Effluent contaminant concentrations were analysed as a function of pore volumes (PV) collected. Clay samples were analysed periodically for changes in resident ion concentrations. The following points summarize the results obtained:

1. pH increased from 6.9 in the leachate to > 8.5 in the effluent from the clay columns.

2. After 5 PV, all heavy metals were effectively retarded. The values of C/Co in the effluent increased slightly to

0.05 for Fe	0.02 for Cu
0.04 for Zn	0.01 for Pb

    The Fe profile within the clay pore liquid decreased from 5 mg.l^{-1} at the clay surface to 0 mg.l^{-1} at a depth of approximately 3 cm.

3. Column effluent concentrations of Na were substantially higher than that of the leachate influent. The values for Ca and K were slightly higher. Cation exchange was apparently taking place.

4. Chloride ion was conservative within the clay but exhibited a dispersed breakthrough curve over approximately 3 PV's.

5. TOC breakthrough paralleled that of Cl until later in the experiments when anaerobic conditions were induced in the columns and biodegradation had been initiated.

These results are similar to those obtained by Griffin et al., 1976 and Farquhar and Constable, 1978. Although there exists a substantial data base from which to work, there are significant limitations when the attenuation capacity of clay liners is relied upon to protect the soil and groundwater adjacent to disposal sites. These include:

1. Some contaminants are poorly attenuated in clay soils. Examples include chloride, monovalent cations, some organic acids (Weinburg et al., 1985) and neutral nonpolar organics (Brown and Anderson, 1983).

2. Adsorption of many ionic species has a capacity limitation and thus, the liner as an adsorbent, can become exhausted with respect to these contaminants and contaminant breakthrough will occur.

3. Predicting adsorptive capacity requires a knowledge of contaminant loading from the leachate and this is very difficult to determine accurately with current information and methods available.

4. Attenuation of some contaminants depends on the presence of others (Griffin, 1983) and thus may be delayed within and even discharged from the liner. Specific testing is required to determine this.

5. Ionic exchange reactions frequently occur in clay soils. Thus, while one ion is removed, another is released to solution and can be discharged from the liner.

6. Predicting liner attenuation requires quantification of all attenuating mechanisms for the unique liner/leachate combination of the site under consideration. Experimentation to provide the necessary data base is either difficult because of the time involved or impossible if no leachate has been produced. Thus reliance upon data from other sources is usually required and this is uncertain at best.

7. Estimating contaminant breakthrough of a soil medium is usually based on plug flow principles and equilibrium attenuation reactions. However, dispersed contaminant fronts, fissure flow, and nonequilibrium and nonlinear adsorption reactions will occur in many cases.

3.2.2 ORGANIC LIQUIDS

The growing body of literature written about interactions between specific organic liquids and natural barriers can be separated into categories based on:

- pure liquids or dilute aqueous solutions of them
- pure clays, mixtures of pure clays or mixtures of clay plus other materials such as cement, fly ash, etc.
- emphasis on either permeability changes or retardation capacity.

The information therefore covers several different applications ranging from liners to contain high strength industrial wastes to vertical barriers to enclose a plume of organics travelling in an aqueous phase. A brief, general summary of information available is presented here.

Reviewing the work of Brown and Anderson (1983) is an important first step to understanding the impact of pure organic liquids on clay barrier permeability. Their work involved representatives from four major organic liquid groups:

Organic Liquid Group	Representative Liquids	Comment on Potential Interaction
organic acids	acetic acid	- multiple potential attack mechanisms - poor adsorption
organic bases	aniline	- rapid adsorption onto clays
neutral polar	acetone, methanol, ethylene glycol	- compete with H_2O to wet clay - reduce fluid viscosity - adsorption inversely proportional to solubility
neutral nonpolar	xylene, heptane	- do not compete with H_2O to wet clay - adsorb poorly onto clay

Major attack mechanisms were identified as "dissolution and piping" in which material at the soil pore wall dissolves and can be released as "fragments" to the flow and "volume changes" caused in part by substitutions in the clay matrix. The long term effect of dissolution and piping is to increase permeability in enlarged flow channels. In the short term however, permeability may reduce due to clogging by soil fragments. Volume increases induced by organics can increase permeability through crack formation. Other mechanisms include: dessication producing cracks which increase permeability, changes to aqueous phase viscosity and surface tension which change flow properties.

Experiments were performed by passing liquids through 4 clay mixtures (> 35% clay minerals) remoulded into rigid wall permeameters and compacted to produce $k < 10^{-7}$ cm.s^{-1}. A minimum of 2 PV's were collected under pressure induced flow. The permeability of all 4 clays increased in response to all organic liquids. The mixed cation kaolinite and the calcareous smectite exhibited less change than the mixed cation, illite and the noncalcareous, smectite.

Aniline increased permeabilities from $< 10^{-8}$ cm.s^{-1} to $> 10^{-7}$ cm.s^{-1}. The mechanism appears to have been due to cracking and pore enlargement.

Acetone first reduced permeabilities but then increased them in excess of 10^{-6} cm.s^{-1} by displacing H_2O from clay and causing it to crack.

Xylene caused structural changes in the form of blocks and cracks through expanded lattice structure. Permeabilities increased from 10^{-8} cm.s^{-1} to greater than 10^{-6} cm.s^{-1}.

Similar results were reported by Anderson et al. (1985) in their work comparing double ringed permeameters with fixed and rigid wall units using bentonite slurry mixtures. Methanol and xylene increased permeability from $< 5 \times 10^{-8}$ cm.s^{-1} to near 10^{-5} cm.s^{-1} for most slurry preparations.

Acar et al. (1985) used a flexible wall permeameter with fluid head gradients up to 100 to test the effects of organic liquids on the permeability of remoulded kaolinite. They found that 100% solutions of acetone and phenol increased permeability slightly while benzene and nitrobenzene actually reduced it. They also found however that 0.1% solutions of all 4 chemicals reduced permeability slightly. The duration of the tests or the PV collected were not reported.

Weinberg et al. (1985) conducted experiments to evaluate retardation of organic compounds by slurry trench materials. They defined adsorption of two types as the main retardation mechanisms: chemisorption including coulombic attraction and ligand exchange which tend not be reversible; physisorption by means of van der Walls attraction which is easily reversed. Equilibrium adsorption isotherms were produced for many organic compounds (mainly pesticides) in aqueous solutions in contact with various slurry trench materials. The materials involved mixtures of bentonite, cement, sand, ground slate, fly ash, polyelectrolytes and flocculents. The results of the work are too extensive for detailed review here. It is sufficient to say that some combinations produced high capacity, nearly irreversible retardation of certain organics, e.g. hexachlorobenzene on fly ash and ground slate. Some materials however such as phenoxy acetic acids were poorly retarded. Adsorption trends were summarized with respect to the organic groups as shown in the table above.

3.2.3   INORGANIC AQUEOUS SOLUTIONS

Studying the impact of inorganic liquids on natural barriers has involved acids, bases and salt solutions. The emphasis has been mostly on changes in permeability. One of the few field investigations reported upon involved a brine storage pond lined with clay (Jones, 1985). The liner consisted of 20-40% quartz, 15-20% illite, 10-15% montmorillinite, 5-10% feldspar and 2-5% kaolinite at a depth of 0.5 m and compacted to 98% standard Proctor. It was placed in 1963 and examined in 1982. Permeability measurements averaged about 10^{-6} cm.s^{-1} in the field and 3×10^{-6} cm.s^{-1} in the lab, both values somewhat less than the 1963 estimates. Seepage rates through the clay in 1982 were estimated to be 0.4 mm. d^{-1} as compared to 3mm. d^{-1} in 1963. Apparently the brine solution (with TDS of 334,000 mg.l^{-1} measured in 1982) had no adverse impact and in fact reduced the permeability of the clay. Unfortunately, the author did not discuss brine leakage through the liner. Assuming an average seepage rate of 1 mm. d^{-1}, theoretical fluid breakthrough of the liner would occur in approximately one and one-half years. Chloride ion however, because it does not normally interact with clay and with the extreme concentration gradient, is likely to have broken through much sooner.

Peterson and Gee (1985) used laboratory columns to evaluate the impact of leachate from acidic mine tailings on the permeability of a clay liner. After 3 years of experimentation and the collection of 30 PV of column effluent, it was found that the permeability of the clay reduced by 2 orders of magnitude. This was attributed to precipitate formation in response to acid neutralization. The authors estimated that, for a 1 m thick liner with k = 10^{-8} cm.s^{-1} and a hydraulic head of 10, it would take approximately 300 years for acid breakthrough to occur. Discussions on the effects of diffusion, nonhomogeneities in the liner or channeling on this estimate were not presented.

Alther and co-workers (1985) used a standard API water loss chamber to examine changes in bentonite permeability induced by exposure to salt solutions. Slurries of bentonite were mixed with various salts in concentrations up to 8% by weight and formed into filter cakes at pressures up to 6.9 x 10^2 kPa. Permeabilities were calculated using Darcy's Law, measured permeate flow and cake area and thickness. Typical results are summarized below:

Salt	Measured k (cm.s^{-1}) For Salt Concentrations:		
	0%	0.5%	5.0%
KCl	3.85×10^{-8}	4.57×10^{-8}	2.0×10^{-7}
NaCl	4.14×10^{-8}	---	1.31×10^{-7}
CaCl$_2$	3.85×10^{-8}	1.63×10^{-7}	1.83×10^{-7}
CaSO$_4$	3.85×10^{-8}	9.4×10^{-8}	9.71×10^{-8}

It can be seen from the results that bentonite permeability is somewhat susceptible to high salt concentrations, the effect increasing with concentration. No mechanisms for this were provided but volume change due to ion substitution was likely taking place. Although permeability did increase, the residual permeability was still acceptable for many sites.

Lentz et al., (1985) studied the effect of acid (HCl) and base NaOH) on the permeability of kaolinite, kaolinite-bentonite and Mg-montmorillinite. Triaxial falling head permeameter tests were conducted with aqueous solution pH's ranging from 1 to 13. The only change in permeability occurred with the Mg - montmorillinite and the pH = 13 solution. In this case it decreased by a factor of 13 due apparently to Mg(OH)$_2$ precipitation in the pores.

3.3   DISCUSSION AND CONCLUSIONS

There is little evidence in the literature reviewed for this paper to show that MSW landfill leachates adversely affect the permeability of natural barriers.   This applies to essentially all soil materials which have been used for barrier construction except for the swelling clays. The same can also be said for dilute aqueous solutions of salts and inorganic acids and bases.   However, the definition of dilute in terms of specific concentrations is dependent on the type of clay.   For nonswelling clays, concentrations up to 1.0% by weight are not likely to have an adverse effect on permeability.   For swelling clays however, salt (especially Ca salts) concentrations on the order of 0.5% by weight have been shown to increase permeability somewhat.

An assessment of the effect of organic materials on the permeability of natural liners is somewhat more complex.   It would appear that dilute aqueous solutions of many organic solvents, acids, and bases on the order of 0.1% by weight have little adverse effect on permeability.   Even higher concentrations are likely to be acceptable but a paucity of data precludes any statement in this regard.   In contrast many pure organic liquids including acetic acid, acetone, aniline, methanol, xylene and benzene seriously increase the permeability of both swelling and nonswelling clays.

It is essential that these conclusions be considered in the context of   the   experimental   conditions   upon   which   they   are   based:

1.   Experimentation has   not been   carried   out long enough to encompass the period for which barriers are required to perform in most cases.

2.   Most of the   information available comes from accelerated laboratory scale tests which do not fully represent and can not effectively be extrapolated to field performance.

3.   Test   equipment,   materials   and   procedures   have   not   been standardized making comparing and normalizing data from different sources difficult.   This includes such components as:

   - permeameter   type,   applied   pressure,   fluid   used   to pressurize, sample size and preparation;
   - liquid type, preparation and mode of application;
   - experimental duration;
   - measurements taken.

Natural barriers can be expected to retard the transport of some contaminants through interaction with the barrier materials or biodegradation. Retardation generally improves when the barrier exhibits increased pH, CEC and organic and hydrous oxide content. Reliance on barrier retardation must acknowledge several limitations:

1. Not all contaminants will be retarded to the extent necessary to meet groundwater quality standards.

2. Retardation may have a capacity limit and thus contaminant loading must be used to design the barrier thickness to avoid breakthrough.

3. Ideal conditions of homogeneous barrier materials, bulk intergranular flow and equilibrium reactions will not occur in the field.

4. Interferences and interactions between contaminants affect retardation and thus leachate/barrier specific data are required to design the system.

Because of these conditions imposed, the conclusions reached must be taken as tentative, subject to confirmation by long term field investigations.

4.   INTERACTIONS WITH SYNTHETIC BARRIER MATERIALS

Synthetic barrier materials encountered in the literature span a broad range of substances. The more common of these include (Haxo, Jr. et al., 1985; Koerner 1986; Reades, 1987):

Plastics: polyvinyl chloride (PVC)
                polyethylene

Rubbers: butyl (BR)
                neoprene
                ethylene propylene diene monomer (EPDM)

Other: asphaltic

Synthetic materials have gained acceptance as barriers at landfills because they exhibit permeabilities $< 10^{-11}$ cm.s^{-1} for many fluids, they are highly resistant to numerous chemicals and in many cases can be installed at lower cost than clay liners. In contrast however, synthetic materials are vulnerable to damage and are difficult to seam in the field, some are susceptible to attack from certain chemicals and field experience with them is limited (Reades, 1987).

Published research on interactions between leachates, organic liquids and synthetic materials as barriers is sparse and, that which is available, has dealt mainly with simulated conditions. The work of Haxo, Jr. et al. (1985) is key in this regard and is summarized briefly below.

4.1 MSW LEACHATE

The research programme was initiated in 1973 and involved the contact of 65 materials as liners at the base of landfill simulators for up to 56 months. The liners including test seams and joints were placed at the base of the simulators, covered with 2.4 m of MSW and exposed to leachate produced through the addition of H_2O on top of the MSW. A total of 64 cm of H_2O were added each year and a leachate head of 30 cm was maintained on the liners. Measurements were taken periodically to determine if leachate was penetrating the liners. At the end of the contact period, the liners were recovered and tested to detect changes in such properties as organic content, puncture resistance, tensile strength and others. The programme also included the immersion of sample sheets of material in landfill leachate and the suspension of sealed pouches filled with leachate in ion free water. After the test period, the sheets were tested as above and the ion free water around the pouches was analysed to detect contaminant leakage.

The results from the landfill simulators showed that no leachate was leaking from any of the liner materials for periods up to 56 months. This is most encouraging in terms of synthetic liner use as barriers at landfills. There were however other results and observations which must be acknowledged in field applications:

1. The tests were short term with respect to landfill leaching life and the results must eventually be compared to longer term field behavior.

2. Changes in material properties did occur even though leachate leakage did not. The impact that these changes might have on longer term failure is not known. Some of the changes are noted below:

- All materials softened after 12 months but PVC, chlorosulphonated polyethylene (CSPE) and chlorinated polyethylene (CPE) hardened somewhat subsequently due to the loss of plasticizers up to 36% for some materials. PVC hardened the most.

- Many materials absorbed organics from the leachate with CSPE, CPE and EPDM absorbing the most.

- Seam strength loss occurred for CPE, CSPF and EPDM. Hot sealed and body solvent welded seams were best.

- LDPE retained its properties best but has poor puncture resistance which impedes field use.

3. Results from the pouch tests showed that leachate contaminants had diffused through the liner materials and into the ion free water. In addition, osmotic pressure forced ion free water through the liner and into the pouch. Transfer rates were not given nor were the contaminant concentrations within the pouches.

Emcon Associates (1983) made similar observations from tests conducted at the Boone County, Kentucky field site. LDPE, CSPE and CPE liners were placed on soil, covered with MSW in a large landfill and in test cells and exposed to the leachate generated for 9 years. Samples of the liners were examined subsequently as was the soil beneath the liners.

The LDPE liner had been exposed to full strength leachate and appeared to have been unaffected by the leachate. Very little swelling and no punctures were evident. Examination of the soils beneath the landfill showed no evidence of leachate having seeped through the liner.

The CSPE sample was in contact with only diluted leachate. However, it was swollen and had softened. The volume and weight of the liner had increased by 57% and 26%, respectively, due to absorption of the leachate. The seams had blisters and these appeared to contain water. No evidence of a loss in permeability was found.

The CPE liner was exposed to full scale leachate within one of the test cells. It exhibited significant absorption with volume and weight increases of 32% and 28% respectively. Analysis of the sand below the liner showed no evidence of leachate leakage.

These results are encouraging for the use of synthetic liners at MSW landfills. However, the conditions imposed upon interpreting the results from the Haxo, Jr. et al. study are also applicable to this work.

4.2 ORGANIC LIQUIDS

Haxo, Jr. et al. (1985) also examined interactions between synthetic materials and specific organic fluids using methods similar to those described in the previous section. The results of this work are extensive and are therefore summarized in general here. Observations by Reades (1987) have also been included.

Chlorinated solvents including chloroform, trichloroethylene, tetrachloroethylene and tetrachloroethane had significant effects on most synthetic materials producing increased permeabilities and swelling.

Polyethylene exhibited high permeation rates for chlorinated solvents but performed well with acids, bases and hydrocarbons in general. It retained its physical properties well in particular its puncture and tear resistance; it also maintained seam integrity.

CPE and CSPE performed nearly as well as HDPE but showed less resistance to solvents and hydrocarbons in general. Swelling in some cases was substantial. Physical property retention was good to excellent except for the puncture resistance of CSPE.

Neoprene demonstrated good resistance to hydrocarbons but was difficult to seam properly and is expensive in comparison to other synthetic materials.

Oil resistant PVC performed well in many respects but weathered poorly, becoming brittle and therefore being of limited use in large applications.

Selection of the proper synthetic material as a barrier is a complex process which must address:

1. liquid/material compatibility,
2. resistance to weathering,
3. ease and quality of seam preparation,
4. physical property requirement,
5. cost.

Chlorinated solvents appear to eliminate most synthetic materials with respect to criterion 1. PVC is eliminated by 2 in many applications. HDPE, CSPE and CPE would appear to be good choices for many cases. However, long term compatibility testing particularly in the field is essential before definitive judgements can be made on the suitability of synthetic materials as envelopes at waste disposal sites.

5.   COMMENT ON FINAL STORAGE

The current data base is not long enough nor does it contain enough contaminant/barrier interaction information to permit confident statements to be made on final storage. There are trends which look good but these require additional long term field study for verification. Since waste storage is inevitable at the present time, use of barriers remain necessary. Their performance must therefore be monitored and backed up with contingency systems in case of failure. Long term study of these systems will provide data essential to the development of rational barrier design principles.

5.1 NATURAL BARRIERS

There is little evidence to show that MSW landfill leachates adversely affect the permeability of compacted clay barriers. A more important problem is to maintain uniform compaction, grading and thickness throughout the site to avoid weak areas in the system. Hydraulic conductivities on the order of 10^{-8} cm s^{-1} are possible with clay soils but this does permit leakage, the amount depending on leachate head and barrier thickness. A 1.0 m thick liner with k=10^{-7} cm.s^{-1} and a leachate head of 0.5 m can produce leakage of several cm.yr^{-1}.

Clay soil barriers attenuate some contaminants well but others poorly. Thus reliance upon contaminant attenuation in the barrier must be limited to specific contaminants if at all. Design for this requires;

1.   a contaminant loading function (mass.time^{-1}) for the life of the site,

2.   a relationship to quantify the attenuation mechanism, its capacity, its kinetics and interferences imposed by other contaminants, and

3.   recognition that capacity limits exist for some mechanisms and that breakthrough can occur.

While much information exists, mainly in the form of equilibrium adsorption isotherms, data on loading functions, removal kinetics, interferences and flow through barriers are sparse. Confident predictions of contaminant attenuation within a barrier are therefore difficult to make.

Contaminants which are poorly attenuated in the barrier will diffuse at velocities which can exceed the fluid velocity by an order of magnitude in clay soils. Predictions require knowledge of specific diffusion coefficients and concentration gradients both of which are difficult to determine accurately with current information.

It also seems likely that inorganic liquids including salts and acids and bases in the pH range of 1 to 13 do not adversely affect the permeability of clay barriers. But contaminant diffusion through the barrier driven by large concentration gradients can be substantial.

Pure organic liquids of several types but solvents in particular can substantially increase the permeability of clay barriers. This situation is obviously one to avoid. Dilute aqueous solutions of many organic liquids do not appear to affect the permeability of clay barriers especially the non-swelling clays. It would appear that concentrations up to 0.1% by weight (total solvents) can be tolerated but knowledge of these interactions is limited.

Attenuation of organics in clay barriers must also be considered. Biodegradable organics should be removed perhaps after an initial breakthrough as biomass is established. Neutral nonpolar organics and organic acids are poorly adsorbed onto clay soils and will not be attenuated. Diffusion will cause them to travel at velocities exceeding the theoretical fluid velocity and eventually to break through the barrier. Again, predictions are hampered by limited data on diffusion coefficients.

5.2 SYNTHETIC BARRIERS

It is likely that MSW landfill leachates do not appreciably increase the permeability of many synthetic flexible membranes. The larger problem is one of maintaining integrity during installation and filling and of achieving proper seals. Hydraulic conductivities in the range of 10^{-11} cm.s^{-1} are possible and this means that leakage rates less than 1 cm.s^{-1} are possible depending on membrane thickness and leachate head. One disturbing trend in the data is the swelling of several membranes (CSPE, CPE, EPDM and PVC) due to organic uptake from leachate. The long term effect of this on permeability is not known. One set of experimental results has shown that contaminants can diffuse through many of the membranes. Determining rates of flux requires further experimentation.

Chlorinated solvents have deleterious effects on many synthetic membranes and contact should be avoided. HDPE, CSPE and CPE appear to perform well with other organics although swelling can occur in these cases as well.

Predicting membrane performance beyond our 12 or so years of experience is not possible.

5.3   COLLECTING FIELD DATA

It is essential that field investigations be continued to measure contaminant and fluid flux across barriers for as long as unacceptable leachate is produced. Documentation with respect to barrier type and design, waste composition, disposal methods, leachate production and composition is needed to provide design principles for the future.

Concentrations in the upflow groundwater, in the leachate and in the clay barrier pore liquid at placement must also be determined for comparison with the barrier seepage composition. Parameters should include as a minimum:

    Mobile Contaminants: Cl, conductivity, TOC, benzene,
                           toluene and xylene
    Special Contaminants: from special wastes
    Barrier Components: clay; Na, HCO_3
                           synthetic; " plasticizers"

6.   EPILOGUE

Those of us involved in the disposal of wastes are obliged to ensure that adequate protection of the environment is provided. Yet we must work with systems whose performance we do not fully understand and can not adequately predict. This dilemma requires that we be conservative in our judgements and vigilant in assessing their impact.

REFERENCES

Acar, Y.B., Hamidon, A.B., Field, S.D., and Scott, L., 1985. "The Effect of Organic Fluids on Hydraulic Conductivity of Compacted Kaolinite". In "Hydraulic Barriers in Soil and Rock", ASTM STP 874, A.I. Johnson, R.K. Frobel, N.J. Cavalli, C.B. Pettersson, Eds., American Society for Testing and Materials, Philadelphia, Pa.

Alther, G., Evans, J.C., Fang, H.Y., and Witmer, K., 1985. "Influence of Inorganic Permeants Upon the Permeability of Bentonite". In "Hydraulic Barriers in Soil and Rock", ASTM STP 874, A.I. Johnson, R.K. Frobel, N.J. Cavalli, C.B. Pettersson, Eds., American Society for Testing and Materials, Philadelphia, Pa.

Anderson, D.C., and Jones, S.G., 1985. "Clay Barrier - Leachate Interaction". Proceedings of National Conference on Management of Uncontrolled Hazardous Waste Sites, Oct. 31-Nov. 2, Washington, D.C.

Anderson, D.C., Crawley, W., and Zobcik, J.D., 1985. "Effects of Various Liquids on Clay Soil: Bentonite Slurry Mixtures". In "Hydraulic Barriers in Soil and Rock", ASTM STP 874, A.I. Johnson, R.K. Frobel, N.J. Cavalli, C.B. Pettersson, Eds., American Society for Testing and Materials, Philadelphia, Pa.

Beine, R.A., and Geil, M., (1985). "Physical Properties of Lining Systems Under Percolation of Waste Liquids and Their Investigations". In "Contaminated Soil", J.W. Assink and W.J. Van Den Brink, Editors, Proceedings of the First International TNO Conference on Contaminated Soil, 11-15 November, Utrecht, The Netherlands, Martinus Nijhoff Publishers, Dorchecht.

Bowders, J.J., Daniel, D.E., Broderick, G.P., and Liljestrand, H.M., 1986. "Methods for Testing the Compatibility of Clay Liners with Landfill Leachate". In "Hazardous and Industrial Solid Waste Testing: Fourth Symposium, ASTM STP 886, J.K. Petros, Jr., W.J. Lacy, and R.A. Conway, Eds., American Society for Testing and Materials, Philadelphia.

Brown, K.W., and Anderson, D.C., 1983. "Effects of Organic Solvents on the Permeability of Clay Soils". US EPA Report EPA-600/52-83-016.

Emcon Associates, 1983. "Field Verification of Liners from Sanitary Landfills". US EPA Report No. EPA-600/2-83-046.

Farquhar, G.J. and Constable T.W., 1978. "Leachate Contaminant Attenuation in Soil". Waterloo Research Institute, Project No. 2123, University of Waterloo, Waterloo, Canada.

Fuller, W.H., and Korte, N., 1976. "Attenuation Mechanisms Through Soil". In "Gas and Leachate from Landfills". E.J. Genetelli and J. Cirello, Eds., US EPA Report EPA-600/9-76-004, Cincinnati.

Gordon, M.E., Heubner, P.M., and Kmet, P., 1984. "An Evaluation of the Performance of Four Clay-Lined Landfills in Wisconsin". In Proceedings of the Seventh Madison Waste Conference, Madison, Wisconsin.

Griffin, R.A., 1983. "Mechanisms of Natural Leachate Attenuation". Notebook for Sanitary Landfill Design Seminar, Department of Engineering and Applied Science, University of Wisconsin, Madison.

Griffin, R.A., Cartwright, K., Shimp, N.F., Steele, J.D., Ruch, R.R., White, W.A., Hughes, G.M., and Gilkeson, R.H., 1976. "Attenuation of Pollutants in Municipal Landfill Leachate by Clay Minerals: Part 1 - Column Leaching and Field Verification". Environmental Geology Notes No. 78 Illinois State Geological Survey, Urbana.

Haxo, H.E., Jr., Haxo, R.S., Nelson, N.A., Haxo, P.D., White, R.M., Dakessian, S., and Fong, M.A., 1985. "Liner Materials for Hazardous and Toxic Wastes and Municipal Solid Waste Leachate". Noyes Publications.

Jones, C.W., 1985. "Effects of Brine on the Soil Lining of Evaporation Pond". In "Hydraulic Barriers in Soil and Rock", ASTM STP 874, A.I. Johnson, R.K. Frobel, N.J. Cavalli, C.B. Pettersson, Eds., American Society for Testing and Materials, Philadelphia, Pa.

Koerner, R.M., 1986. "Use of Flexible Membrane Liners for Industrial and Hazardous Waste Disposal". In "Hazardous and Industrial Solid Waste Testing and Disposa: Sixth Volume". ASTM Special Technical Publication 933, Philadelphia, Pa.

Lentz, R.W., Horst, W.D. and Uppot, J.O., 1985, "The Permeability of Clay to Acidic and Caustic Permeants". In "Hydraulic Barriers in Soil and Rock", ASTM STP 874, A.I. Johnson, R.K. Frobel, N.J. Cavalli, C.B. Pettersson, Eds., American Society for Testing and Materials, Philadelphia, Pa.

Peterson, S.R. and Gee, G.W., 1985. "Interactions Between Acidic Solutions and Clay Liners: Permeability and Neutralization". In "Hydraulic Barriers in Soil and Rock", ASTM STP 874, A.I. Johnson, R.K. Frobel, N.J. Cavalli, C.B. Pettersson, Eds., American Socity for Testing and Materials, Philadelphia, Pa.

Quigley, R.M., and Rowe, R.K., 1986. "Leachate Migration Through Clay below a Domestic Waste Landfill, Sarnia, Ontario, Canada: Chemical Interpretation and Modelling Philosophies". In "Hazardous and Industrial Solid Waste Testing and Disposal: Sixth Volume". ASTM Special Technical Publication 933, Philadelphia, Pa.

Reades, D., 1987. "Liner Selection - Leachate Collection System Design". Second Annual Workshop on Waste Management. University of Toronto, Canada.

Weinberg, R., Heinze, E., and Forstner, U., 1985. "Experiments on Specific Retardation of Some Organic Contaminants by Slurry Trench Materials". In "Contaminated Soil", J.W. Assink and W.J. Van den Brink Editors, Proceedings of the First International TNO Conference on Contaminated Soil,11-15 November, Utrecht, The Netherlands. Martinus Nijhoff Publishers, Dorchecht.

Wuellner, W.W., Wierman, D.A., and Koch, H.A., 1985. "Effect of Landfill Leachate on the Permeability of Clay Soils". Proceedings of the Eighth Annual Madison Waste Conference, Sept. 18-19, 1985, Wisconsin.

Yong, R.N., Warith, M.A., Boonsinsuk, P., 1986. "Migration of Leachate Solution Through Clay Liner and Substrate". In "Hazardous and Industrial Solid Waste Testing and Disposal: Sixth Volume". ASTM Special Technical Publication 933, Philadelphia, Pa.

CHEMICAL EFFECTS ON CLAY FABRIC AND HYDRAULIC CONDUCTIVITY

Fritz T. Madsen[1] and James K. Mitchell[2]

[1] Swiss Federal Institute of Technology Zurich, Division
of Foundation Engineering and Soil Mechanics
[2] University of California, Berkeley, Department of Civil
Engineering, Geotechnical Engineering

ABSTRACT

Hydraulic conductivity and its susceptibility to changes with time or
exposure to chemicals are major factors in selection of clay for use
in waste containment barriers. Available concepts of clay-chemical
interactions and data permit development of conclusions useful for
prediction of clay barrier performance in waste containment
applications. Among the most important conclusions are that **(1)** the
influences of the many factors that can cause changes in hydraulic
conductivity can be understood from the perspective of their effects
on the soil fabric, **(2)** the influences of chemicals on high water
content clays, such as in slurry walls, are likely to be much greater
than on lower water content compacted clays, **(3)** the effects of
inorganic chemicals are consistent with their effects on particle
surface double layers, their effects on surface and edge charges, and
on pH, and **(4)** the effects of organic chemicals are influenced
primarily by their water solubility, their dielectric constant, their
polarity, and whether the clay is exposed to the pure chemical or a
dilute solution. The type of test used may have a very significant
effect on the values of hydraulic conductivity that are measured. In
almost all cases pure organic liquids will interact adversely with
clays by causing some shrinking and cracking, with concurrent large
hydraulic conductivity increases; however, dilute solutions of
organics have essentially no effect.

INTRODUCTION

The hydraulic conductivity and its susceptibility to changes with time
or exposure to chemicals are the major factors in the selection of
clay for use in waste containment barriers. Such barriers are usually
in the form of landfill liners and covers, lagoon liners, and slurry
walls. Much has been written about the influences of chemicals in

permeants on the hydraulic conductivity of the permeated clay. There has also been much debate about the most suitable tests for measurement of the hydraulic conductivity for use in hazardous waste problems.

The purpose of this paper is to synthesize the available information so as to permit the development of conclusions about chemicals and clay hydraulic conductivity that may be useful for evaluation and prediction of clay barrier performance in waste containment applications. The following subjects are addressed:

1. Factors affecting hydraulic conductivity

2. Laboratory measurement methods

3. Inorganic chemical effects on hydraulic conductivity

4. Organic chemical effects on hydraulic conductivity

5. Conclusions relative to clays in waste containment applications

6. Comments on "final storage" and "final storage quality"

FACTORS AFFECTING HYDRAULIC CONDUCTIVITY

Of the soil properties needed for solution of most geotechnical problems, none varies over so wide a range or is so difficult to determine reliably as the hydraulic conductivity. A given soil may exhibit large variations in hydraulic conductivity as a result of changes in fabric (particle arrangements), density, and water content. This variation is clearly indicated by the test results in Fig. 1, from Mitchell et al. (1965), in which contours of equal hydraulic conductivity are shown for a silty clay compacted to different water contents and dry densities. To understand how variations of such magnitude can occur within such small ranges of density and water content, and also, how different chemicals can influence the hydraulic conductivity of fine grained soils requires consideration of the importance of soil fabric and the factors which influence it. Fabric-

hydraulic conductivity relationships are considered in detail
elsewhere; e.g., Michaels and Lin (1954), Lambe (1954), Olsen (1962),
and Mitchell (1976).

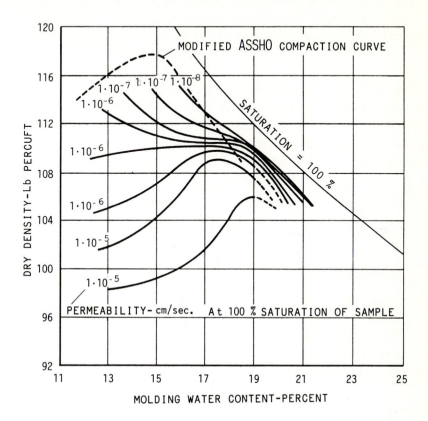

FIG.1: COUNTOURS OF EQUAL HYDRAULIC CONDUCTIVITY FOR SAMPLES
OF SILTY CLAY PREPARED BY KNEADING COMPACTION
(from Mitchell et al., 1965)

The soil fabric is strongly influenced by the repulsive force between
the clay particles. According to van Olphen (1977) a large repulsive
force will cause a dispersed soil fabric, whereas, a small repulsive
force tends to give a more flocculated fabric.

The reason for the repulsive force between clay particles is their net
negative charge which is caused by irregularities in the clay crystal.
The negative charge is balanced by cations outside the clay particle.
The distribution of ions adjacent to a clay surface for a clay
particle in an electrolyte can be described according to the diffuse
double layer theory as developed by Gouy (1910), and Verwey and

Overbeek (1948). The diffuse double layer consists of the negative charged clay particle surface and the ions surrounding it as shown in Fig. 2. According to the theory, the overlap of double layers of individual clay particles is the source of the interparticle repulsion which controls the flocculation-deflocculation behavior of clay suspensions and the swelling-shrinking behavior of soils (Bolt, 1955, 1956).

The double layer is sensitive to variation of the dielectric constant (ε) of the fluid, the temperature (T), the electrolyte concentration (n) in the pore water, and the cation valence (ν). An approximate quantitative indication of the influence of these factors can be seen in terms of the "thickness" of the double layer as given by

$$\frac{1}{\kappa} = \sqrt{\frac{\varepsilon \cdot k \cdot T}{8\pi n e^2 \nu^2}}$$

where k is the Boltzmann constant and e is the electronic charge.

From this relationship it may be noted that the thickness of the double layer, and thus also the magnitude of the repulsive force between the clay particles, varies directly with the square root of

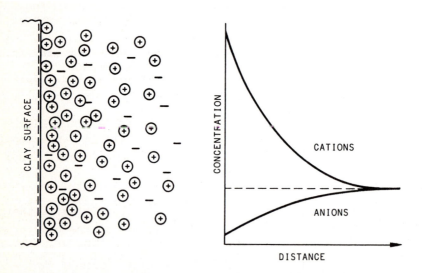

FIG. 2: DISTRIBUTION OF IONS ADJACENT TO A NEGATIVE CHARGED SURFACE ACCORDING TO THE DOUBLE LAYER THEORY

the dielectric constant and temperature, and inversely as the valence and the square root of the electrolyte concentration. As shown by Mitchell (1976), the influence of the temperature is very small because a change in temperature will cause a change of the dielectric constant in a way that the product $\varepsilon \cdot T$ remains nearly constant.

Beside the above mentioned variables some other factors as the size of the cations in the double layer, pH of the fluid, and anion adsorption on the clay particles may also influence the clay behavior and thus the soil fabric.

The smaller the hydrated cation in the double layer, the closer it can approach the clay surface resulting in a smaller thickness of the diffuse layer and a tendency toward a more flocculated fabric.

The pH influences the dissociation of the hydroxyl groups (OH^-) exposed on the clay crystal surfaces and edges; the higher the pH, the greater the tendency for the H^+ to go into solution, and the greater the negative charge of the clay particle; e.g.

$$SiOH \xrightarrow[\text{high pH}]{H_2O} SiO^- + H^+$$

In addition, alumina, which is exposed at the edges of the clay crystals, is amphoteric and ionizes positively at low pH and negatively at high pH. A low pH then promotes a positive edge to negative surface interaction of clay particles leading to a flocculated fabric. A high pH promotes a negative charge on the edges and leads to a more dispersed fabric.

Attraction of anions and negatively charged radicals to the positively charged edges of the clay particles, particularly anionic groups such as phosphate, arsenate, and borate which have about the same molecular size and geometry as the silica tetrahedron, appear well established. Phosphates in particular seem strongly attracted, and polyphosphates are among the most effective dispersing (deflocculation) agents for soil suspensions. After adsorption of the anions, the clay particles will be all over negatively charged and result in a more dispersed fabric.

The effect of the various parameters influencing the double layer thickness, and thus the soil fabric and the hydraulic conductivity, are summarized in Table 1.

TABLE 1

EFFECT OF CHANGES IN PORE FLUID PARAMETERS ON THE DOUBLE LAYER
THICKNESS, SOIL FABRIC, AND HYDRAULIC CONDUCTIVITY

Pore Fluid Parameter	Change in Parameter	Change in Double Layer Thickness	Fabric Change Tendency	Usual Effect on Hydraulic Conductivity
Dielectric Constant	increase	increase	dispersed	decrease
	decrease	decrease	flocculated	increase
Electrolyte Concentrat.	increase	decrease	flocculated	increase
	decrease	increase	dispersed	decrease
Cation Valence	increase	decrease	flocculated	increase
	decrease	increase	dispersed	decrease
Cation Size	increase	increase	dispersed	decrease
	decrease	decrease	flocculated	increase
pH	increase	increase	dispersed	decrease
	decrease	decrease	flocculated	increase
Anion Adsorption	increase	increase	dispersed	decrease
	decrease	decrease	flocculated	increase

From Evans et al. (1985), modified.

The hydraulic conductivity will be almost totally determined by the flow rates through the largest flow channels. Three levels of fabric may be important when dealing with fine-grained soils as barriers to the flow of chemicals. These fabric levels are well illustrated by the photomicrograph of Tucson silty clay obtained by Collins and McGown (1974) and shown in Fig. 3.

The microfabric consists of the regular aggregations of particles and the very small pores between them through which little fluid will flow.

The minifabric contains these aggregations and the interassemblage pores between them. Flow through these pores will be much greater than through the intraaggregate pores, since the hydraulic conductivity varies with the square of the pore radius.

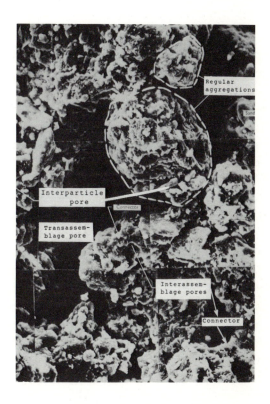

FIG. 3: FABRIC OF TUCSON SILTY CLAY
(from Collins and McGown, 1974)

On a larger scale there may be a macrofabric that contains cracks, fissures, root holes, laminations, etc., corresponding to the "transassemblage pores" in Fig. 3, through which the flow rate is so great as to totally obscure that through the other pore space types. The effects of clay-chemical interactions are generally much more pronounced in the case of high water content systems such as slurry walls, because particles are less constrained from movement in response to chemically induced interparticle force changes than in compacted clays. As a result larger changes in fabric are likely in the wetter clay.

Much attention has been directed at the large differences between the laboratory and field measured values of hydraulic conductivity of compacted clay liner material; e.g., Daniel (1984), with actual liner conductivities reported to be as much as 1000 times the value for laboratory compacted or undisturbed small samples. Such large differences can only be explained by differences in the mini- and macrofabrics.

An additional consideration of great importance when evaluating the effects of chemicals on the hydraulic conductivity of clays is that the composition of the fluid used to prepare samples for testing or for compaction in the field can influence the fabric significantly, as was demonstrated clearly by Michaels and Lin (1954) in connection with permeability measurements on samples of kaolinite consolidated in different pore fluids.

Once the hydraulic conductivity is viewed from the perspective of the fabric of the soil being permeated, then the influences of the myriad of factors that can influence it can be readily understood. It becomes necessary to consider only how the variable influences the fabrics at the various levels. Important factors in this regard include: **(1)** water content, **(2)** density, **(3)** method and effort of compaction and their relation to layering, break down of clods and uniform dispersion of particles, **(4)** sample size relative to fabric features, **(5)** hydraulic gradient and the influences of seepage forces, **(6)** anisotropy and flow direction, **(7)** physical and chemical properties of the permeant, and **(8)** thixotropic or other time effects.

At least two other factors may be important also; namely, the degree of saturation, which controls the space available for fluid flow, and biological processes which may lead to pore clogging as a result of the growth of microorganisms.

LABORATORY MEASUREMENT METHODS

Although the suitability of laboratory hydraulic conductivity measurements for quantitative prediction of the conductivity in the field remains a subject for debate, laboratory testing does have a very important and useful place in hazardous waste projects. In the laboratory it is possible to complete a relatively large number of tests in a reasonable time, thereby making it possible to investigate

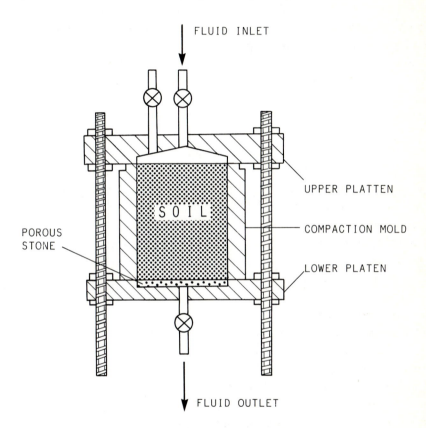

FIG. 4: RIGID WALL PERMEAMETER
(from Day, 1984)

the influences of several variables. Laboratory tests are particularly well suited for determination of compatibility between chemicals and clay under different conditions, thus enabling early decisions about whether specific soils are useful at all for containment of specific wastes. They enable study of the influences of water content, density, and saturation that would be difficult, time consuming and expensive in the field.

Nonetheless, laboratory tests must be carried out with great care, as there are many variables that must be controlled relating to sample preparation, apparatus, and testing procedure (Dunn and Mitchell, 1984). Three test apparatus types are being used; namely (1) rigid wall permeameter, Fig. 4, (2) flexible wall permeameter, Fig. 5, and (3) consolidometer permeameter, Fig. 6. Each of these test types has advantages and limitations relative to each other (Daniel et al., 1985), and none represents field conditions exactly. A complete discussion of each of these test types is beyond the scope of this paper. However, in most cases use of rigid wall permeameters will usually lead to a very large increase in measured hydraulic

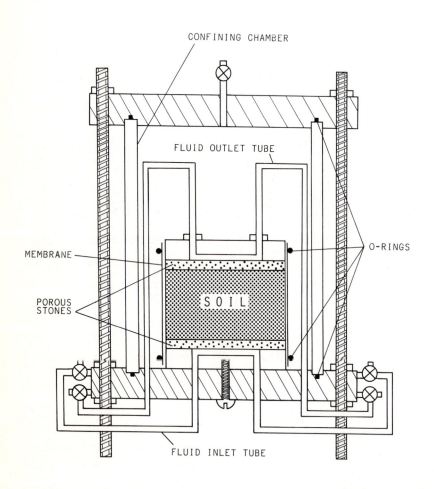

FIG. 5: FLEXIBLE WALL (TRIAXIAL) PERMEAMETER
(from Day, 1984)

conductivity as a result of sidewall leachage in cases where chemical-clay interactions lead to shrinkage and cracking. Flexible wall tests, on the other hand, may underestimate these effects, owing to the closing of cracks and fissures as they form.

It is the authors' opinion that the consolidometer permeability test is potentially the most useful of the three test types because of the flexibility it offers for testing specimens under a range of confining stresses and for accurate determination of changes in sample thickness as a result of both seepage forces and chemical influences on the soil structure. Furthermore, the thinner samples relative to the other test types means that the pore fluid replacement can be achieved in a shorter time for a given hydraulic gradient.

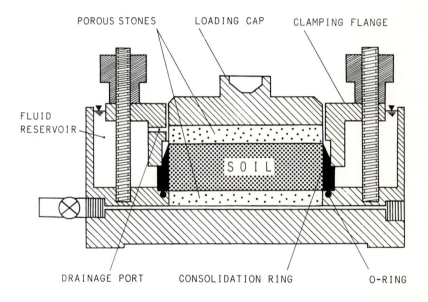

FIG. 6: CONSOLIDOMETER PERMEAMETER
(from Day, 1984)

In a number of recent studies the hydraulic conductivity to water has, in reality, been determined using 0.01N calcium sulfate as the permeant. In view of the significant effects even small differences in chemistry of the permeant can have on the hydraulic conductivity; as shown, for example by Fig. 7, from Dunn and Mitchell (1984), it would seem more reasonable the permeant be of the same composition as that to which the clay will be subjected in the field during compaction.

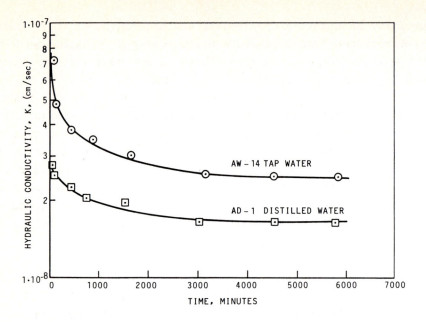

FIG. 7: COMPARISON OF HYDRAULIC CONDUCTIVITY OF ALTAMONT
SOIL WITH TAP WATER AND DISTILLED WATER PERMEANTS
(from Dunn and Mitchell, 1984)

INORGANIC CHEMICAL EFFECTS ON HYDRAULIC CONDUCTIVITY

Inorganic chemicals may affect the hydraulic conductivity of clays
through their effects on fabric and pore space relationships.
Alterations may result from changes in ion concentrations, ion
exchange, anion adsorption, and dissolution.

In general, increases in concentration and cation valence cause
flocculation of clay particles and limit the swelling of expansive
minerals. Acids tend to cause flocculation and may attack the crystal
lattice of clays, especially the octahedral sheets of the minerals.
Bases tend to disperse and attack mainly the silica tetrahedral sheets
of the minerals.

Salts

The results of tests on a silty loam by Quirk and Schofield (1955)
clearly showed that reduction in electrolyte concentration led to a
substantial decrease in hydraulic conductivity as a result of swelling

and deflocculation. Tests on low clay content soils intended to simulate some types of aquitard settlements by Hardcastle and Mitchell (1974) yielded similar results, and also showed that reversible changes in hydraulic conductivity were possible owing to changes in electrolyte concentration. The reduction in conductivity as a result of clay dispersion and swelling accompanying reduction in NaCl concentration in the pore water is shown in Fig. 8, and the subsequent increase in conductivity as a result of reintroduction of a higher NaCl concentration is shown in Fig. 9. In these figures the cumulative throughput refers to the number of pore volumes of electrolyte solution passed, and the relative hydraulic conductivity is the ratio of the value at any given throughput to that at the start of pore fluid exchange.

Other data to illustrate the effects of inorganic salts on hydraulic conductivity are given by Olsen (1962) and Dunn and Mitchell (1984). Alther et al. (1985) present photographs of cracking patterns in bentonite after mixing with different concentrations of KCl which show clearly the pronounced effects that increasing salt concentrations can have on the fabric of high water content clays.

Acids

A soil-bentonite backfill and a slurry wall filter cake were tested by D'Appolonia (1980) using a 5 percent hydrochloric acid solution, and increases in conductivity up to one order of magnitude were measured. Tests on a compacted clay containing carbonates using sulfuric acid at a pH of 1.5 showed no change in hydraulic conductivity (Gordon and Forrest, 1981). Calcium sulfate precipitate was formed in addition to carbon dioxide and water.

The passage of six pore volumes of HCl at pH values of 1, 3, and 5 had no effect on the hydraulic conductivity of consolidated specimens of kaolinite, kaolinite-bentonite mixtures, and magnesium montmorillonite (Lentz et al., 1985). Compacted natural clays were tested in a rigid wall permeameter using a 5 percent solution containing equal parts of hydrochloric, sulfuric, and nitric acid by Simons et al. (1984). Testing extended over 300 days, with an increase in hydraulic conductivity of about half an order of magnitude after 120 days. Some dissolution of clay minerals, especially kaolinite, was detected.

Taken together the results of these studies suggest that the effects of acids will be significant if the soil particles are not constrained from movement, there is solutioning and removal, solution and reprecipitation, or gross shrinkage and cracking, if there is sufficient time for these reactions to occur.

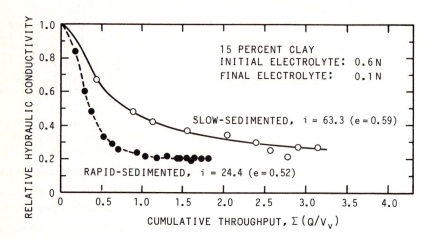

FIG. 8: HYDRAULIC CONDUCTIVITY REDUCTIONS AS
A RESULT OF CLAY SWELLING
(from Hardcastle and Mitchell, 1974)

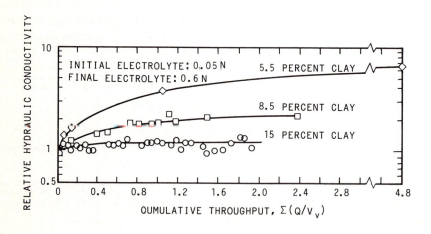

FIG. 9: HYDRAULIC CONDUCTIVITY INCREASES FOR
REFLOCCULATED DISPERSED MIXTURES
(from Hardcastle and Mitchell, 1974)

Bases

The effects of sodium hydroxide at pH values of 9, 11, and 13 were
determined by Lentz et al. (1985) on the same clays that were used in
their tests of the effects of acids.   Little effect on the hydraulic
conductivity was measured at pH 9 or 11.   Conductivity decreases by up
to a factor of about 10 were measured for the pH 13 permeant, as may
be seen in Fig. 10. These decreases were interpreted to have resulted
from the use of tap water rather than distilled water for preparation
of the permeants.   The tap water contained 300 mg/liter calcium
carbonate, and the resulting formation of calcium hydroxide may have
blocked some of the pores.   These test results also illustrate the
significant effects that even small concentrations of different
chemicals may have and the need for complete knowledge of permeant
composition in studies of this type.

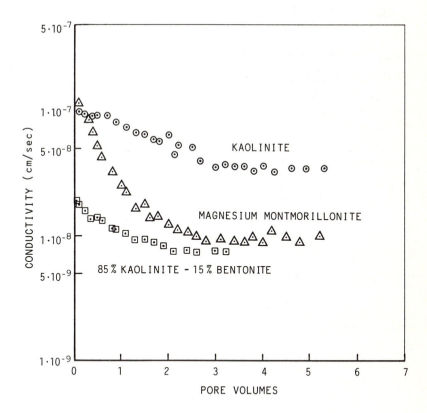

FIG. 10: EFFECT OF SODIUM HYDROXIDE AT pH 13 ON
THE HYDRAULIC CONDUCTIVITY OF CLAYS
(from Lentz et al., 1985)

Contrary results were obtained by D'Appolonia (1980), who found that when a slurry wall filter cake material and soil-bentonite backfill were permeated with a 5 percent sodium hydroxide solution, the hydraulic conductivity increased by a factor of 2 to 10. The different initial states of the clays: i.e., high water content, low density in D'Appolonia's tests and lower water content, higher density in Lentz et al.'s tests, and the different test types; i.e. flexible wall in the Lentz et al. tests and rigid wall in D'Appolonia's tests, can account for the different results obtained. In addition, the 5 percent NaOH solution was probably strong enough to cause flocculation in these samples; whereas, in the tests by Lentz et al. at the lower values of pH, the dilute sodium solution could cause dispersion.

ORGANIC CHEMICAL EFFECTS ON HYDRAULIC CONDUCTIVITY

The movement of organic chemicals, especially solvents, through clay barriers and aquitards is a major concern as regards groundwater and environmental protection, because many of these compounds are considered to adversely affect biological systems. Accordingly, there has been extensive study of the effects of organics on the hydraulic conductivity of both natural fine-grained soils and processed clays that are useful in formation of seepage barriers. The results of these studies have been both conflicting and confusing, which is perhaps not surprising, since tests of different types have been used, and concentrations in the permeants have ranged from 0.1 percent to pure compounds.

Those classes of organic compounds that have been investigated relative to their effects on hydraulic conductivity are shown in Table 2.

Clay minerals interact with organic chemicals by adsorption, intercalation, and cation exchange. Adsorption seems to be the most important from the standpoint of effects on the hydraulic conductivity. Water molecules in the double layer can be replaced by a variety of organic molecules. Some of the organic molecules more or less strongly solvate the cations, with the others being associated between the solvation shells. The bond strength between the double layer cations and solvent molecules varies widely (Lagaly, 1984). The lower dielectric constant of most organic compounds compared to water

TABLE 2

CLASSES OF ORGANIC COMPOUNDS

Type of Compound		Functional Group	Specific Example	
Saturated Hydrocarbons	(alkanes)	$\equiv C-C\equiv$	H_3C-CH_3	ethane
Unsaturated Hydrocarbons	(alkenes)	$=C=C=$	$H_2C=CH_2$	ethylene
	(alkynes)	$-C\equiv C-$	$HC\equiv CH$	acetylene
	(polyolefins)	$=C=C-C=C=$	$H_2C=CH-CH=CH_2$	butadiene
Aromatic Hydrocarbons				benzene
Alcohols		$-OH$	CH_3CH_2-OH	ethyl alcohol
Phenols		$-OH$		1-phenyl-ethanol
Ethers		$-O-$	$CH_3CH_2-O-CH_2CH_3$	ethyl ether
Aldehydes		$\overset{}{\underset{H}{>}}C=O$	$CH_3-\overset{O}{\overset{\|}{C}}-H$	acetaldehyde
Ketones		$>C=O$	$CH_3-\overset{O}{\overset{\|}{C}}-CH_3$	acetone
Organic Acids (Carboxylic Acids)		$-C\overset{O}{\underset{OH}{<}}$	$CH_3-\overset{O}{\overset{\|}{C}}-OH$	acetic acid
Organic Bases (Amines)		$-NH_2$	CH_3-NH_2	methylamine

TABLE 3

PROPERTIES OF ORGANIC CHEMICALS USED IN HYDRAULIC CONDUCTIVITY TESTING OF CLAYS

Class of Compound	Compound	Formula	Solubility in Water (g/L)	Dielectric Constant	Dipole-Moment (Debye)	Density (g/cm^3)
Hydrocarbons and Related	Heptane	C_7H_{16}	<0.3	1.9	0	0.684
	Cyclohexane	C_6H_{12}	<0.3	2.0	0	0.779
	Benzene	C_6H_6	0.7	2.3	0	0.879
	Xylene (Dimethyl-benzene)	C_8H_{10}	<0.3	2.27 para 2.37 meta 2.57 ortho	0 0.62	0.880
	Tetrachloro-methane (Carbontetra-chloride)	CCl_4	0.8	2.2	0	1.594
	Trichloro-ethylene (TCE)	C_2HCl_3	1	3.4	0	1.464
	Nitro-benzene	$C_6H_5NO_2$	2	35.7	4.22	1.204
Alcohols and Phenols	Methanol	CH_3OH	∞	33.6	1.70	0.791
	Ethanol	C_2H_5OH	∞	25.0	1.69	0.789
	Ethylene-glycol	$C_2H_6O_2$	∞	37.7	2.28	1.119
	Phenol	C_6H_5OH	86	13.1	1.45	1.072
Ethers	1.4-Dioxane	$C_4H_8O_2$	∞	2.2	0	1.034
Aldehydes and Ketones	Acetone	C_3H_6O	∞	21.5	2.9	0.79
Organic Acids	Acetic Acid	$C_2H_4O_2$	∞	6.15	1.74	1.049
Organic Bases	Aniline	$C_6H_5NH_2$	36	6.89	1.55	1.02
Mixed Chemicals	Xylene Acetone	C_8H_{10} C_3H_6O	∞ in acetone	2.27-2.37 21.5	0-0.62 2.9	0.880 0.79
	Sodium Acetate Glycerol Acetic Acid Salicylic Acid	CH_3COONa $C_3H_8O_2$ $C_2H_4O_2$ $C_7H_6CO_3$	∞ ∞ Slightly	42.5 6.15	1.74	1.261 1.049 1.443
	5% Ammonia Copper Tetra-mine, Nickel Hexamine	NH_3		17.3		

may limit swelling or induce a tendency towards flocculation if particles are not restricted from movement.

Intercalation of distinct organic compounds in kaolinite has been studied by Weiss (1963). Molecules enter the interlayer spaces and split apart the unit layers. The alkali salts (K^+, $NH4^+$, Rb^+, Cs^+) of short chain fatty acids, in particular acetic and propionic acid, are of interest. Ammonium acetate, for example, would change the basal spacing of kaolinite from about 7 to 14 Angstroms. Formamide, dimethylsulfoxide (DMSO), and urea could have similar effects. The resulting swelling of the kaolinite particles should cause a decrease in the hydraulic conductivity; however, this decrease has not yet been demonstrated.

Interlayer adsorption of acids or bases from aqueous solutions is accompanied by a cation exchange. For the bases, particularly the amines, the protonated species exchange interlayer cations (Vansant and Uytterhoeven, 1973). The influence on the hydraulic conductivity then depends on the valence of the cations and the pH. The latter may influence the edge and surface charges of the clay particles.

According to Huang and Keller (1971) dilute organic acids may be able to dissolve clay minerals. Depending on the clay mineral type and the specific acid, silica or alumina will be preferentially dissolved. This could be a long term problem for clay liners; however, no long term investigations appear to have been made.

Available data on the effects of organic chemicals on the hydraulic conductivity of clays have been reviewed within the framework of the compound classifications given in Table 2. The chemicals used for hydraulic conductivity testing and some of their properties are listed in Table 3. The results of a number of investigations are summarized in Table 4, and some of them are illustrated by the figures that follow.

Hydrocarbons and Related Compounds

There have been a number of laboratory studies, but only a few welldocumented field measurements of the effects of hydrocarbons on the hydraulic conductivity of clays.

220

The effects of the hydrocarbons, heptane and xylene, were studied by Anderson, Brown, and Thomas (1985) using a rigid wall compaction mold permeameter. These hydrocarbons are nonpolar, have low dielectric constants of about 2, and are lighter than water. Gradients of 361.6 were used for tests on smectite clay soils, and 61.1 were used for illitic and kaolinitic soils. The results for both chemicals were similar. Breakthrough and conductivity curves as a function of pore volumes of heptane passed for four soils are shown in Fig. 11. In each case the hydraulic conductivity increased by two to three orders of magnitude but then remained approximately constant.

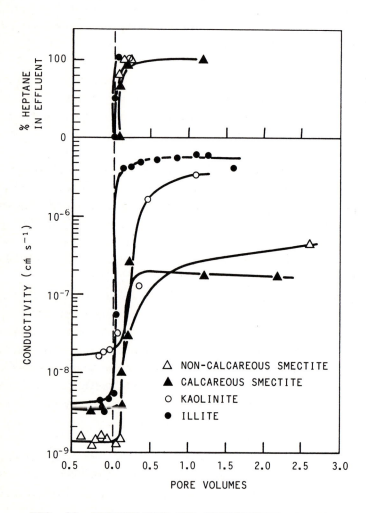

FIG. 11: BREAKTROUGH AND CONDUCTIVITY CURVES
FOR FOUR SOILS TREATED WITH HEPTANE
(from Anderson et al., 1985)

Tests with heptane as a permeant both in concentrated form and at its solubility limit in water (53 mg/liter), were done by Bowders (1985) in both flexible and rigid wall permeameters and gave quite different results. The dilute heptane had no effect at all. Pure heptane in flexible wall tests caused a decrease in conductivity owing to sample shrinkage, but in the rigid wall tests the conductivity increased by 250 to 660 times that of water. The large increases in conductivity in the case of rigid wall tests and concentrated hydrocarbons were attributed to the observed formation of cracks and macropores by both Anderson et al. (1985) and Bowders (1985).

Tests on a compacted illitic-smectitic soil in a rigid wall permeameter using benzene, cyclohexane, and o-xylene under a hydraulic gradient of 500 were done by Fernandez and Quigley (1985). These tests followed initial permeation with 0.01N calcium sulfate solution. The results were similar in each case, and those for benzene are shown in Fig. 12. There was essentially no effect on the hydraulic conductivity.

Analyses of the sample pore fluid after the test showed that only 8 percent by weight was benzene, the rest being water. Fig. 12 shows that the effluent concentration reached 50 percent benzene after passage of only 0.28 pore volumes and reached 100 percent by 2 pore volumes. This is strong evidence that the flow of solvent was primarily through interconnected large pores, with little invasion of the micro-fabric units. Due to the immiscibility of the non-polar solvents with water and the large interfacial tension at the entrance to water filled micro-pores, this result is to be expected.

When tests were done using sequential permeation of water followed by ethanol and then the non-polar hydrocarbons (Fernandez and Quigley, 1985), quite different results were obtained, as can be seen in Fig. 13. In this case the solubility of polar ethanol in water and non-polar hydrocarbon benzene in ethanol meant that greater penetration of the pores by benzene was possible. The lower dielectric constant of the organics results in a decrease in double layer thickness and increase in the sizes of both mini- and macropores, thus accounting for the large increases in conductivity.

Fernandez and Quigley (1985) also did tests in the reverse direction, with samples initially molded using benzene then permeated by ethanol followed by water. The results shown in Fig. 14 show sequential

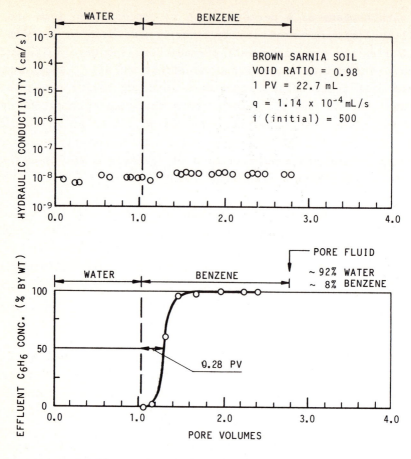

FIG. 12: HYDRAULIC CONDUCTIVITY AND EFFLUENT
CONCENTRATIONS OF COMPACTED ILLITE-
SMECTITE SOIL PERMEATED WITH BENZENE
(from Fernandez and Quigley, 1985)

decreases in conductivity as the dielectric constant of the permeant
increases, as would be expected, because of the increased double layer
thickness and accompanying tendencies for swelling and dispersion.
The large difference in initial hydraulic conductivities between
samples compacted in water (Fig. 13) and in benzene (Fig. 14) should
also be noted. This difference is a direct consequence of the
influence of the fluids on the initial fabric.

Tests using dilute (0.1 percent) and concentrated benzene as permeants
through compacted kaolinite by Acar et al. (1985) using a flexible
wall permeameter showed that there was little effect of the dilute
solution, but a decrease of three orders of magnitude in the hydraulic

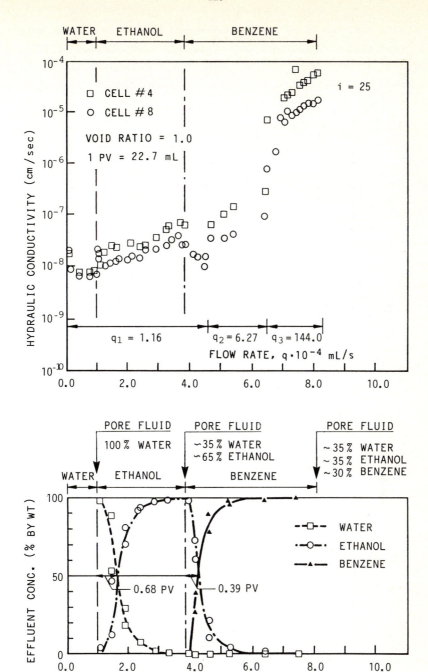

FIG. 13: HYDRAULIC CONDUCTIVITY AND EFFLUENT
CONCENTRATIONS FOR SEQUENTIAL PERMEA-
TION BY WATER, ETHANOL, AND BENZENE
(from Fernandez and Quigley, 1985)

conductivity was measured with concentrated benzene. When a bentonite slurry was tested by Anderson, Crawley, and Zabcik (1985), using concentrated xylene in a double ring, rigid wall permeameter, a two to three order of magnitude increase in hydraulic conductivity was

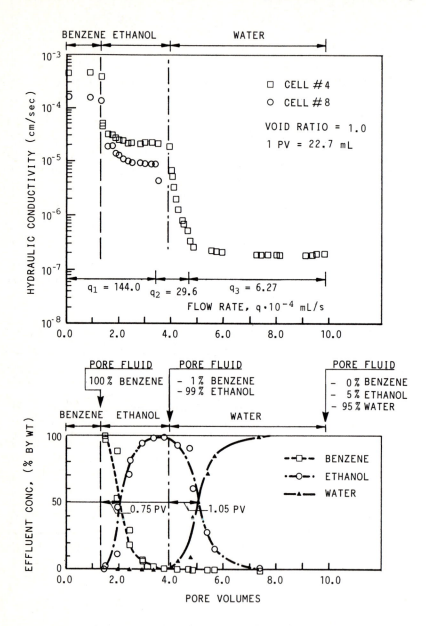

FIG. 14: HYDRAULIC CONDUCTIVITY AND EFFLUENT
CONCENTRATIONS FOR SEQUENTIAL PERMEA-
TION BY BENZENE, ETHANOL, AND WATER
(from Fernandez and Quigley, 1985)

measured. The apparent discrepancy between the results of these two investigations can be accounted for by the difference in test types, flexible wall vs. rigid wall, and initial conditions of the samples, compacted vs. slurry.

Outdoor tests on three soils in large diameter (1.8 m) cells were done by Brown et al. (1986) using xylene and a hydraulic gradient of 7, a value that is much more representative of actual field conditions than the higher values usually used for laboratory tests. In the large cell tests the hydraulic conductivity increased by two to three orders of magnitude in the presence of xylene; whereas, that of the same soil investigated in the laboratory increased by three to four orders. The initial values of the laboratory samples with water as a permeant were about half an order of magnitude less than the outdoor values, however. It was observed that the xylene moved through preferential pathways such as cracks rather than uniformly through the soil mass. An additional significant observation in these tests was that from time to time rain water leaked into the field test cells. When this happened the xylene floated, and as the water permeated the soil, the hydraulic conductivity decreased. After most of the water had passed through, the conductivity increased again.

Hydraulic conductivity increases of two orders of magnitude were measured when soil-bentonite slurry wall material was permeated with concentrated carbontetrachloride in a flexible wall permeameter (Evans et al., 1985). When dilute CCl_4 solution (720 mg/liter) was used the conductivity was unaffected. Bowders (1985) found that dilute (1.1 g/liter) trichloroethylene (TCE) had no effect on the hydraulic conductivity in either flexible or rigid wall tests. Pure TCE caused a large increase, by a factor of 140 to 500 times, in rigid wall tests, but a decrease in flexible wall tests.]

Finally, nitrobenzene, a nitrated aromatic hydrocarbon with a density greater than water, a solubility of about 2 g/liter, a large dipole moment, and a dielectric constant of 35 was tested by Acar et al. (1985). Hydraulic conductivity measurements were made using a 0.1 percent aqueous solution and pure nitrobenzene and three soil types and flexible wall permeameters. Slight decreases in hydraulic conductivity were measured with the dilute solution, and larger decreases were measured with the pure chemical. The results of several investigations (Table 4) considered together lead to the following general conclusions about the effects of hydrocarbons on the hydraulic conductivity of clay soils.

1. Dilute solutions; i.e., concentrations at or below the solubility limit have no effect irrespective of the test method.

2. Permeation with pure hydrocarbons leads to a decrease in hydraulic conductivity when flexible wall permeameters are used.

3. Permeation with pure hydrocarbons can lead to hydraulic conductivity increases of up to three orders of magnitude when rigid wall permeameters are used.

4. Flow of concentrated hydrocarbons through fine grained soils is predominantly through the cracks and macro-pores. Most of the water in the micro-pores remains in the soil.

5. In the case of sequential permeation by mutually soluble organics; e.g., water followed by ethanol followed by benzene, the results are quite different. A much greater proportion of the water in the pores was replaced, and a large increase in the hydraulic conductivity developed.

6. Subsequent replacement of the organic compounds by water causes a hydraulic conductivity decrease.

Alcohols and Phenols

The simple alcohols whose effects on hydraulic conductivity have been determined are lighter than water, are totally soluble in water, have dipole moments similar to that of water (about 1.7 Debye), and dielectric constants in the range of 25 to 36. They should be able to replace the double layer water in clays and reduce the double layer thickness owing to their lower dielectric constant.

The hydraulic conductivity of compacted kaolinite to methanol was determined by Bowders (1985), and the values obtained are shown in Fig. 15. Conductivity values greater than that for water were obtained only when the methanol concentration reached 80 percent.

Anderson, Brown, and Thomas (1985) measured the hydraulic conductivity of compacted clay soils to pure methanol in rigid wall permeameters, and Anderson, Crawley, and Zabcik (1985) tested bentonite slurries in a double ring permeameter, also using pure methanol. Large increases were measured, and the samples developed large pores and cracks.

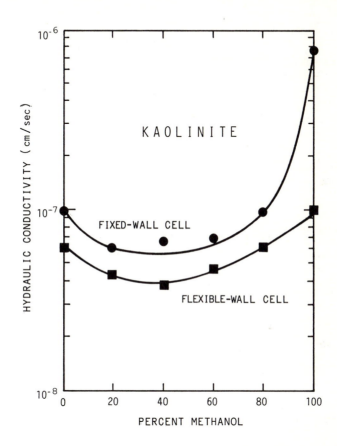

FIG. 15: HYDRAULIC CONDUCTIVITY OF COMPACTED
KAOLINITE TO AQUEOUS SOLUTIONS CONTAINING
METHANOL

(from Bowders, 1985)

Fernandez and Quigley (1985) did tests on compacted illitic-smectitic soil with both ethanol and methanol in a rigid wall permeameter under an initial gradient of 500. An increase in hydraulic conductivity from 10^{-8} to 10^{-7} cm/sec was measured in each case. The results for methanol are shown in Fig. 16.

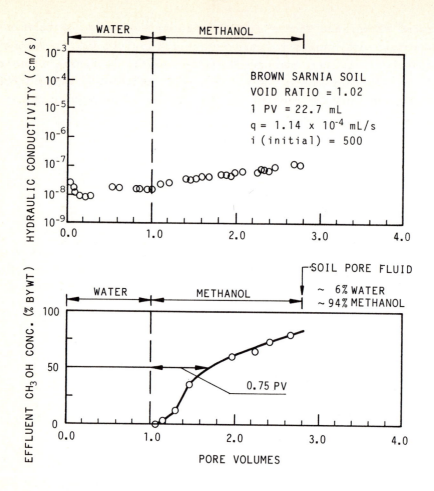

FIG. 16: HYDRAULIC CONDUCTIVITY AND EFFLUENT
CONCENTRATIONS DURING PERMEATION OF
COMPACTED CLAY BY METHANOL
(from Fernandez and Quigley, 1985)

After passage of about 2.8 pore volumes the effluent concentration had reached about 80 percent methanol and the pore fluid in the sample consisted of 94 percent methanol and 6 percent water. Thus the alcohols replace water throughout the soil in contrast to the non-polar hydrocarbons which remove water only from fissures and large pores.

Tests using ethylene glycol, which is slightly heavier than water, is infinitely soluble in water, has a strong dipole moment (2.28 Debye), and a dielectric constant of about 38, were done by Anderson, Brown, and Thomas (1985) on four clay soils in a compaction mold permeameter.

Their results are shown in Fig. 17. The trends indicate increases in hydraulic conductivity for the smectite (after an initial decrease in the case of the calcareous smectite), but decreases, at least initially, for the kaolinite and illite. The causes of the initial decreases are not fully understood, but may have been caused by swelling due to an increase of double layer thickness at low concentration of glycol.

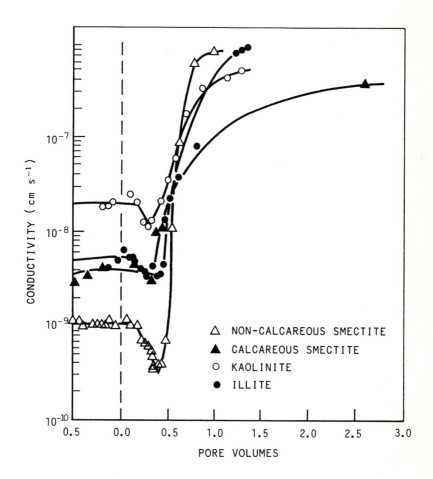

FIG. 17: EFFECT OF ETHYLENE GLYCOL ON THE HYDRAULIC
CONDUCTIVITY OF FOUR SOILS. ETHYLENE GLYCOL
INTRODUCED AT PORE VOLUME = 0.0
(from Anderson, Brown, and Thomas, 1985)

Phenol is less soluble (86 g/liter) in water than the simple alcohols, has a lower dielectric constant (9.8), and a dipole moment of 1.45 ·Debye.

Tests on compacted kaolinite using 0.1 percent solution and pure phenol were done by Acar et al. (1985) in a flexible wall permeameter. The pure phenol caused a conductivity increase by a factor of about two, but the dilute solution resulted in a slight decrease.

The results of tests with alcohols and phenols that are summarized in Table 4 indicate that:

1. Water soluble alcohols replace double layer water, and permeation causes essentially complete replacement of the pore water.

2. Alcohol concentrations less than about 80 percent have little effect on hydraulic conductivity.

3. Pure alcohols and phenol may cause hydraulic conductivity increases up to three orders of magnitude in rigid wall permeability tests.

4. Pure alcohols and phenol may cause a slight hydraulic conductivity increase in flexible wall permeability tests.

5. Subsequent replacement of alcohol by water causes a hydraulic conductivity decrease.

Ethers

The only information that could be found for the effects on ethers on hydraulic conductivity were the tests by Michaels and Lin (1954) using pure 1,4-dioxane. A consolidometer type permeameter was used, and the test specimens were obtained by consolidating pure kaolinite from a suspension in distilled water. Permeation by dioxane, which is totally soluble in water, caused a permeability increase of 20 to 30 percent. The non-polar dioxane is slightly heavier than water, and its low dielectric constant of 2.2 caused a decrease in double layer thickness and a tendency towards flocculation, which accounts for the increase in permeability. Although test data are not available, it would be expected that water soluble ethers would cause a large increase in the permeability of expansive soils owing to shrinkage and cracking.

Aldehydes and Ketones

Although there apparently have not been any measurements of the effects of aldehydes on hydraulic conductivity, a number of tests have been done using the ketone acetone. Acetone is lighter than water, has a dielectric constant of about 22, and a strong dipole moment of 2.9 Debye. Acetone is soluble, it can readily replace the water in clay double layers, and its lower dielectric constant relative to water can limit swelling and induce flocculation. The influence of a range of acetone concentrations on a compacted micaceous soil in rigid wall permeameters was determined by Brown et al. (1983, 1984) using a hydraulic gradient of 91.

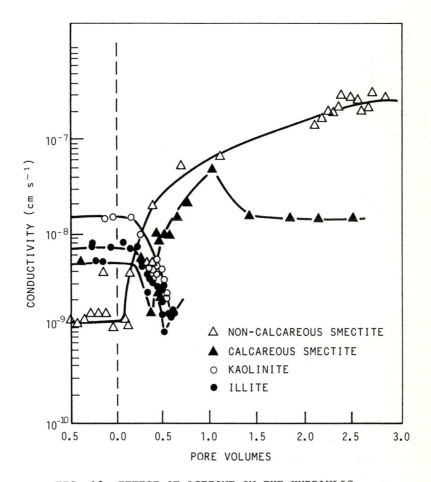

FIG. 18: EFFECT OF ACETONE ON THE HYDRAULIC
CONDUCTIVITY OF FOUR SOILS.
ACETONE INTRODUCED AT PORE VOLUME = 0.0
(from Anderson, Brown, and Thomas, 1985)

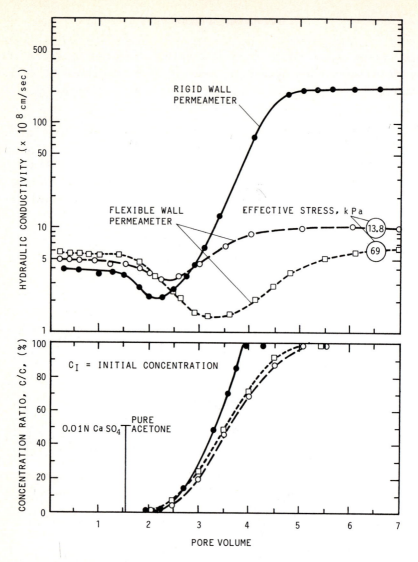

FIG. 19: EFFECT OF ACETONE ON THE HYDRAULIC CONDUCTIVITY
OF COMPACTED KAOLINITE IN RIGID WALL TESTS AND IN
FLEXIBLE WALL TESTS AT TWO EFFECTIVE STRESSES
(from Acar et al., 1985)

Substantial increases in hydraulic conductivity were measured when the
acetone concentration was 75 percent or higher. Tests by Anderson,
Brown, and Thomas (1985) on four soils in rigid wall permeameters gave
the results shown in Fig. 18, which show large increases in
conductivity after an initial decrease. This behavior was similar to
that found in tests with ethylene glycol. After permeation by pure
acetone the soil showed extensive shrinkage and cracking. The initial

hydraulic conductivity decrease may have been the result of swelling at low concentrations of acetone, as shown for montmorillonite by Griffin and Roy (1985).

A 0.1 percent solution of acetone caused a small decrease in the hydraulic conductivity of compacted kaolinite in flexible wall tests (Acar et al., 1985). Pure acetone caused an increase by a factor of about two in flexible wall tests, but much greater increases were measured when rigid wall permeameters were used, as may be seen in Fig. 19. The curves in Fig. 19 indicate also the effects of effective confining stress in flexible wall tests.

The influence of pure acetone on mixtures of sand, kaolinite and montmorillonite in both flexible wall permeameter and consolidometer were investigated by Acar et al. (1987). The increase in hydraulic conductivity -in maximum a factor of about 10- was found to be dependent on the activity of the clay-sand mixture. Samples with high activity (more montmorillonite) showed a smaller increase in hydraulic conductivity as samples with low activity (more kaolinite).

In outdoor tests on 1.8 m diameter, 15 cm thick compacted clay samples by Brown et al. (1986) it was found that the conductivity to acetone was about one order of magnitude greater than values with water in laboratory rigid wall tests. Examination of the permeated samples after the tests showed that the acetone moved as a uniform front through the compacted clay.

Taken together the results of tests using the water soluble ketone acetone as permeant indicate that:

1. Permeation causes essentially complete replacement of the pore water.

2. Acetone solutions at concentrations less than about 75 percent have little effect on hydraulic conductivity regardless of the testing method.

3. Pure acetone can cause an increase in hydraulic conductivity of up to three orders of magnitude in rigid wall tests.

4. Pure acetone can cause a slight increase in hydraulic conductivity in flexible wall tests.

Organic Bases

The weak base aniline has been tested as representative of the organic
bases. It has a water solubility of 36 g/liter, a dielectric constant
of 6.9, and a dipole moment of 1.55 Debye. Tests on soil-bentonite
slurry in a flexible wall permeameter using both aniline solutions in
water and pure aniline were done by Evans et al. (1985). The water
solutions had no effect; however, the pure aniline caused an increase
in the hydraulic conductivity of about four orders of magnitude.

Tests by Anderson, Brown, and Thomas (1985) on four compacted clays in
rigid wall permeameters using pure aniline showed that pure aniline
moved through cracks and macropores rather than uniformly through the
soil. The hydraulic conductivity increased by up to two orders of
magnitude relative to the values for the intact structure permeated by
water. Reintroduction of water caused a hydraulic conductivity
decrease of about one order of magnitude, but the original water
conductivity was not reached owing to the irreversible structural
changes caused by the aniline. The results of these few investigations
using aniline suggest the following possible consequences of
permeation of clay soils by simple organic bases.

1. Water solutions have no large influence on the hydraulic
    conductivity.

2. Pure bases can cause a large increase in the hydraulic
    conductivity.

3. Weak bases may not cause solutioning of clay particles.
    The effects of strong organic bases are not known.

4. The effects of the base-soil interactions are, to some extent,
    reversible when water is again permeated through the soil.

Organic Acids

The effect of various concentrations of acetic acid on the hydraulic
conductivity of a kaolinite and an illite-chlorite in both flexible
wall and rigid wall tests was determined by Bowders (1985). The

conductivity decreases in all cases except for rigid wall tests on kaolinite using pure acid, where an increase by a factor of two was measured. The mechanism for this decrease is thought to be that some of the soil constituents are dissolved by the acid. Then as the acid flows further, it begins to be buffered by the soil. As the pH increases the dissolved constituents start to precipitate, thereby clogging small pores. Bowder's measurements showed that the pH of the effluent was substantially higher than in the influent. The tests lasted 3 to 4 months, and after this time it was found that some of the sample had not yet been exposed to the acid.

Dilute acetic acid (pH 1) was found by Evans et al. (1985) to cause a small but steady increase in the conductivity of soil-bentonite slurry with time over a period of almost 70 days in flexible wall tests, during which about eight pore volumes of acid had flowed through the soil.

The effects of acids and the difficulty of testing soils that contain calcium carbonate or iron oxide were shown by Anderson et al. (1986). Initial decreases in the hydraulic conductivity are postulated to be due to solutioning and subsequent migration of soil particle fragments that then lodge in conducting pores. The calcareous smectite and the iron oxide bearing kaolinite both showed continuous decreases in hydraulic conductivity. The acid-treated calcareous smectite generated a cream colored foamy effluent of dispersed calcium salts and carbon dioxide, and the kaolinite effluent was dark red due to dissolved iron oxides.

Simons et al. (1984) permeated three natural clays with a dilute (5 percent total) mixture of acetic and propionic acid in rigid wall permeameters. The variations of hydraulic conductivity, electrical conductivity, and pH of the illitic clay over the 300-day testing period are shown in Fig. 20. Although the pH decreased continuously from the start of the tests, no change was observed in the hydraulic conductivity until about 100 days, at which time the effluent pH had decreased to about 4. Chemical analysis showed that only about 0.2 percent of the aluminum by weight and 0.08 percent of the silicon was washed out of the kaolinitic soil during the 300-day period. In similar tests using a 5 percent inorganic acid solution, about 10 percent of the aluminum and 0.2 percent of the silicon were removed due to solution of the clay.

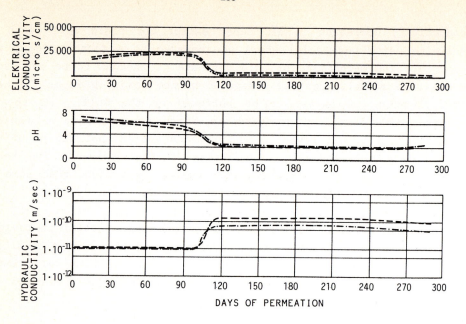

FIG. 20: ELECTRIC CONDUCTIVITYL AND pH OF THE EFFLUENT AND
HYDRAULIC CONDUCTIVITY OF A SOIL DURING PERMEATION
WITH 5% ORGANIC ACID

(from Simons et al., 1984)

The results of several investigations (Table 4) of the effects of organic acids on hydraulic conductivity lead to the following general conclusions.

1. Dissolution of carbonates and iron oxides, if present in the soil, buffers the acid, and the resulting rise in pH leads to precipitation of salts in the pores.

2. The hydraulic conductivity of soils containing carbonates or iron oxides may decrease due to pore clogging by precipitates and particle migration.

3. In the absence of carbonates or iron oxides, organic acids may cause an increase in hydraulic conductivity of up to an order of magnitude owing to solutioning and removal of small amounts of clay.

4. The long term effects of organic acids are not well established, but it is possible that after carbonates and iron oxides are

finally removed by solution an increase in hydraulic
conductivity will occur due to formation of more void space.

5. Although the organic acids cause some solutioning of the clay,
   it appears to be less than caused by inorganic acids.

Mixed Organic Chemicals

Landfill leachates and leaking tanks may expose soils to permeation by
mixtures of more than one chemical. Very little is known about the
effects of mixed chemical solutions on the hydraulic conductivity.

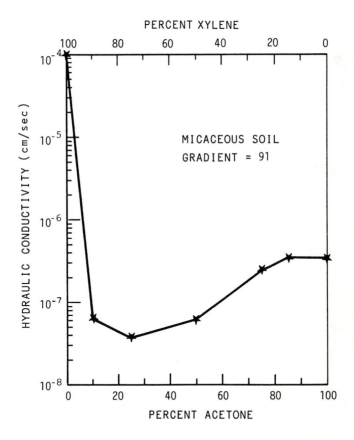

FIG. 21: HYDRAULIC CONDUCTIVITY OF COMPACTED
MICACEOUS SOIL PERMEATED BY MIXTURES
OF XYLENE AND ACETONE
(from Brown et al., 1984)

Brown et al. (1984) tested a compacted, partly saturated soil in a rigid wall permeameter using mixtures of xylene and acetone. The results are shown in Fig. 21.

Simons et al. (1984) tested "synthetic leachates" in rigid wall tests on natural clays. The acid leachate consisted of sodium acetate, glycerol, acetic acid, and salicylic acid. The basic leachate was a 5 percent ammonia solution containing copper tetramine and nickel hexamine complexes. The effects on the hydraulic conductivity of an illitic and a smectitic clay were small, whereas, that of a kaolinitic clay increased less than one order of magnitude.

Brunelle et al. (1987) investigated the influence of a natural leachate on a compacted clay in both flexible wall permeameter, rigid wall double ring permeameter, and consolidation cell. Results indicated no signifificant difference between the hydraulic conductivity of water (0.01N $CaSO_4$) and leachate in any one type of the three permeameters. Permeabilities in the fixed wall permeameters were higher than in the flexible wall devices, although there was less than an order of magnitude difference in these measurements.

Summary of the Effects of Organic Chemicals

Several general conclusions can be stated relating to the effects of organic chemicals on the hydraulic conductivity of fine-grained soils.

1. The hydraulic conductivity is invariably influenced when permeation is by an organic chemical in its pure or concentrated form. The effect depends on the method of testing that is used. The hydraulic conductivity increases very significantly in rigid wall tests, but usually decreases in flexible wall tests.

2. Solutions of compounds having a low solubility in water, such as the hydrocarbons, have no large effect on the hydraulic conductivity.

3. Water soluble organics, such as the simple alcohols and ketones, have no effect on hydraulic conductivity at concentrations less than about 75 to 80 percent.

4. Hydraulic conductivity increases caused by permeation by organics are partly reversible when water is reintroduced as the permeant.

5. Concentrated hydrophobic compounds permeate soils through cracks and macro-pores. Water remains within mini- and micro-pores.

6. Hydrophilic compounds permeate the soil more uniformly, as the polar molecules can replace the water in the hydration layers of the cations and are more readily adsorbed on the particle surfaces.

7. Organic acids can dissolve carbonates and iron oxides. Buffering of the acid can lead to downstream precipitation and pore clogging. After long times and removal of all carbonates and oxides, the conductivity may increase.

8. Pure bases can cause a large increase in the hydraulic conductivity, whereas concentrations at or below the solubility limit in water have no influence.

9. Organic acids do not cause large scale dissolution of clay particles. Weak organic bases appear to have no effect on the clay particles, and strong bases have not been tested.

CONCLUSIONS RELATIVE TO CLAYS IN WASTE CONTAINMENT APPLICATIONS

Based on the findings from the above review it is appropriate to ask: "What does it all mean?" If clays and clay soils are to be used in a particular waste containment application, if intact and durable clay barriers in the form of liners, covers, and slurry walls are to be constructed, and if the potential consequences of exposure of an aquitard to chemicals in the ground are to be assessed, then detailed knowledge of the effects of chemicals on hydraulic conductivity is needed. From the considerations in the preceding pages it is possible to state general principles and conclusions which can help in anticipating probable effects and for selection of methods for proper evaluation of the consequences of clay-chemical interactions. Acceptably low values of hydraulic conductivity can be obtained only if mini- and macropores can be essentially eliminated from the clay barrier. This means that liner compaction and slurry wall construction must be done so that there will not be either large aggregates and clods or fissures and cracks. For compacted liners and covers heavy sheepsfoot or tamping rollers and water contents above optimum are needed.

Each of the three types of laboratory tests that are used for hydraulic conductivity measurement has inherent advantages and limitations. The rigid wall permeameter system will overestimate in any case where clay-chemical interactions cause shrinkage and cracking. The flexible wall system will lead to an underestimate of the hydraulic conductivity. The consolidometer permeameter system offers the best means for quantitative assessment of chemical interaction induced volume changes and for testing clays under confining stresses that are most representative of those in the field. If possible, compaction or mixing of clay samples or soil-bentonite slurry samples for hydraulic conductivity testing should be done using water of the same composition as will be used in the field. The influences of chemicals on the hydraulic conductivity of high water content clays such as in slurry walls may be much greater than on lower water content compacted clays owing to the greater particle mobility and easier opportunity for fabric changes in the high water content system.

The effects of inorganic chemicals on hydraulic conductivity are consistent with **(1)** their effects on the double layer and interparticle forces that promote flocculation, dispersion, shrinkage, and swelling, **(2)** their effects on surface and edge charges on particles and the influences of these charges on flocculation and deflocculation, **(3)** the effects of high and low pH. Acids promote solutioning of carbonates, iron oxides, and the alumina octahedral layers of clay minerals. Bases promote solutioning of the silica tetrahedral layers and, to a lesser extent, alumina octahedral layers of the clay minerals. Removal of the dissolved material can cause increases of hydraulic conductivity; whereas, precipitation of this material can cause pore clogging and conductivity decrease.

The most important factors controlling the influences of organic chemicals on hydraulic conductivity are **(1)** the water solubility, **(2)** the dielectric constant, **(3)** the polarity, and **(4)** whether the clay is exposed to the pure organic or a dilute solution. The different classes of organics have been described herein, and available information from hydraulic conductivity research using chemicals from each class has been summarized. Summaries are given at the end of each section, so they are not repeated here.

It seems clear, however, that in almost all cases pure organic liquids will interact adversely with clays by causing some shrinkage and cracking, with concurrent large increase in hydraulic conductivity. In practice exposure of clay barriers to water-insoluble pure or concentrated organics is likely only in the case of spills, leaking tanks, and with organic "sinkers" that accumulate at low spots above liners.

Contrary to the inorganics, dilute solutions of organics have essentially no effect on the hydraulic conductivity of clays. This does not mean that they will not be transported through liners or covers. Unless adsorbed by the clay itself or by organic matter in the clay, they will migrate with the seepage water by advective transport.

Finally, the data base so far available is comprised almost totally of test results for simple monochemical systems and test programs that have extended only over very short time periods relative to the active life and post-closure security periods for most waste containment and storage facilities.

More research on the influence of natural waste leachates, mixed chemical systems and the effects of time on the hydraulic conductivity of clays is needed. This, and investigations on various Swiss clays and specially treated clays which -depending on the treatment- would be able to adsorb hazardous parts of the waste leachate, will be a field of research in the Division of Foundation Engineering and Soil Mechanics at the ETH Zurich in the immediate future.

COMMENTS ON "FINAL STORAGE" AND "FINAL STORAGE QUALITY"

Looking at the present situation of waste deposits in Switzerland it seems that -in a practical sense- nearly all waste sites are a kind of final storage sites. This is because it is more or less impossible to remove the waste from a site for any reason. In "Bundesamt fur Umweltschutz (1986)" the ideas of final storage are quite different from this actual situation. According to "Bundesamt fur Umweltschutz

(1986)" a final storage is a waste site whose content combined with a proper liner system only allows escape of leachates which are harmless to the environment and not requiering treatment. The difference between the actual situation -especially for hazardous waste- and the ideas of how it should look in the future seems quite large and it may take many years to develop suitable systems. This also because the necessary amount of hazardous waste treatment and combustion plants are missing.

It therefore seems that in the near future the content of the waste sites will be more or less the same as it is today. This means that we will have to deal with a mixture of very different substances and chemicals which are not totally immobilized. It is the authors' opinion that the liner system for those sites can and should be designed to encapsulate the waste. For a hazardous waste site a combination of plastic and clay liners with a leak detection system guarded by monitoring wells inbetween will be required. The inflow of water into the waste site and the amount of leachate should be minimized. The clay liner should be designed in a way that it will be able to adsorb as many hazardous parts of the leachate as possible. At the present and for the near future this may require a rather thick sandwich of different clays or differently treated clays.

In addition future waste sites will need a properly designed liner system to encapsulate the waste. Even if only treated and immobilized wastes that have been solidified and of "final storage quality" will be disposed in "monosites," a clay liner able to adsorb possible hazardous parts of the leachate should be used. These liners may be less complicated and less expensive than for waste sites with mixed content of more or less unknown chemicals.

ACKNOWLEDGEMENTS

In addition to the financial support for this study provided by the Environmental Institute for Waste Management Studies, University of Alabama, Tuscaloosa, of which the second author is a member, support was provided by the Swiss Federal Institute of Technology Zurich, and by the California Water Resources Center under Agreement No. W-689, "Mechanics of Organic Solvent Movement in Aquitards". This support is acknowledged with appreciation.

TABLE 4   RESULTS OF HYDRAULIC CONDUCTIVITY TESTING OF CLAYS USING DIFFERENT ORGANIC CHEMICALS

Class of Compound	Compound	Soil Type	Soil State	Concentration	Permeameter*	Effect on Conductivity	Effects on Samples, Remarks	Reference
	Heptane	Natural soils: Non-calcareous smectite Calcareous smectite Kaolinite, Illite	Compacted	Pure	RW	10^2 to 10^3X increase	Blocky structure formed. Conductivity and break-through curves in Fig.11.	Anderson, Brown & Thomas (1985)
		Kaolinite, Georgia Natural soil: Illite-chlorite	Compacted	Pure 53 mg/L	FW RW FW RW	10^2X decrease 250 to 660X increase None None	Large pores & cracks formed. None None	Bowders (1985)
	Cyclohexane	Natural soil: Illite-smectite	Compacted	Pure	RW	None		Fernandez & Quigley (1985)
	Benzene	Natural soil: Illite-smectite	Compacted	Pure	RW	None	See Fig.12.Only 8% of pore water replaced.See Figs.13 & 14 & text for sequential replacement effects.	Fernandez & Quigley (1985)
Hydrocarbons and Related		Kaolinite, Georgia	Compacted	Pure 0.1%	FW	10^3X decrease Slight decrease		Acar et al. (1985)
		Natural soils: Non-calcareous smectite Calcareous smectite Kaolinite, Illite	Compacted	Pure	RW	10^2 to 10^3X increase	Blocky structure formed.	Anderson, Brown & Thomas (1985)
	Xylene (Dimethyl-benzene)	Calcareous smectite-soil-bentonite mixture	Slurry	Pure	RW,DR	10^2 to 10^3X increase	9% bentonite	Anderson, Crawley & Zabcik (1985)
		Natural soils: Kaolinite, Mica, Bentonite	Compacted	Pure	RW	10^2 to 10^3X increase	Outdoor tests on 1.8m diameter samples. Gradient = 7. Cracks formed	Brown, Thomas & Green (1986)
		Natural soil: Illite-smectite	Compacted	Pure	RW	None		Fernandez & Quigley (1985)

*RW-Rigid Wall; RW,DR-Rigid Wall, Double Ring; FW-Flexible Wall; C-Consolidometer.

TABLE 4   RESULTS OF HYDRAULIC CONDUCTIVITY TESTING OF CLAYS USING DIFFERENT ORGANIC CHEMICALS

Class of Compound	Compound	Soil Type	Soil State	Concentration	Permeameter*	Effect on Conductivity	Effects on Samples, Remarks	Reference
Hydrocarbons and Related	Tetrachloro-Methane (Carbontetrachloride)	Sand-bentonite mix	Consolidated Slurry	Pure 720 mg/L	FW	10^2X increase None	7% bentonite	Evans et al. (1985)
	Trichloroethylene (TCE)	Kaolinite, Georgia Natural soil: Illite-chlorite	Compacted	Pure 1100 mg/L	FW FW FW RW	10^2 to 10^3X decrease 140 to 500X increase None None	No flow through the samples. Cracks formed.	Bowders (1985)
	Nitro-benzene	Kaolinite, Georgia	Compacted	Pure 0.1%	FW	10^3X decrease Slight decrease		Acar et al. (1985)
		Kaolinite, Georgia Natural soil: Illite-chlorite	Compacted	Pure 20%,40% 60%,80%	FW FW FW RW	None 7.5 to 44X increase None None	See Fig. 15.	Bowders (1985)
	Methanol	Natural soils: Non-calcareous smectite Calcareous smectite Kaolinite, Illite	Compacted	Pure	RW	10^3X increase	Large pores & cracks formed.	Anderson, Brown & Thomas (1985)
		Calcareous smectite-soil-bentonite mixture	Slurry	Pure	RW,DR	10^3X increase	9% bentonite	Anderson, Crawley & Zabcik (1985)
Alcohols and Phenols		Natural soil: Illite-smectite	Compacted	Pure	RW	10X increase	See Fig.16. 94% of the pore fluid replaced.	Fernandez & Quigley (1985)
	Ethanol	Natural soil: Illite-smectite	Compacted	Pure	RW	10X increase	See Fig.13, 10^3X increase when permeated with benzene after ethanol.	Fernandez & Quigley (1985)
	Ethylene-glycol	Natural soils: Non-calcareous smectite Calcareous smectite Kaolinite, Illite	Compacted	Pure	RW	10^2X increase	See Fig.17. Initial decrease of conductivity	Anderson, Brown & Thomas (1985)
	Phenol	Kaolinite, Georgia	Compacted	Pure 0.1%	FW	2X increase Slight decrease		Acar et al. (1985)

*RW-Rigid Wall; RW,DR-Rigid Wall, Double Ring; FW-Flexible Wall; C-Consolidometer

TABLE 4 (Contd.) RESULTS OF HYDRAULIC CONDUCTIVITY TESTING OF CLAYS USING DIFFERENT ORGANIC CHEMICALS

Class of Compound	Compound	Soil Type	Soil State	Concentration	Permeameter*	Effect on Conductivity	Effects on Samples, Remarks	Reference
Ethers	1.4-Dioxane	Kaolinite	Consolidated suspension	Pure	C	20 to 30% increase	Samples prepared initially in dioxane showed much higher conductivities.	Michaels & Lin (1954)
		Micaceous soil	Compacted	Pure 75% 2%,12.5% 25%,50%	RW	10^2X increase 10X increase None		Brown, Thomas & Green (1983)
		Kaolinitic soil	Compacted	Pure	RW	10^2X increase		Brown, Green & Thomas (1984)
Aldehydes and Ketones	Acetone	Kaolinite, Georgia	Compacted	Pure Pure 0.1%	RW FW	10X increase 2X increase Slight decrease	See Fig. 19.	Acar et al. (1985)
		Natural soils: Kaolinite Mica, Bentonite	Compacted	Pure	RW	10X increase 10^3X increase	1.8m diameter outdoor tests, gradient = 7. Laboratory tests.	Brown, Thomas & Green (1986)
		Natural soils: Non-calcareous smectite Calcareous smectite Kaolinite Illite	Compacted	Pure	RW	10 to 10^3X increase	Shrinkage & cracking, initial decrease of conductivity. See Fig.18.	Anderson, Brown & Thomas (1985)
		Mixtures of: Na-Montmorillonite Kaolinite, and Sand	Compacted	Pure	FW C	2 to 10X increase	Increase dependent on the activity of soil mixture.	Acar et al. (1987)
Organic Bases	Aniline	Sand-bentonite mix	Consolidated Slurry	Pure 30 g/L 15 g/L	FW	10^4X increase None None		Evans et al. (1985)
		Natural soils: Non-calcareous smectite Calcareous smectite Kaolinite, Illite	Compacted	Pure	RW	10^2X increase	Large pores and cracks formed, reintroduction of water caused a 10X decrease.	Anderson, Brown & Thomas (1985)

*RW-Rigid Wall; RW,DR-Rigid Wall, Double Ring; FW-Flexible Wall; C-Consolidometer.

TABLE 4 (Contd.) RESULTS OF HYDRAULIC CONDUCTIVITY TESTING OF CLAYS USING DIFFERENT ORGANIC CHEMICALS

Class of Compound	Compound	Soil Type	Soil State	Concentration	Permeameter*	Effect on Conductivity	Effects on Samples, Remarks	Reference
Organic Acids	Acetic Acid	Kaolinite, Georgia Illite-chlorite Natural soil	Compacted	Pure	FW FW FW RW	Slight decrease 2X increase Slight decrease Slight decrease	Pore clogging by precipitates. Effluent pH > influent pH.	Bowders (1985)
		Sand-bentonite mix	Consolidated Slurry	Dilute, pH 1	FW	Slight continuous increase	Effluent pH > influent pH.	Evans et al. (1985)
		Natural soils: Non-calcareous smectite Calcareous smectite Kaolinite, Illite	Compacted	Pure	RW	10 to 10^3X decrease	Dissolution of calcium carbonate and iron oxide, pore clogging by soil particle fragments.	Anderson, Brown & Thomas (1985)
		Natural soils: Kaolinite Illite Smectite	Compacted	5%	RW	10X increase	Acetic acid + propionic acid mixture, 0.2% aluminum and 0.06% silicon washed out after 300 days	Simons et al. (1984)
	Xylene Acetone	Micaceous soil	Compacted	Pure Xylene Acetone: 12.5%,25%, 50%,75%, 90%,100%	RW	10^4X increase 10 to 15X increase	See Fig. 21.	Brown, Thomas & Green (1984)
Mixed Chemicals	Sodium Acetate Glycerol Acetic Acid Salicylic Acid	Natural soils: Kaolinite Illite Smectite	Compacted	Dilute	RW	10X increase None None	"Synthetic leachate acid".	Simons et al. (1984)
	5% Ammonia Copper tetramine, Nickel Hexamine	Natural soils: Kaolinite Illite Smectite	Compacted	Dilute	RW	10X increase None None	"Synthetic leachate basic".	Simons et al. (1984)
	Type II Landfill leachate	Natural soil (25% < 2 μm)	Compacted	Dilute	FW RW C	None None None	Hydraulic conductivity RW and C higher than in FW	Brunelle et al. (1987)

*RW-Rigid Wall; RW,DR-Rigid Wall, Double Ring; FW-Flexible Wall; C-Consolidometer.

REFERENCES

Acar, Y.B., Hamidon, A., Field, S.D., and Scott, L. (1985). "The Effect of Organic Fluids on Hydraulic Conductivity of Compacted Kaolinite," Hydraulic Barriers in Soil and Rock, ASTM STP 874, pp. 171-187.

Acar, Y.B.and D'Hollosy, E. (1987). "Assessment of Pore Fluid Effects Using Flexible Wall and Consolidation Permeameters," Geotechnical Practice for Waste Disposal '87, University of Michigan, Ann Arbor, June 15-17, 1987. ASCE Geotechnical Special Publication No. 13.

Alther, G., Evans, J.C., Fang, H-Y., and Witmer, K. (1985). "Influence of Organic Permeants upon the Permeability of Bentonite," Hydraulic Barriers in Soil and Rock, ASTM STP 874, pp. 64-73.

Anderson, D.C., Brown, K.W., and Thomas, J.C. (1985). "Conductivity of Compacted Clay Soils to Water and Organic Liquids", Waste Management and Research, Vol. 3, No. 4, pp. 339-349.

Anderson, D.C., Crawley, W., and Zabcik, J.D. (1985). "Effects of Various Liquids on Clay Soil: Bentonite Slurry Mixtures," Hydraulic Barriers in Soil and Rock, ASTM STP 874, pp. 93-103.

Bolt, G.H. (1955). "Analysis of the Validity of the Gouy-Chapman Theory of the Electric Double Layer," Journal of Colloid Science, Vol. 10, pp. 206-219.

Bolt, G.H. (1956). "Physico-Chemical Analysis of the Compressibility of Pure Clays," Geotechnique, Vol. 6, No. 2, pp. 86-93.

Bowders, J.J. (1985). "The Influence of Various Concentrations of Organic Liquids on the Hydraulic Conductivity of Compacted Clay," Geotechnical Engineering Dissertation GT85-2, The University of Texas at Austin, 218 pp.

Brown, K.W., Green, J.W., and Thomas. J.C. (1983). "The Influence of Selected Organic Liquids on the Permeability of Clay Liners," Land Disposal of Hazardous Waste, Proc. of the 9th Annual Research Symposium, EPA-600/9-83-018, Ft. Mitchell, Kentucky, pp. 114-125.

Brown, K.W., Thomas, J.C., and Green, J.W. (1984). "Permeability of Compacted Clays to Solvent Mixtures," Land Disposal of Hazardous Waste, Proc. of the 10th Annual Research Symposium, EPA-600/9-84-007, Ft. Washing-ton, PA, pp. 124-137.

Brown, K.W., Thomas, J.C., and Green, J.W. (1986). "Field Cell Verification of the Effects of Concentrated Organic Solvents on the Conductivity of Compacted Soils," Hazardous Waste and Hazardous Materials, Vol. 3, No. 1, pp. 1-19.

Brunelle, T.M., Dell, L.R., and Meyer, C.J. (1987). "Effects of Permeameter and Leachate on a Clay Liner", Geotechnical Practice for Waste Disposal '87, University of Michigan, Ann Arbor, June 15-17, 1987. ASCE Geotechnical Special Publication No. 13.

Bundesamt fur Umweltschutz (1986). "Leitbild fur die Schweizerische Abfallwirtschaft". Schriftenteihe Umweltschutz Nr. 51. Herausgegeben vom Bundesamt fur Umweltschutz, Bern, Juni 1986.

Collins, K. and McGown, A. (1974). "The Form and Function of Microfabric Features in a Variety of Natural Soils," Geotechnique, Vol. 24, pp. 223-254.

Condon, F.E. and Meislich, H. (1960). Introduction to Organic Chemistry, Holt, Rinehart, and Winston, Inc., New York, 841 p.

Daniel, D.E. (1984). "Predicting Hydraulic Conductivity of Clay Liners," Journal of Geotechnical Engineering, ASCE, Vol. 110, No. 2, pp. 285-300.

Daniel, D.E., Anderson, D.C., and Boynton, S.S. (1985). "Fixed-Wall Versus Flexible Wall Permeameters," Hydraulic Barriers in Soil and Rock, ASTM STP 874, pp. 107-126.

D'Appolonia, D. (1980). "Soil-Bentonite Slurry Trench Cutoffs," Journal of Geotechnical Engineering, ASCE, Vol. 106, No. GT4, pp. 399-417.

Day, S.R. (1984). "Field Permeability Test for Compacted Clay Liners," M.S. Thesis, The University of Texas at Austin, May 1984, 105 p.

Dunn, R.J. and Mitchell, J.K. (1984). Fluid Conductivity Testing of Fine-Grained Soils," Journal of Geotechnical Engineering, ASCE, Vol. 110, No. GT11, pp. 1648-1665.

Evans, J.C., Fang, H-Y., and Kugelman, I.J. (1985). "Organic Fluid Effects on the Permeability of Soil-Bentonite Slurry Walls," Proc. National Conf. Hazardous Waste and Environmental Emergencies, May 14 - 16, 1985, Cincinnati, OH.

Fernandez, F. and Quigley, R.M. (1985). "Hydraulic Conductivity of Natural Clays Permeated with Simple Liquid Hydrocarbons", Canadian Geotechnical Journal, Vol. 22, pp. 205-214.

Gordon, B.B. and Forrest, M. (1981). "Permeability of Soils Using Contaminated Permeant," Permeability and Groundwater Contaminant Transport, ASTM STP746, pp. 101-120.

Gouy, G. (1910). "Sur la constitution de la charge electrique a la surface d'un electrolyte," J. Physique, 9, pp. 457-468.

Griffin, R.A. and Roy, W.R. (1985). "Interaction of Organic Solvents with Saturated Soil-Water Systems," Open File Report, Environmental Institute for Waste Management Studies, University of Alabama, Tuscaloosa.

Hardcastle, J.H. and Mitchell, J.K. (1974). "Electrolyte Concentration-Permeability Relationships in Sodium Illite-Silt Mixtures," Clays and Clay Minerals, Vol. 22, pp. 143-154.

Hart, H. and Schuetz, D. (1966). A Short Course in Organic Chemistry, Third Ed., Houghton Mifflin Co., Boston, 346 p.

Huang, W.H. and Keller, W.D. (1971). "Dissolution of Clay Minerals in Dilute Organic Acids at Room Temperature," The American Mineralogist, Vol. 56, pp. 1082-1095.

Lagaly, G. (1984). "Clay-Organic Interactions," Phil. Trans. Royal Society of London, A311, pp. 315-332.

Lambe, T.W. (1954). "The Permeability of Fine-Grained Soils," ASTM STP 163, pp. 56-67.

Lentz, R.W., Horst, W.D., and Uppot, J.O. (1985). "The Permeability of Clay to Acidic and Caustic Permeants," Hydraulic Barriers in Soil and Rock, ASTM STP 874, pp. 127-139.

Madsen, F.T. and Mitchell, J.K. (1987). "Chemical Effects on Clay Hydraulic Conductivity and their Determination," Open File Report, Environmental Institute for Waste Management Studies, University of Alabama, Tuscaloosa, 70 p.

Michaels, A.S. and Lin, C.S. (1954). "The Permeability of Kaolinite," Industrial and Engineering Chemistry, Vol. 46, pp. 1239-1246.

Mitchell, J.K. (1976). Fundamentals of Soil Behavior, John Wiley & Sons, New York, 422 p.

Mitchell, J.K., Hooper, D.R., and Campanella, R.G. (1965). "Permeability of Compacted Clay," Journal of the Soil Mechanics and Foundations Division, ASCE, Vol. 91, No. SM4, July 1965, pp. 41-65.

Mitchell, J.K. and Madsen, F.T. (1987). "Chemical Effects on Clay Hydraulic Conductivity," Geotechnical Practice for Waste Disposal '87, University of Michigan, Ann Arbor, June 15-17, 1987. ASCE Geotechnical Special Publication No. 13, pp. 87-116.

Olsen, H.W. (1962). "Hydraulic Flow Through Saturated Clays," Proc. Ninth National Conf. on Clays and Clay Minerals, pp. 131-161.

Quirk, J.P. and Schofield, R.K. (1955). "The Effect of Electrolyte Concentration on Soil Permeability," Journal Soil Science, Vol. 6, No. 2, pp. 163-178.

Simons, H., Hansel, W., and Reuter, E. (1984). "Physical and Chemical Behavior of Clay-Based Barriers under Percolation with Test Liquids," Proc. Int. Symposium on Clay Barriers for Isolation of Toxic Chemical Wastes, May 28-30, 1984, Stockholm, p. 117. Expanded version in German, pp. 118-127.

van Olphen, H. (1977). An Introduction to Clay Colloid Chemistry, Wiley Interscience, New York, pp. 260-293.

Vansant, E.F. and Uytterhoeven, J.B. (1973). "The Adsorption of Aromatic, Heterocyclic and Cyclic Ammonium Cations by Montmorillonite," Clay Minerals, Vol. 10, pp. 61-69.

Verwey, E.J.W. and Overbeek, J.Th.G. (1948). Theory of the Stability of Lyophobic Colloids, Elsevier, Amsterdam, New York, London, pp. 22-76.

Weiss, A. (1962). "Ein Geheimnis des Chinesischen Porzellans," Angew. Chem., 75 Jahrg, Nr. 16/17, pp. 755-762.

GROUP REPORT: MATERIAL TRANSPORT AND PROPERTIES OF REACTOR ENVELOPES

Objectives

Envelope is necessary around a reactor landfill to control emissions (leachate and gas).

If a landfill has reached final storage quality liners have not to be effective any more.

The discussion whether it is possible to achieve final storage quality is not an objective of our group.

As long as the liner system is in operation it has to be observed and maintained and the top liner system has to be repaired if necessary.

Basics

The geological and hydrogeological situation is not considered in our group work, assuming that careful field investigations and modeling are done.

No recipes but requirements and recommendations are given.

Landfill liners (bottom and top) are not impermeable. They will release - although leakage rates might be very low - both leachate and gas into the environment.

Because of time and background limitations we have reduced in our considerations the variety of liner materials to clay (or clay mixtures) and high-density polyethylene (HDPE). These are the main materials that are suitable for the desired purposes. If a single liner is installed only, then preference should be given to clay.

Slurry walls do not meet liner quality criteria and are, therefore, excluded.

Strategy of envelope installation

A bottom liner system has to be installed before landfill operation starts.

When the main biological degradation processes are terminated, a surface liner (surface cap) has to be installed.

Unless the landfill is stabilized (see groups B) it should not be used for public purposes. The landfill has to be controlled, and installations have to be maintained.

The bottom liner has to be placed with a significant distance above the groundwater table. Above-ground landfills are favoured for several reasons. In any case, bottom-liner area should be minimized.

QUESTIONS

1. Which "indicator parameters" should be used to describe transport processes across envelopes ?

1.1. Models

Existing commonly used transport models are based on Darcy's and Fick's law. It is assumed that Darcy's law is valid for low hydraulic conductivities too.

In those transport models attenuation can be incorporated in a simplistic form. The various processes for attenuation (e.g. sorption, biodegradation, precipitation/dissolution etc.) are not well understood. More research is needed.

The transport through the liner is dependent on:

      a) the liner material properties and

      b) the leachate components.

1.2 Indicator parameters of liner materials

1.2.1. Natural materials

Pure clays:

a) swelling clays

b) non-swelling clays;

Clayey sediments:

a) rocks (pulverized)

b) unconsolidated.

Upgraded materials (e.g. with bentonite).

Transport-relevant liner properties:

a) mineralogical composition;

b) physical properties of the compacted material, e.g. porosity/density, moisture content, structure (flocculated - disperse); appropriate testing methods should be standardized;

c) Hydraulic conductivity/intrinsic permeability;

d) Adsorption/desorption is dependent on:

      - Clay content and type of clay;

      - Specific surface area and cation exchange capacity (CEC);

      - Content of organic matter;

e) Special treatment (e.g. hydrophobic clays).

1.2.2. Synthetic materials

HDPE is presently widely recommended as a liner material. Properties describing the liner material are density, permeability, tensile strength, elongation, and weathering.

1.3. Transport rates

Transport rates may be influenced by:
- material characteristics
- waste/material interactions
- biological processes (clogging in clays or degradation in synthetic materials);
- weathering (temperature, ultraviolet (UV) light, ozone, decompositions through loss of plasticizers)

1.4 Indicator substances

Due to molecular diffusion certain substances pass the liners faster than others ("forerunners"). Examples are chloride, bromide, certain organic solvents.
Due to sorption heavy metals can effectively be retarded in clay-liner materials.

2. With artificial and/or natural envelopes, is tailor-made fluid control possible ?

"Tailor-made" fluid control can be achieved to a certain extent by synthetics and/or clayey materials including an effective water collecting system.

Whether the lifetime of the system components or their performance can be guaranteed over the period required is uncertain and can hardly be predicted.

3. What is the lifetime of liner materials ?

Most available data about the performance of liner materials are from a) lab simulations with strong solutions and high pressure gradients acting on small samples, thus not reflecting real site conditions; and b) from field experience limited in time (less than 15 years). Therefore, no confident long-term prediction is possible. However, some conclusions can be drawn:

a) Clayey materials: Stable under given pressure/temperature -conditions and "natural" fluid composition for geological time periods; unstable if exposed to strong bases and acids.

b) HDPE: May change its physical properties on short terms by organic solvents, on long terms by microbiological attack to an unknown extent.

c) Asphalt: Cases of biodegradation and material incompatibilities were found for asphalt liners (Haxo, 1985). As the precise conditions are poorly known, further investigations are required.

The low confidence level in prediction is asking for further field observations and testing as part of the monitoring of real landfills as long as final storage quality is not reached.

4. The influence of variations in design and configuration on liquid and contaminant transport through liners

Considerations of this issue have involved:

- clays, synthetic materials and combinations of the two;
- control of leachate and gases and structural stability;
- bottom liners plus sidewalls and top liners or caps;
- other factors.

In particular the following objectives were taken into account:

- bottom liners plus sidewalls must reduce leachate seepage, contaminant discharge and gas migration (sidewalls);
- top liner must minimize infiltration of precipitation, control gaseous exchange with the atmosphere, support adequate vegetation, control vectors and enhance appearance.

4.1 Clay liners

Specific properties have been presented in response to question no.1 and consideraton of these allows selection of clay composition, compaction, porositiy and hydraulic conductivity. These have a direct influence on leachate flux and contaminant transport through diffusion. Additional design features must also be considered however.
If possible avoid the use of high-plasticity clays because it is difficult to compact them.
Clay liner thickness reduces the seepage rate of the liquid and delays contaminant breakthrough. Increased depth helps to reduce the effect of inhomogeneities produced at placement and the impact of settlement. However, very large thicknesses may lead to stability problems related to slippage and shear failure. In Germany a minimum thickness of 60 cm in combination with synthetics is required, while in some states of the US minimum thicknesses in excess of 1 m are required. A maximum thickness constraint requires a site specific geotechnical and hydrogeological evaluation.
It is also necessary to protect the clay during and after placement against desiccation. Care must be taken to avoid damage by equipment. Proper methods for placement of clay liners must be used and reference to this from other sources should be made.

4.2 Synthetic liners

Properties related to the performance and selection of synthetic materials used for liners have been given in response to question no.1. Consideration of these will provide input on stability to leachate, permeability, contaminant diffusion and other essential design features. Membrane thickness is important in limiting leachate seepage,

but very thick materials impede on-site handling and seaming. A thickness of 2.0 - 3.0 mm is recommended. Protection against UV rays, ozone,
and mechanical disruption from equipment during construction is necessary. A fine-grained granular blanket possibly with geofabrics is recommended. Methods for preparation of the supporting base of the liner,
seaming, anchoring on the side wall, quality assurance testing and
other factors must be addressed but are not considered here in detail.

Some sites may be unsuitable for synthetic liners because of geological
and geometric constraints. For either clay or synthetic liners, it is
essential to account for sedimentation and stability in the liner
subgrade.

4.3 Design features of the leachate collection system

The objective of the collection system is to remove the leachate from
the landfill and to reduce hydraulic head on the liner at the same
time. It is essential that the drainage blanket extends over the entire
area of the liner. It includes a granular layer and pipes with
geofabrics in some cases. The granular layer should exclude limestone
and also materials which might adversely affect the liner system. The
drainage slopes should be at least 2 % to permit gravity flow. Pipes
should extend through the liner to external storage tanks thus
eliminating internal manholes and pumping and reducing mounding in the
landfill if clogging occurs. It is recommended that appr. 2 m of
uncompacted refuse be placed above the drainage blanket to provide protection against equipment, to enhance drainage, and to provide a
buffer against suspended contaminant transport.

The following features are important in collection system design:

- Pipe diameters should be sufficient   to allow inspection   and
  cleaning;

- materials should adequately resist corrosion and applied loads;

- siphons are required for gas-transfer control;

- provision of a protective layer to minimize clogging is desirable but not easily accomplished.

It should be borne in mind that the leachate collection system can be the weak part in the system because of clogging and pipe failure. Measurements of head build up and frequent pipe maintenance are essential.

4.4 Landfill cap

The final landfill cap on cover should not be placed until settlement is near completion. This will reduce damage to the final cover and will allow moisture entry to enhance biodegradation. However, an interim equalization layer and handling is needed. Gas extraction and handling is required as soon as the landfill is in Operation. Grading of the finished surface must allow for enhanced runoff but should not encourage erosion. The final cover should consist of the following layers A to E:

A: a growth support layer including topsoil and sufficient subsoil to match local pedogenesis and prevent freeze-up of lower layers;

B: a granular drainage layer;

C: a low hydraulic conductivity layer consisting of clay and/or synthetic material;

D: a granular gas drain to reduce the pressure against layer C;

E: the equalization layer already in place. Synthetics should be discontinuous to avoid rupture during cover movement and must be shingled to impede infiltration. This layer will also function for root control.

4.5 Combination liners

Consideration should be given to the use of a multiliner system. There are many possibilities.

5. Emplacement criteria for liner systems, control and maintenance

The right emplacement of a liner may in practice have more effect
on the functioning of a liner than the liner material itself. It
is essential to control the quality of the installation according
to specification.
Clay has to be adjusted to the required water content and quality,
if necessary by means of separate preparation steps (additives,
mixing, water addition, separation of sand, etc.).

6. Influence of solid waste quality and landfill operation on liners

Enhancement of biological degradation processes is mostly favoured for
the following reasons:

- decrease in organic leachate concentrations which results in
  less reactivity with liner materials;

- minimizing the time period in which the main settling takes
  place;

- minimizing the time period in which the main gas production
  takes part.

Emplacement of the waste should take place in each landfill section
most evenly so that significant weight differences on the liner can be
avoided.

The following materials should be excluded because of the interference
with liner systems: Waste containing considerable amounts of strong
acids and bases, pure organic solvents (in particular chlorinated hydrocarbons), and sludges/slurries which might migrate to collection
systems and clog.

THE LANDFILL AS A FINAL STORAGE

SCIENTIFIC AND TECHNICAL CRITERIA
FOR THE FINAL STORAGE QUALITY

Chairman: Adrian Pfiffner (CH)
Lecturers: Eckehard Bütow (FRG)*
                        Peter Huggenberger (CH)
                        Klaus Stief (FRG)
                        Josef Zeyer (CH)
Troublemaker: Thomas Lichtensteiger (CH)
Rapporteur: Jan van Stuijvenberg (CH)
Discussion participants: Markus Egli (CH)
                        Donald H. Gray (USA)
                        Anders Lagerkvist (SW)
                        Auguste Zurkinden (CH)

SPECIFIC QUESTIONS:

1) Which factors determine the geological, geotechnical, chemical and biological quality of a final storage?

2) How is the pollution potential of sanitary landfills defined in practice?

3) What kind of selection and pretreatment is most suitable in practice to accelerate the transformation of wastes into a final storage quality?

4) What is the influence of the site selection on geotechnical criteria?

5) What is the approximate "lifetime" of geotechnical properties in landfills?

6) Which hydrogeological criteria are suitable to determine final storage quality?

7) Are there ecotoxicological methods to evaluate the pollution potential of substances put in a landfill?

INTRODUCTION

SCIENTIFIC AND TECHNICAL CRITERIA FOR FINAL STORAGE QUALITY

O. Adrian Pfiffner

Geological Institute, University of Bern

1. Final storage

Final storage according to the Swiss concept of waste management (EKA 1986) involves waste disposal with fluxes of substances into the environment (air, water and soil) that do not significantly alter this environment and transformation of waste into forms that do not require any treatment in the long- and short-term. In this concept, final storage includes the use of appropriate barriers to reach the desired quality. The concept aims at outlining feasable solutions for operating final storage facilities by the year 1995-2000, rendering Switzerland independent in its waste management.

Strategies followed to attain final storage quality depend strongly on the type of waste considered. This is a result of the different types of processes encountered in different waste types. In particular, interactions among chemically different waste compounds must be considered (e.g. the change in solubility due to changes in ph-conditions). As a consequence, so-called monowastes consisting of a group of sub-stances with a specific chemical composition are envisaged in the Swiss concept. The advantage of monowastes is the minimization of complex interactions between various chemical compounds and the potential for their re-use as resources for future gene-rations.

2. Landfill

Landfills are engineered disposal systems of solid wastes. For the purpose of final storage, one must consider the effects that waste disposal can have on the groundwater environment. These effects stem from the interaction between percolating water and the waste, which leads to the formation of a leachate. The objectives are to reduce the concentration of hazardous compounds in such a leachate to a minimum and to minimize the contamination of groundwater by this leachate. The first objective concerns the nature of the waste body itself, i.e. the chemical and physical properties of the waste. The second objective pertains to the interface between the waste body and the natural environment, namely the geosphere. This interface is commonly referred to as barrier.

Land disposal of waste entails the use of both natural and artificial barriers. The role of these barriers is both containment and retention or attenuation of hazardous waste compounds. The barriers must therefore meet certain standards with regard to permeability and attenuation-capacity and to the possible deterioration of these factors with time.

Artificial barriers commonly used beneath and beside the waste body (landfill bottom liners) as well as barriers put on top of the waste (caps or landfill surface liners) include synthetic membranes and clay liners. In final storage landfills the waste body itself must be considered as barrier too. Landfill bottom liners have the role to prevent leachate percolation into the bedrock and groundwater.

Synthetic membranes (textiles) contain, but do not attenuate and are vulnerable to defects and thus to losing their containment capacity.

Clay liners both contain and attenuate. Attenuation mechanisms include precipitation and filtration, sorption and exchange of polar and ionic solutes on clay minerals, degradation of organic compounds to inert forms, and neutralization of excess acidity. Attenuation capacity can be significantly affected by interactions between leachate and the clay liner. Such interactions may increase the hydraulic conductivity and thus diminish the containment effect. Besides they may as well reduce the attenuation effect, the leachate largely bypassing the clay liner.

Caps have the role of sealing the waste from precipitation so as to avoid leachate formation, to prevent emission of odours into the environment and to meet other legal environmental standards. The cap must obviously be less permeable than the least permeable barrier around the waste in order to prevent leachate overflow (bath-tub effect).

The first natural barrier encountered by the leachate is the bedrock, which is required to have a high attenuation capacity, to be relatively impermeable, and to be stable from an engineering point of view. Rock types which best meet these requirements are shales (sedimentary rock with an appreciable content of clay minerals) and marls (sedimentary rocks consisting chiefly of clay and carbonate minerals). The next and last "barrier" should garantee dilution and dispersion of the remaining contaminants in the natural groundwater. This requires that substantial groundwater circulation occur, e.g. within unconsolidated Quaternary sediments (e.g. fluvial and glacial deposits). However such groundwater is also extensively exploited as drinking water in Switzerland.

3. Scientific and technical criteria

Scientific and technical criteria to evaluate final storage quality must rely on the chemical, physical and biological processes which go on within and around the waste. These processes encompass

a) chemical and biological reactions and transformation of various waste compounds;
b) interaction of these products and reactants with the synthetic or natural liners which separate the waste from the natural environment (the geosphere);
c) transportation of diluted contaminants in groundwater.

Since criteria for final storage depend upon all these processes, a basic understanding of the processes themselves is of utmost importance for final storage quality evaluation and extrapolation of final storage quality with time (risk-assessment).

Final storage is often associated with the notion of the waste being rendered "inert" (i.e. non-reactive). This is implied when the waste is said to be similar to the earth's crust (rock-, ore- or soil-like), or when requirements are made for its transformation into a "mineralized form" (EKA 1986). The latter designates solely organic compounds which are transformed into H_2O, CO_2 and alkaline salts of halogens by burning. In earth sciences minerals are defind as naturally occurring inorganic crystalline substances of a particular homogeneous composition.

When considering even short geological time-spans of the order of 100 to 10'000 years, the notion of an "inert" waste body needs to be carefully assessed. It is known from many mining areas that massive ore bodies (aggregates of minerals) within solid rocks undergo severe chemical weathering or alteration as soon as the ore body finds itself in an oxidyzing environment (e.g. Baumann, 1976). This weathering may lead to voluminous transportation of metals and to a chemical zonation of the ore body. In an upper zone of oxidation (also called the gossan), extensive leaching of metal oxides occurs. These metals are then enriched by precipitation under more alkaline conditions in the lower zone of cementation. The leached metals precipitate as sulphates onto grains of sulphides of less noble metals of the ore body, thereby "cementing" the ore body. In a next step, the less noble metal is then dissolved as sulphate while

the formerly leached noble metal reacts to sulphide. An example of a commonly observed reaction is:

$$4\ FeS_2 + 4\ CuSO_4 \longrightarrow 2\ Cu_2S + 4\ FeSO_4 + 3\ S_2$$

In the geological past, such chain reactions produced considerable concentrations of
noble metals in the earth's crust, where their stability is maintained by the prevailing
alkaline conditions.

It is evident from the discussion above that there is no clear distinction between
"reactor landfills" (as discussed in first half of this book) and "final storage landfills". Rather, there is a gradual transition between the two types. The situation is
explained in Fig. 1, in which the future development of a specific chemical compound
within a landfill is sketched by plotting a "toxicity index" as a function of time. The
"toxicity index" could represent for example, the total amount of the hazardous compound leached out within one month. For waning biological activities, this "toxicity
index" might level off with time. For chemical reactions (e.g. due to changing pH-Eh
conditions), such a "toxicity index" would level off in time only if the waste body became thermodynamically stable. Fluctuations in the nature of the percolating leachate
(e.g. changing pH-Eh-conditions) could result in an irregular course of the "toxicity
index", whereby final storage quality could exceed acceptable standards during certain periods.

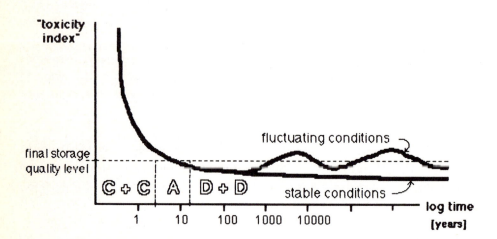

Fig. 1: Evolution in time of the "toxicity index" for a specific chemical compound
and strategies to be followed (C+C: "concentrate-and-contain", A: "attenuate",
D+D: "dilute-and-disperse"). The "final storage quality level" represents the maximum acceptable "toxicity index" in final storage.

An important stage in the assessment of final storage quality is the "risk analysis". Risk (R) is defined (Wynne 1987) as the product of the degree of harm or "consequences" (C) of a given event and its probability of occurrence (P):

$$R = P \times C$$

The degree of harm is related to hazard, which describes the damage a landfill could cause in a "worst-case scenario". Hazard assessment relies on ecotoxicological criteria, while risk analysis incorporates variable situational qualifications and so is based on additional scientific and technical criteria.

It should be remembered however, that the same risk factor may describe very different situations and potential consequences (e.g. severe consequences and low probability of occurrence as opposed to slight consequences and high probability of occurrence).

4. Strategies in depositing solid wastes

Strategies for achieving final storage quality depend on the type of emission of waste from the landfill. Toxic emissions require a strategy which is often called "concentrate-and-contain". Here "concentrate" refers to creating a high concentration of hazardous material (waste) in a restricted area (the landfill). With "contain", one endeavors to keep the waste isolated from the natural environment by means of barriers. It must be kept in mind that absolute permanent containment is not possible. The concentration of the inevitable toxic emissions can be reduced by a "dilute-and-disperse" or "dilute-and-attenuate" strategy. "Dilute" simply refers to the dilution of the various compounds in the natural environment, namely the groundwater, and "disperse" refers to the dispersion of the various dissolved compounds by means of groundwater flow. In the "dilute-and-attenuate" strategy, attenuation (or retention) is obtained by sorption, i.e. reaction between a solid phase (barriers and geosphere) and a fluid phase (leachate, groundwater). Most important are clay minerals which can adsorb ions by cation exchanges.

The various strategies and their use in practice are discussed in the paper by K. STIEF. The practical solutions are illustrated by extracts from government regulations from the U.S., and from recently published reports describing strategies envisaged in the U.K. and the F.R.G. STIEF proposes a "multi-barrier strategy" consisting of a combination of "dilute-and-attenuate" on one hand and "containment" on the other. The concept requires an appropriate site for the landfill, a bottom liner, a cap and waste quality equal or close to final storage quality as defined by EKA (1986).

A combination of these strategies could work as follows: in a first phase of "reactor-type" landfill the waste could obtain final storage quality (by e.g. biodegradation) with a contain strategy. In subsequent phases, the strategies would be attenuation and then dilution and dispersion.

5. Hydrogeological and geological aspects

Ideally, site-selection for a landfill is constrained alone by the geological structure and the hydrological conditions of the site. But in a densely populated country such as Switzerland political and social-psychological factors will become increasingly important as well. The hydrological conditions are intimately related to the geological structure by the permeability of the various rock types and the distribution and geometry (i.e. structure) of these rock-types. For containment and, to some extent, attenuation, the geological structure of the immediate vicinity is of utmost importance. Consideration of the geological structure of a greater region becomes critical for a dilute-and-disperse strategy, for attenuation and for biodegradation.

The structure of the site depends largely on the geological events which produced the association of rock-types. In many cases, these events produce very complex geometries of rock bodies and so render predictions and modelling difficult.

For Switzerland, complex geometries are the rule rather than the exception. This stems from the fact that many of the potential landfill sites lie within fluvial, glacial, lacustrine and fluvio/glacial deposits related to the Quaternary glaciations. A discussion of these aspects is given in the paper by P. HUGGENBERGER.
Risk-assessment of the pollution-potential involves extensive modelling and this, in turn, requires numerical data concerning permeabilites and a basic understanding of flow patterns. Geological and hydrogeological criteria are treated in the paper by P. HUGGENBERGER, whereas the paper by E. BüTOW & H.-P. LüHR concentrates on theoretical aspects of leachate-transport models.

Insight into the pollution potential of the metals in a landfill can be gained by studying the processes inside and around mine tailings (i.e. spoil heaps of washed or milled ore). P. HUGGENBERGER summarizes the situation in the Coeur d'Alène District, where mining activities started 85 years ago.

6. Ecotoxicological aspects

Ecotoxicology encompasses the impacts of selected chemical compounds on a gene, a cell or an organism and a population or a whole ecosystem, together with the evolution and fate of these chemical compounds in the environment. As discussed earlier, chemical and biological activity must be low in final storage landfills.

For the long term, slow continuous emissions must be envisaged. Consequently long-term exposure to hazardous chemicals at low concentration levels are more relevant to the final storage concept than short term emissions with acute toxicity. These aspects and their bearing on ecotoxicological tests aimed at assessing hazard and risk are dealt with in the paper by J. ZEYER & J. MEYER.

7. References

Baumann, L., 1976, Introduction to ore deposits. – Scottish Academic Press.

EKA (Eidg. Komm. für Abfallwirtschaft), 1986, Leitbild für die Schweizerische Abfall-wirtschaft. – Schriftenreihe Umweltschutz (BUS), Nr. 51.

Wynne, B., 1987, Risk management and hazardous waste: implementation and the dia-lectics of credibility. – Springer Verlag.

STRATEGY IN LANDFILLING SOLID WASTES
Different Solutions in Practice

Klaus Stief
Fachbereich Abfall- und Wasserwirtschaft
Umweltbundesamt, Berlin

Abstract

There are two main strategies for landfill design and operation:
1) the 'dilute-and- attenuate strategy'
2) the 'containmment strategy'.

The dilute-and-attenuate strategy depends for its effectiveness on controlled and reliable chemical, physical, and biological processes which
operate within the wastes themselves, and on the predictable attenuation
processes in the underlying rock strata. On a site- by- site basis, detailed and professional site assessment, followed by a high quality, engineered landfill design are necessary. Relevant properties of the strata
which are important to know include particle size distribution, clay content, cation exchange capacity, unsaturated zone thickness, and carbonate
content.

Main goals of the containment strategy are: minimization of leachate generation in the landfill and removal of leachate from the landfill. Controlled removal from the landfill can be achieved by installation of an
effective bottom liner system. Infiltration of water into the landfill
can be minimized by effective run-on control systems and by capping the
filled parts of the landfill. Neither limited lifetime of flexible membrane liners nor (low) permeability of clay liners are disputed. However,
attenuation properties of the site, and waste quality within the landfill
body appear to be of less importance than bottom liner and cap effectiveness.

The implementation of a 'multi-barrier strategy', a combination of the
dilute-and-attenuate strategy and the containment strategy, could serve
to obtain better landfills. The main barriers to be considered are : the
site, the bottom liner system, the waste, the cap, and the long-term monitoring. The main idea of the multi-barrier strategy is: at every landfill all above mentioned barrieres are required. If it can be proven that
the wastes disposed of have 'final storage quality' according to the attenuation capacity of the underlying strata, the landfill could be clas-

sified as 'Low Maintenance Landfill', equivalent to 'Final Storage Facility'. If not, depending on how much the final storage quality standards
are exceeded, the landfill would have to be classified as 'Medium or High
Maintenance Landfill'.

1. Introduction

To landfill wastes meant and means that one intends to get rid of the
wastes forever. Every landfill therefore is designed to perform as a 'final storage facility'.
A final storage facility must be distinguished from an 'intermediate storage facility'. In an intermediate storage facility waste is only stored
for a limited time, e.g. because appropriate treatment facilities are not
available.

In the past preferred disposal sites were often those pieces of land that
were deemed to be worthless, e.g. wetlands or abandoned gravel pits, clay
pits, quarries etc. This was ongoing even when in the seventieth dumps
developed into controlled tips, sanitary landfills or Deponien.

In designing sanitary landfills, i. e. landfills for municipal solid wastes,
two strategies have been adopted by the responsible authorities:
    1) the 'dilute-and-attenuate strategy'
    2) the 'containment strategy'.

The dilute-and-attenuate strategy is in particular accepted in the U.K.
as appropriate to protect human health and the environment. A lot of research was done to investigate the attenuation processes inside landfills and within underlying strata.
Landfills existing in many other countries landfills may also be classified as 'dilute-and-attenuate-landfills', in particular landfills for
demolition material, but also sanitary landfills. However, little is
known about the specific design and operation which is necessary for a
landfill to qualify as a dilute-and-attenuate landfill according to the
U.K. approach.

The containment strategy, which requires to establish a bottom liner and
a cap, is widely accepted for hazardous waste landfills all over the world.
For sanitary landfills a bottom liner is required in many countries ,
but rather a surface cover than a cap. Currently, when designing sanitary
landfills , caps are considered to be necessary to minimize leachate generation and landfill-gas migration.

In the following chapters both strategies will be explained in more detail. For this purpose from recent expert-publications will be quoted, concisely describing the basics of the various strategies in landfilling solid wastes.

Finally conclusions will be drawn from the weak points of these two strategies, to find out what a better, a 'final-storage strategy' should be look like.

2.Dilute-and-Attenuate STRATEGY

In the U.K. the dilute-and-attenuate strategy is widely accepted by the responsible authorities. This acceptance is mainly based on results of research done by HARWELL on behalf of the UK Department of Environment. Robinson (1987) explained and discussed this strategy:
  "The dilute and attenuate philosophy of landfill design depends for its effectiveness on chemical, physical, and biological processes which operate within the wastes themselves, and in the underlying rock strata. In the UK many landfills continue to operate safely as the results of these principles, and significant case histories of major ground or surface water pollution are rare, despite great increases in monitoring activities in the last 10 years."
  "..., what continues to be required, on a site by site basis, is detailed and professional site assessment, followed by a high quality, engineered landfill design."
  "If a dilute and attenuate landfill is to be constructed, it must be possible to demonstrate with a reasonable degree of confidence that the likelihood of a detectable adverse impact on local ground water resources is minimal and acceptable."
  "To allow this [to landfill according to the dilute- and-attenuate strategy], relevant properties of the strata must be measured and assessed. These include particle size distribution, clay content, cation exchange capacity, unsaturated zone thickness, capacity for buffering of pH-values, as carbon dioxide and fatty acids migration from the wastes tend to lower these to values which may be hostile to beneficial micro-organisms."
  "A significant further benefit of an unsaturated zone is in controlling the seepage of leachate into the saturated zone, such that the residual contaminant load is balanced by the continuing supply of oxygen and dilution which is commonly available from ground water underflow.
  It is important to remember that there are two aspects to consider in site design. One is the minimization of leachate quantity, by good

landfill practice, and the other the safe attenuation of that leachate which is generated and released."

"The wide range of different problems which these [wide range of geological formations in Britain] pose means that there is no single landfill design which can be universally applied. What is essential in each
instance is detailed and professional site assessment, followed by a
high quality, site specific engineered landfill design. This must be
good enough to demonstrate with a high degree of confidence that the
site will not have an adverse impact on the environment."

"Results from the site investigations and assessments continue to demonstrate the remarkable attenuation capacity of many natural materials
(e.g. ...) and the ability of physical, chemical, and biological processes (including dilution) both within and below landfills, to reduce
levels of contaminants. It is, however, more of a problem to quantify
and assess the efficiency and reliability of these processes, and allow
decisions to be made regarding the assessment of potential landfills in
varying geological conditions."

"Unsaturated zone calculations, together with estimates of flow rates,
must therefore be coupled with determinations of supply of oxygen, dilution and additional attennuation available within the saturated zone,
to allow a total estimate to be made of the overall degree of protection afforded to adjacent aquifers."

"..., leachate/rock interactions are complex and variable, so that there
can be no simple means of assessing the degree of attenuation which
will occur at a specific site.
In spite of this, the following properties are important, and can be
determined accurately by field or laboratory tests:

a) Particle Size Distribution. Medium to fine grained strata, where
flow is primarily intergranular, have high surface areas for reactions
to take place, and increase the potential for filtration. Major fissure
flow systems are clearly disadvantageous.

b) Clay Content. Clays have very large surface areas, reduce permeability
(increasing contact time for reactions to occur), and different clay
minerals have significantly different exchange capacities and hence
potential for attenuation.

c) Cation Exchange Capacity. This is a major attenuating mechanism, for
such ions as ammonium (ammonium acetate is in fact used for the laboratory determination of CEC).

d) Carbonate Content. Neutral or slightly alkaline conditions to reduce
the mobility of many pollutants, for example several metals, and provide

an optimum environment for both aerobic and anaerobic micro-organisms. It has been demonstrated (...) that landfill gases diffuse into the unsaturated zone beneath landfills in advance of a leachate front. This has two consequences: first that all subsequent processes will occur under essentially anaerobic conditions, and second that carbon dioxide in the gas will dissolve in pore waters to form carbonic acid (H_2CO_3). This is a weak acid, but in the absence of sufficient carbonate minerals and at typical partial pressures of carbon dioxide beneath landfills, pH-values as low as 5 can occur, and can act to inhibit beneficial micro-organisms (...).

Carbonates are not only means by which acids can be neutralised, but they are generally much more important than other (usually surface) processes.

e) Oxides of Iron. can also reduce pollutant mobility.

f) Unsaturated Zone Thickness has a direct relationship with the attenuating capacity of a given stratum, and influences the contact time between it and leachate during which reactions can take place.

g) Groundwater Underflow can provide a constant supply of well-oxygenated ground water, allowing aerobic degradation processes to take place subsequently, and giving potential for dilution of more-resistant contaminants.

With a knowledge of these characteristics at a given location, obtained as parts of a professional and thorough site investigation, it will be possible to estimate a likely minimum degree of attenuation, which can be used in conjunction with a detailed landfill design to arrive at a sound assessment of pollution risk. A vital factor to remember is that operation of an acceptable attenuation landfill design depends on matching the overall flux of contaminants to the available attenuation capacity. Good landfill practice, such as cellular tipping, progressive restoration, or even removal of leachate as it arises, all assist in reducing the quantity of leachate entering the underlying strata and requiring to be attenuated.

h) Biological Degradation of Contaminants. The unsaturated zone underlying a landfill generally becomes anaerobic within a matter of months in the regions where attenuation will occur. Production of an anaerobic microbial population will therefore depend on a supply of nutrients (principally organic material), pH-values, redox potential, temperature, and an inoculum, which form the wastes themselves. Unlike the earlier parameters, biodegradation rates cannot be measured in advance, and although laboratory studies (...) can help in predictions, the most useful data come from detailed monitoring of existing sites. Studies at the Stangate East Landfill, described below, demonstrate what is possible, and how the fund of useful knowledge can be increased."

3. The Containment Strategy

The containment strategy is most developed for hazardous waste landfills in the USA on the basis of RCRA and US EPA regulations. Therefore it makes sense to look into the US regulations to explain and understand what 'containment' is, and how it should be performed.
The US EPA hazardous waste landfill containment strategy is also of importance for all sanitary landfills, which require at least bottom liners, and perhaps caps, as the principles are the same.

In the proposed rules for 'Liners and Leak Detection for Hazardous Waste Land Disposal Units'(May 29, 1987) EPA explains its 'Liquids Management Strategy', to help to understand the relationship of the proposal. It is cited below, as it will help to understand the containment strategy in general.

"EPA believes that in order to protect human health and the environment, a fundamental goal of RCRA regulations must be to minimize, to the extent achievable, the migration into the environment of hazardous constituents placed in land disposal facilities. One element of EPA's strategy for achieving this goal is the liquids management strategy: The minimization of leachate generation in the unit and the removal of leachate from the unit.

First, the generation of leachate is minimized through the use of design controls and operational practices such as run-on control systems capable of preventing the flow of liquid onto the active portion of the unit, the placement of a cap on the unit at closure, and the restriction of liquid waste in landfills. Second, the removal of leachate is maximized by requiring leachate collection and removal systems above (for landfills and waste piles) and between the liners.
....
The Agency views leachate collection and removal systems as principal means of removing liquids from units. Although a liner is a barrier to prevent migration of liquids out of the unit, no liner can be expected to remain impervious forever. As a result of waste interaction, environmental effects and the effects of construction processes and operating practices, liners eventually may degrade, tear, or crack and may allow liquids to migrate out of the unit (47 FR 32284, July 26, 1982). Because generation of leachate cannot be eliminated completely

during the active life and postclosure care period of a land disposal facility, leachate removal is essential to prevent subsurface migration (47 FR 32313, July 26, 1982). For example, in a double liner system measures must be taken to remove liquid that migrates through the top liner, thereby preventing hazardous constituents from migrating through the bottom liner and into the environment.

......

Today's proposed rule, therefore, helps to implement the liquids management strategy. The land disposal system elements function in an integrated independent manner along with a construction quality assurance program to prevent leachate migration out of the unit by maximizing its collection and removal. The liners serves as a barrier to leachate migration and facilitate its collection and removal; the leachate collection and removal system (LCRS) above the top liner in landfills minimizes buildup of liquid pressure on the top liner; the LCRS system between the liners serves to reduce the buildup of head on the bottom liner; and the leak detection system notifies the owner or operator of leakage through the top liner, which may in turn require the owner or the operator to implement certain response actions to prevent migration of hazardous constituents from the unit."

In the draft RCRA Guidance Document 'Landfill Design, Liner Systems and Final Cover`, to be used with RCRA Regulations Sections 264.301 (a) and 264.310 (a) issued in July 1982, the cap (final cover) design has been described. The section on final covers is still valid. EPA wrote:

" 1. Regulation
The cap or final cover must be designed to minimize infiltration of precipitation into the landfill after closure. It must be no more permeable than the liner system. It must operate with minimum maintenance and promote drainage from its surface while minimizing erosion. It must also be designed so that settling and subsidence are accommodated to minimize the potential for disruption of continuity and function of the final cover.

2. Guidance
(a) The cap or final cover should be placed over each cell as it is completed
(b) The cap (final cover) should consist of the following as a minimum:
(1) A vegetated top cover.......
(2) A middle drainage layer,.....; and
(3) A low permeability bottom layer.......
....

3. Discussion

The guidance calls for placing the final cover at closure of each cell, or preferably, as filling of the cell progresses.

The Agency believes that a three layer final cover (cap) will adequately minimize infiltration of precipitation, which is the primary purpose of the final cover.

By minimizing infiltration, the generation of leachate will also be minimized, thereby reducing the longterm discharge of pollutants to the ground water to a bare minimum. To prevent the "bath-tub effect", i.e., to prevent the landfill from filling with leachate after closure when the leachate collection system ceases to function, the final cover must be no more permeable that the most impermeable component of the liner system (or of the underlying soils). In this way, no more precipitation is allowed to infiltrate the cell than can escape through the bottom liner. Prevention of the "bath-tub effect" is important to eliminate the possibility of surface overflow or migration through porous surface strata. The latter phenomenon is largely the cause of the problems at Love Canal in New York. Other functions of the final cover include prevention of direct contact with hazardous wastes by people and animals straying into the site.

......

Owners and operators using different final slopes should determine that an alternate slope will not be beset with erosion problems and that it will promote efficient drainage. The U.S. Department of Agriculture Universal Soil Loss Euation (USLE) is recommended as a tool for use in evaluating erosion potential. The USLE predicts average annual soil loss as the product of six quantifiable factors. The equation is:

$$A = R \times K \times L \times S \times C \times P$$

where

A = average annual soil loss, in tons/acre
R = rainfall and run-off erosivity
K = soil erodibility factor, tons/acre
L = slope-length factor
S = slope steepness factor
C = cover/management factor
P = practice factor

The data necessary as input to this equation is described in Evaluating Cover Systems for Solid and Hazardous Waste (SW-867) September 1980, U.S. EPA.

......

The function of the low permeability layer is to reject fluid transmission, thereby causing infiltrating precipitation to exit through the

drainage layer. It should consist of at least two components. While the regulations do not specify that the final cover <u>prevent</u> infiltration, the requirement that it be no more permeable than the bottom liner, as a practical matter, necessitates the use of a synthetic membrane. ....
......

Even with protection from damage, the synthetic cap will not last forever. At some point in the future, the synthetic membrane will degrade. At that time, the function of minimizing infiltration will fall to the second component, a 2-foot minimum clay soil-cap with a maximum hydraulic conductivity of 1×10^{-7} cm/s. Although some small amount of precipitation will seep through this secondary cap, the amount of leachate generated will be small and escape to ground water should be minimal. Unless damaged or affected by differential settlement, the secondary soil layer should remain intact and effective into the distant future. One source of damage, that can disrupt the continuity of the impermeable layer is frost heaving. For this reason, the impermeable layer should be wholly below the average depth of frost penetration in the area. ...... One of the more difficult problems associated with designing final covers for landfills is how to allow for settlement especially differential settlement. Settlement occurs a a result of natural compaction and consolidation and biological degradation of organics. Differential settlements unevenly distributed can cause disruption in continuity of the final cover.
......"

4. Requirements For Underlying Rock Strata

Drescher (1987) described and discussed investigations of clayey rock strata of potential landfill sites with respect to the adsorption capacity. The investigations were carried out in Lower Saxony (Niedersachsen), Federal Republic of Germany. Parts of his publication, which probably are most important for the 'final-storage-quality' discussion, will be quoted in the following (translation by the author):

" A complete permanent containment of leachate constituents is economically possible only for sorptive, degradable groups of constituents. For all the other constituents the "safety" is delayed release into the biosphere. This is combined with a reduced flux of pollutants released. Both are dependent on adsorption, and the thickness of the underlying rock strata.

Adsorption of Pollutants in Soil
Some pollutants will be fixed to a certain extent by clay minerals as well as by organic soil particles, usually available in soil. These pre-

cipitated and /or adsorbed constituents at some point in future may be
remobilized and will migrate due to change of conditions in the unsaturated and saturated zones and/or concentrations in soil water and
leachate.
Barriers against pollutants also do only delay migration of non-perma-
nent fixable pollutants. However, this time limited fixation results in
lower concentrations of non-degradable constituents. What order of magnitude the reduction of pollutant concentrations will be, cannot be
answered according to the current knowledge. Petrographical-mineralo-
gical and geochemical composition of rock and the general hydrogeological condition of the site are critical for the aptitude of the geological setting as barrier against pollutants.

Although clay rocks have a high buffering and precipitating capacity,
and as in particular smectite material has sufficient adsorption properties, some types of waste must be banned from landfills. All types of
waste mobilizing toxic substances should not be allowed to be landfilled. Usually such constituents will not be precipitated and only
partly adsorbed. Constituents precipitated and adsorbed will be mobilized.

If these constraints are considered, and wastes are only landfilled in
clay rock with mainly smectite clay minerals, with sufficient thickness,
and low permeability (bedrock permeability as well as clay permeability)
the underlying soil will have a proper attenuation capacity to guarantee
long term protection of the soil around the landfill site."

" Conclusions
The results of the investigations and their evaluation are summerized
in tables 1 and 2.
For requirements the geological setting should fullfill with respect
to the attenuation capacity see table 1.

NO.	CRITERIA/RANKING	POOR	GOOD	EXCELLENT
1	THICKNESS OF CLAY	5-50 m	50-100 m	> 100 m
2	STRUCTURE OF DISIN-TEGRATION PLANES	dominant	small	negligible
3	DISTURBANCIES	pervious	impervious	none

NO.	CRITERIA/RANKING	POOR	GOOD	EXCELLENT
4	PETROGRAPHIC COMPOSITION	claystone changing with permeable layers	claystone, cobble layers (developed in beds of Lower Cretacious in Lower Saxony)	clay and claystone
5	DIAGENETIC SOLIDIFICATION	very high	medium	small
6	MAIN MINERALS	Quartz, Kaolinite	Quartz, Kaolinite Muskovite-Illite	Quartz, Smectite
7	CLAY FRACTION < 2 um (%-W)	< 20	20 - 50	> 50
8	SWELLING CAPACITY	small	medium	high
9	PLASTICY (DIN 18 196)	low	medium	high
10	SPEC. SURFACE (m^2/g) (GASSORPTION)	> 15	> 20	> 40
11	PH OF ROCK	< 7	7	> 7 , < 9
12	CARBONATE (%-weight)	< 3	> 3	> 6
13	ORGANIC MATTER(%-weight)	< 1	> 1	> 3
14	ADSORPTION CAPACITY	small	medium	high
15	BUFFERING CAPACITY	small	medium	high
16	HEAVY METAL PRECIPITAITION	small	medium	high
17	SOIL PERMEABILITY (m/s)	$< 10^{-8}$	$< 10^{-9}$	$< 10^{-10}$
18	ROCK STRATA (IN-SITU) PERMEABILITY (m/s)	$< 10^{-6}$	$< 10^{-7}$	$< 10^{-8}$

Table 1: Requirements to Underlying Rock Strata With Respect to
Pollutant Retardation Capacity

All criteria listed in table 1 are of importance directly or indirectly
for the travel time of pollutants in the rock strata (retardation),
with different weight.
In table 2 criteria has been comprised to three subcriteria according
to their weight.

SUBCRITERIA / RANKING	o	+	+ +
Thickness of homogeneous plastic clay present	< 10 m	10-20 m	> 20 m
specific groundwater discharge $(m^3/m^2.s)$	$<5 \times 10^{-11}$	$<5 \times 10^{-11}$	$<5 \times 10^{-12}$
pollutant retardation	poor	medium	high

Table 2: Criteria for Rock Strata Evaluation

This evaluation does not say anything about the suitability for a specific group of leachate constituents. It is open whether additional measures, e.g. liners, will be necessary. This has to be decided "case by
case"."

5. Above Ground Landfills

In the Federal Republic of Germany as well as in Austria there is a trend
towards above-ground landfills, so that leachate can flow to a surface
water by gravity, and does not have to be pumped off forever. In addition
it is generally accepted that landfill bottoms must be at least 1 m above
the highest possible ground water level. (NRW, 1987; Lechner et.al.,1986)

6. Conclusions

6.1 Comparison of Strategies

The 'dilute-and-attenuate strategy', as widely accepted in the U.K., requires landfill operation according to the attenuation capacity of the
unsaturated zone beneath and the saturated zone downstream the site. Bottom liners are not only seen as unnecessary, but also as disadvantageous.
Covering or capping are accepted as means to control attenuation processes.

The larger the volume of waste disposed of in a landfill, the larger the flux of released pollutants will become, and the more likely it is to overload the attenuation capacity available. As a consequence bottom liners are rather to minimize releases of pollutants according to the attenuation capacity than to prevent any release into the unsaturated zone beneath the landfill bottom. However, with simple clay liners or simple FML-liners only a reduction of leachate release into ground water can be achieved. This had to be learnt from bad experiences in the past.

The 'containment strategy' is widely misunderstood as a means to dispose of wastes on land irrespective of the geological setting beneath the site. Actually, an appropriate landfill site selection includes an assessment of the geological setting, in particular hydrogeological conditions in the underlying rock strata.

It should be strongly emphasized that bottom liners and caps do not last forever. Therefore restrictions are to be made regarding the types of waste acceptable for landfilling.

The weak point of the containment strategy is that those responsible are much fonder of increasingly sophisticated multi-layer-liners, incl. leak detection devices, rather than improving the quality of wastes by pre-treatment.

At this point it becomes obvious that standards for 'final storage quality' (for 'Deponierbarkeit' in German) of wastes are necessary.

The crux of the matter is:
   - high standards will make current landfills unacceptable;
   - low standards will not at all improve minimization of the release
     of contaminants in the long term.

6.2 A 'Multi-Barrier Strategy'

A good idea might be to combine the 'dilute-and-attenuate strategy' and the 'containment strategy'. The combination could be addressed as a 'multi-barrier strategy' (Stief, 1986).

The following are considered to perform as barriers:

   - the site
   - the bottom liner system
   - the landfill body (the waste)
   - the surface liner system (the cap)

- the controlled afteruse of the closed landfill area
- the long term monitoring and control of the landfill behavior.

From the environmental protection point of view each of the barriers should, independent of the others, be permanently effective. However, the 'man made barriers' are likely to have a limited lifetime. It is therefore necessary to know in which period of time their effectiveness is most important, and what the probable lifetime will be.

It is very important to stress that every 'barrier' has to be selected and or constructed according to the best demonstrated available technology standards (BDAT).

Most severe environmental impacts are caused by the unpredictable mixture of organic and inorganic components in leachate and in landfill gas. Consequently a main objective of the 'multi-barrier-strategy' is to ban landfilling of organic waste whenever possible. The second objective is to dispose of the inorganic waste in a way that prediction of leachate composition will be possible. This can be achieved by operating so called mono-landfills or quasi-mono-landfills.

To obtain maximum stability of the landfill body, the waste has to be solidified by pretreatment and/or to be compacted as much as possible during operation. Liquid wastes must be banned. The watersoluble parts of the landfilled waste should be low, less than 10 %. More soluble wastes should be disposed of in underground salt mines or salt caverns.

The landfill site itself will be the main barrier against movement of pollutants in the long term. Attenuation processes in the unsaturated and saturated zone under the landfill bottom and downstream from the site are necessary to achieve acceptable ground water quality. Acceptable ground water quality should be achieved within a short distance from the landfill site as the result of a simulation of a worst case release of pollutants . This simulation of leachate discharge into ground water is to make sure that the attenuation capacity of the site will be considered. For simulation purposes the bottom liner and the cap should be taken as ineffective, however, immobilisation of pollutants within the waste body has to be assumed as effective.

Attenuation has to be simulated at the time of designing the landfill, before one has complete information about the contents of the landfill, as it is important to assess the environmental impact when designing a landfill. The mathematical model therefore involves a number of assumptions which may later need modification, e.g. about the 'true' leachate

composition. The model therefore has to be improved during the phase of operation of the landfill, and during the post-operation phase.
At the time of planning it is usually unknown what the main pollutants will be, and what quantities will be in the landfill body at the time the landfill is closed. Therefore it is rather impossible to make a reliable risk assessment.
However, at the very end, if one is sure about the leachate composition, it will be possible to leave for future generations a remedial action plan for effective hydraulic measures, in case future generations will not accept the contamination we left.

It is necessary to submit future generations as much information as possible about the landfills we are constructing, filling, and restoring, as we know they will have some impacts on the environment, and as we are not sure if our environmental risk assessment will be acceptable for our descendants. We have to help them to cope with the abandoned landfill sites we leave.
The documentation possible, useful, and necessary will be a regular 'Declaration-of-Landfill-Behavior' which encompasses at a minimum:

- amount and types of wastes, location in the landfill
- leachate, amount and composition
- gas, amount and composition (not in final storage facilities)
- landfill density, waste water permeability
- settlement and subsidence.

By this Declaration-of-Landfill-Behavior, given at least once a year and based on reliable measurements, it will be possible to have data available to predict the long term behavior of a landfill.

From the data declared, one will have quite reliable information about types and amounts of the main pollutants, including what mass has been placed into the landfill, and what has left the landfill via leachate or landfill gas. The pollutants remaining in the landfill body, described with regard to their mobility or immobility, are paramount for the long term risk potential of the landfill site.

7. Closing Remarks

All of our recently and currently operated landfills are designed to perform as 'final storage facilities', as the removal of the waste from the site is not planned at all. Waste disposed of is to be left there forever. However, there is some doubt, if our current landfill sites are designed and operated to cause in the long term only acceptable environmental impacts, even if the bottom liner system and the surface cover (the cap) will become ineffective, even if in the long term due to physical, chemical, and biological reactions within the landfill body leachate with unacceptable 'quality' will be generated and be discharged into the ground water.

It might be helpful to find a new word to make clear what the quality of waste should be to be accepted at a 'perfect' landfill site. Perhaps that waste quality criteria could be described as 'final storage quality'. And 'perfect' landfills could be called 'final storage quality facilities'. The problem will be that the final storage quality criteria will become tougher in the years to come, and the final storage facilities will become more and more acceptable, according to the knowledge on how to treat the waste before landfilling, and according to the availability of the treatment facilities. That means 'final storage quality' as well as 'final storage facilities' are shifting definitions.
Nevertheless, it is reasonable to distinguish landfills in which physical, chemical , and biological reactions will be going on for an unknown period of time from those landfills in which these processes are negligible. According to the Swiss approach, the former could be called 'reactor landfills', and the latter 'final storage facilities'.(EKA,1986)
Final storage facilities should need very little maintenance to control environmental impacts. Reactor landfills of course will need much maintenance as long as the physical, chemical, and biological processes have come to an end. Whether further maintenance will be necessary will depend on the 'final product', the landfilled waste within the reactor has been converted to. Some experts are very optimistic that municipal solid waste will be converted to a harmless mass, even if hazardous wastes have been co-disposed of , others are rather pessimistic.
However, there is little doubt, that the hazardous waste disposal sites we have generated within the last decades all over the world will need high maintenance to keep the environmental impacts acceptable. Some of them are already on the various national prioritiy lists for remedial actions. If we are following the final storage quality approach, we will need some more money for treatment and disposal of our waste today, but much less money for remedial action in some years.

8. References

EKA (Eidg. Kommision für Abfallwirtschaft), 1986: Leitbild für die Schweizerische Abfallwirtschaft, Schriftenreihe Umweltschutz Nr. 51,herausgegeben vom Bundesamt für Umweltschutz Bern, Juni 1986

Drescher, J., 1987: Schadstoffadsorptionspotential des Deponieuntergrundes - Ergebnisse des 9000 ha Programmes des Landes Niedersachsen, in: Fehlau/Stief (Hrsg.): Abfallwirtschaft in Forschung und Praxis, Band 19 'Fortschritte der Deponietechnik 1987',Erich Schmidt Verlag Berlin, 1987, ISBN 3 503 02683 5

Lechner, P., Pawlick, R., et. al., 1987, Richtlinienentwurf für Mülldeponien, TU Wien Institut für Wassergüte und Landschaftswasserbau, Abtlg. Abfallwirtschaft, O. Univ. Prof. Dr. W. Kemmerling, Wien, Sept. 1987

NRW, 1987: Entwurf einer Richtlinie über die Untersuchung und Beurteilung von Abfällen, Teil 2: Empfehlungen zur Beurteilung der Ergebnisse von Abfalluntersuchungen - Beseitigung von Abfällen durch Ablagern unter besonderer Berücksichtigung wasserwirtschaftlicher Gegebenheiten, Landesamt für Wasser und Abfall des Landes Nordrhein-Westfalen, Düsseldorf,1987

Robinson, H., 1987: Studies On the Unsaturated Zone, International Symposium "Process, Tecnology, and Environmental Impact of Sanitary Landfill", Sardinia, Cagliari, October 1987 (Publication by ISWA in prep.)

Stief, K.: Das Multibarrierenkonzept als Grundlage von Planung, Bau, Betrieb und Nachsorge von Deponien. Müll und Abfall, 18. Jg. (1986), Heft 1, Seite 15 - 20, Erich Schmidt Verlag Berlin

Stief, K., 1987: The Multi-Barrier Concept - A German Approach, International Symposium "Process, Technology, and Environmental Impact of Sanitary Landfill", Sardinia, Cagliari, October 1987 (Publication by ISWA in prep.)

USEPA, 1987: Liners and Leak Detection for Hazardous Land Disposal Units; Notice of Proposed Rulemaking, Friday May 29, 1987, Federal Register (USA), Environmental Protection Agency 40 CFR Parts 260, 264, 265, 270, and 271

HYDROGEOLOGICAL CRITERIA FOR
FINAL STORAGE QUALITY

Peter Huggenberger
Institut de Géologie
Genève

Abstract

Disposal of hazardous non-radioactive wastes is a major problem because of our incomplete understanding of subsurface processes. We are forced to simplify with models and thus to overlook some of the natural defects in real storage sites. The aim of this paper is to provide a conceptual framework and to discuss some examples of previously overlooked processes leading to leakage in storage sites. In the late seventies the hydrogeologic methodology and the governmental regulatory guidelines were insufficient to prevent contamination of groundwater and atmospheric pollution. Frequently, engineers dealt with the bulk properties of rocks and grossly underestimated the influence of the 3-D flow system due to the variable interconnections between permeable zones and the heterogeneous mineralogical and sedimentological character of the media under consideration.

Analysis of different geological environments shows that under natural conditions we mostly have to deal with associations of different rock-types belonging to the same genetic system. Predicting the lateral extent of a particular rock-type -e.g. of very low permeability- is very difficult and essentially depends on our understanding of the depositional environment or of the tectonic and igneous processes.

In addition, new concepts for final repositories should provide for a site selection process that includes specific safety analysis prooving that a particular site will satisfy stipulated safety requirements. The following conditions are required to make a proposed repository site acceptable for waste disposal: (1) a good knowledge of the regional hydrogeological regime, (2) a detailed understanding of the geological, geochemical, and hydrogeological conditions of the site, (3) a capacity for accurate predictions of the migration of contaminated waters through the hydrogeological system (4) an understanding of microbiological processes and of biological accumulations.

1. Introduction

Current practice of handling a great number of types of hazardous non-radioactive wastes has created problems at each stage of the waste-chain, from the treatment of the wastes to their disposal or final storage.

The impact of land disposal on the environment is often difficult to determine because of the complexity of the subsurface. The present situation is unsatisfactory as is demonstrated by the closing of most disposal sites to hazardous wastes. We do not understand the governing processes well enough to be able to describe and monitor them adequately. Groundwater contamination associated with land-disposal has most often been detected only by chance.

This unsatisfactory situation leads to a broad controversy about production, recycling, treatment and disposal of domestic and industrial wastes. At the present time, it seems that the concept of separating different wastes prevails. This concept considers two basic disposal types: (1) the reactor and (2) the final repository. The latter is the subject of this paper. It contributes to the geological and hydrogeological aspects of the discussion about the storage of hazardous waste.

First of all, (section 2) current concepts are examined in order to point out problems which are often overlooked. Concepts which have been recently proposed by the Swiss Environmental Protection Agency (BUS) [Oggier, 1987] are used as a guideline in the establishment of hydrogeological criteria for final storage quality.

Section 3 examines a case study of a general geological/hydrological setting in the NE part of the molasse basin and the southern part of the Jura mountains. It illustrates the effect of flow paths within different geological formations.

Section 4 compares the physical and chemical processes of natural ore deposits and shows how these differ from processes affecting mineralized wastes.

Section 5 examines the properties of backfill materials.

Section 6 discusses some problems that may arise from monitoring concepts for the storage of hazardous waste.

Section 7 stipulates the requirements of the final storage sites.

2. Concepts of waste disposal in geological formations

2.1 Previous practices in Switzerland

Marls and clays have generally been selected as the bed-rock or host-rock for the disposal of non-radioactive waste (e.g. Jäckli, 1974) mainly because laboratory analyses show that the bulk hydraulic conductivities range from 10^{-7} to 10^{-11} cm/sec (Freeze and Cherry, 1979) which are low permeability values. In the molasse formations, two units -marls and clays of the lower and the upper fresh-water molasse- are generally used as bed-rock materials. From the late seventies onwards hazardous organic and inorganic waste have been deposited in the "non-permeable marls" of ancient clay pits like the one in Kölliken (Canton Aargau). But the actual permeability of clay and marl formations in field situations has often turned out to be several orders of magnitude higher than that predicted on the basis of laboratory sample tests. The existence of flow paths was illustrated by the migration of hazardous chemical compounds out of the disposal site within a relatively short time.

Risk estimates for release and transport of contaminants out of a specific waste disposal site was frequently based on the hydrogeological data of only a few borehole measurements. Still now a days most of the heterogenities due to deformations or sedimentological structures cannot be recognized because the measurement techniques have been developed mainly for the determination of water quantities. This has led to the present groundwater problems near sanitary landfills, waste disposal sites or uncontrolled waste dumps. Consequently, research is now more and more directed towards the problems of contaminant transport in different geological environments. The literature shows a large number of specific studies on hydrogeological-, geochemical, chemical- and geotechnical processes in fine-grained sediments. They demonstrate the significance of processes such as the variation of attenuation capacity for different chemicals or effects of chemical reactions and the flow of groundwater in fractured rocks together with dispersion of contaminants (e.g. Fahrquar and Rovers, 1975; Griffin et al., 1976,1977;

Goodall and Quigley, 1977; Cartwright et al., 1981; Overbeck, 1981; Gillham and Cherry, 1982; Farlane et al., 1983; Cherry et al., 1984; Cederberg et al., 1984; Lun, 1985; Al-Azawi et al., 1986; Höhn and Santschi, 1987; Siegentaler, 1987).

The creation of regulations for the final storage of radioactive wastes has resulted in an increasing number of publications about groundwater circulation in fractured sedimentary and cristalline rocks. These publications describe conceptual and mathematical models of groundwater flow in fractured rocks (e.g. Wilson & Witherspoon, 1970; Narasimhan, 1982; Long et al., 1985; Tsang, 1986, Kiraly et al. 1985), or deformations -(fractures, shear-zones, joints etc.)- in specific rock-materials or regions (e.g. Huber and Huber, 1984; Gautschi et al., 1986; Guilette, 1986).

The Nagra (national cooperative company, acting as a central coordination and planning institution for research on radwaste disposal in Switzerland) has shown how difficult it is to bring together the results of different research fields in order to fulfill the stipulated safety requirements of a particular site (Nagra, (NGB 85-01 - 85-08, 1985; HSK, 1986;1987).

2.2 New proposed concepts

According to the proposals of the Swiss Environmental Protection Agency [BUS] (Oggier, 1987), material fluxes from a final storage repository to the environment should not exceed specific limits in a range of tens to hundreds of years, so that no post-treatment of the migrating leachate would be necessary. (Final storage strategy reflects the intention of the authorities in charge to find a final solution for a civilisation problem rather than to leave it to future generations.) The BUS further proposes that wastes falling into the category of hazardous non-radioactive wastes should be segregated and deposited in the most suitable and unsoluble form -the so-called mono-disposal-sites- which would contain one particular chemically-defined group of wastes (Oggier 1987).

This basic premise on segregation is used as a guideline in our discussion on hydrogeological criteria for final storage quality. It leads directly to some major points which will be the topics of many future controversies: (1) the definition of acceptable limits of pollution, (2) the criteria for grouping particular wastes into categories, (3) the site selection strategy for the final repository.

If this proposition is accepted as a basic premise, the following requirements are needed to make a proposed repository site acceptable for waste disposal: (1) a good knowledge of the regional hydrogeological system and subsurface structure, (2) a detailed understanding of the geological, geochemical and hydrogeological conditions of the site, (3) a capacity for accurate predictions on the migration of contaminated water through the hydrogeological system to the biosphere, taking different scenarios of leachate migration into consideration, (4) an understanding of microbiological processes and possible biological accumulation (5) a specific performance assessment of the site for the disposal of wastes.

Formations considered for final storage of non-radioactive wastes:

In order to minimize the risk of migrating contaminants reaching the exploited groundwater systems in high concentrations, it is generally accepted that final storage repositories for hazardous wastes should be placed in geological formations of very low permeability. In Switzerland, three main groups of host-rock or bed-rock materials can be considered among existing geological formations:

1. practically non-permeable consolidated and non-consolidated sediments (clays and marls)
2. crystalline rocks (granites, gneisses)
3. anhydrite, gypsum

3. Case study of a geological/hydrogeological system in the molasse basin the southern part of the Jura Mountains

3.1 Generalities

The northeastern part of the molasse basin and the southern part of the Jura mountains provide a good geological setting in which to illustrate how current concepts failed in terms of waste disposal sites. To avoid repetition, most topics in this section are discussed using the example of only one particular formation. The final conclusions, however, should be integrated into the repository site selection process.

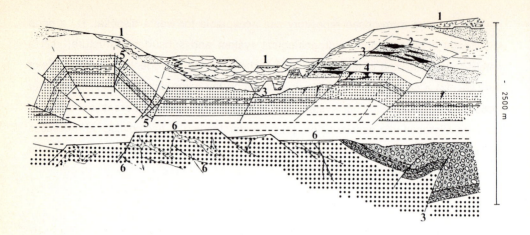

Figure 1: Schematic geological profile of the NE-part of Switzerland
(12-15 x vertical exaggeration)

 Quarternary deposits (mostly unconsolidated gravels, glacial tills, moraines and lake sediments)

Molasse

Upper Freshwater Molasse (coarse detritic-, fluvial- and laccustrine sediments)
Upper Marine Molasse (shallow marine- , tidal flats and delta deposits
Lower Freshwater Molasse (red silty-, sandy shales, fluvial and conglomeratic sediments)
Lower Marine Molasse (shallow marine to brackish or continental detritus)

Mesozoic and Permo-carboniferous rocks

Dogger, Malm (limestones, marls)

Liassic and Triassic formations (limestones, shales, marls, evaporites)

Permo-Carboniferous Units (mostly detritic sediments)

Crystalline basement rocks

 (metamorphic rocks, granites, gneisses)

1 Near-surface flow systems

3 Faults and fracture-zones allowing water exchange between different formations

5 Tectonically induced fractures in competent units of Jurassic formations

2 Highly permeable sands in fractured marls and clays

4 Karst-systems in Jurassic limestones

6 Fractured flow systems in crystalline rocks (fault zones, kakirites, shear-zones, dyke-walls)

The geological framework of the northern margin of the molasse basin -a hilly country with some lakes and a few larger plains- is illustrated by the schematic geological profile of Fig. (1). The valley bottoms lie at 350-600m amsl, and the hills in between lie a few hundred meters higher. The subsurface geological sequence can be roughly divided into four major units:

A   Quarternary deposits: mostly valley fills; consisting of unconsolidated sediments (e.g. gravels, lake deposits, tills, moraines, eskers)

B   Sediments of the molasse basin (marine, lacustrine and fluvial sands   and shales becoming more conglomeratic close to the Alps)

C   Mesozoic and permo-carboniferous sedimentary rocks underlying the   molasse sediments (limestone units, shales, evaporites, marls, sandstones,   conglomerates and coal beds)

D   Crystalline basement, comprising a heterogeneous mixture of gneisses, metamorphic rocks and granitic intrusives.

In all these units practically non- permeable layers or zones alternate with aquifers of varying sizes, chemical compositions, and ages (Balderer 1985). Because of the complex sedimentary, erosional and tectonic history, water exchange can take place between different aquifers. Before applying a risk analysis, it is important to understand the flow path within the different kinds of rocks. In the case of accidental release from a final storage or waste disposal site, chemical compounds could very rapidly be transported and dispersed over long distances.

The key to a better understanding of subsurface processes lies in the analysis of different geological environments.

3.2 Molasse and Quarternary deposits

Rocks of low permeability in the molasse basin:

In the northeastern part of Switzerland several outcrops of the lower fresh-water (Bachs valley (Fig. 2), Eglisau) and the upper fresh-water molasse (Lufingen (Fig 3)) give some indication of the palaeo-environment. Here the molasse deposits consist of alternating

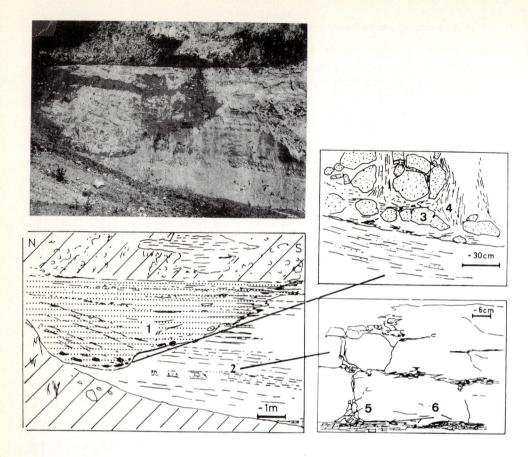

Fig. 2　Lower freshwater molasse (Bachs valley, NE-Switzerland) (1) Channel cut into a sequence of alternating marls, clays and silty sands (2) and subsequent fill and abandonement. Sedimentary structures within the sands only barely recognizable. Blocks of cemented sand (3) in a sandy matrix without cohesion at the channel base (4). Fractures and brecciation parallel (6) and perpendicular to bedding (5).

Fig. 3   Outcrop of meandering stream deposits (upper freshwater molasse,
           Lufingen, NE-Switzerland)
           (1) Alternation of red and grey shales with varying contents of fine sand.
           (2) Horizontal sandstone beds alternating with shales and marls (1).
           (3) Sandy channel fill (notice lateral accretion structure on the left side of
           the channel.

layers of shales, marls and more silty or sandy-layers with varying clay content. Typical
alternations of reddish to greyish colors are not necessarily restricted to certain lithologies.
They sometimes reflect different hydrogeological conditions (periods with low vs. high
water levels) recorded by different oxidation stages of the iron. Fracture systems (Fig. 2)
caused by diagenetic processes or by alpine tectonic are common. One example shows a
fracture set that is parallel to the bedding associated with another fracture set that is

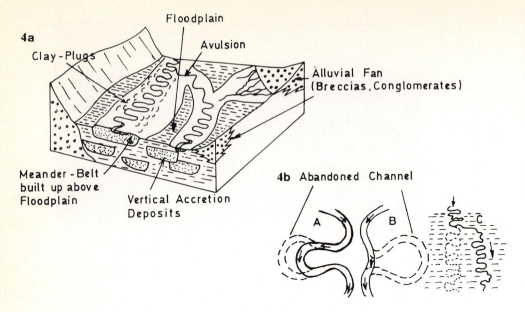

Fig. 4a   Block diagram of meandering stream deposits showing linear sand bodies
oriented down the depositional slope. These porous facies are laid down by
the meander belts, and are enclosed within floodplain shales. (redrawn
after Cant, 1982).

4b   Modes of channel shifting in meandering systems. (A) Chute cut-off, (B)
neck cut-off, (C) development of a new meander belt following avulsion. Old
course is dashed line in each case (redrawn after Allen, 1965c)

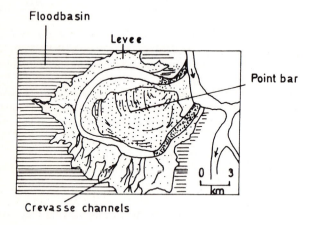

Fig. 5   Levee, crevasse and crevasse splay topography preserved around an
ox-bow lake caused by a neck cut-off (redrawn after Fisk 1947)

perpendicular to the bedding. Other fracture systems do not show any preferred orientation.

Channels have also been cut into these layers and have been filled with trough-shaped sand bodies, showing point-bar lateral accretion structures (Lufingen, Fig. 3). The troughs of the Bachs valley outcrop have different internal structures. Sand layers filling the troughs are still partly cemented by calcite whereas part of them have been strongly altered by diagenetic processes or by subsequent deformation and solution due to circulating groundwaters. The altered sands (medium-grained) are practically without cohesion and might have a considerable permeability. The ratio of the lateral extent of sand bodies over the vertical one is much smaller in the outcrop of the lower fresh-water molasse.

These sequences represent meandering river systems (fig. 4a,b). Such systems are more stable than braided river systems (see chapter Quarternary valley fills), because they contain thicker, more heavily vegetated, cohesive floodplain deposits. Abandoned mud-filled channels are a common feature of most meandering river floodplains (Fig. 5). The presence of clay plugs on a floodplain restrict any tendency towards lateral migration of the meandering belt. The channel migrates freely within the dominant sandy meanders (Fig. 4a). The varying positions of the trough-shaped sand bodies at different levels are typical of an aggrading river-system. As a river aggrades, it builds up above the level of its own floodplain. In some cases, the break of a levee during a flood results in the abandonment of the existing meander belt and in the establishment of a new one. This process is known as "avulsion" (Fig. 4a) and leads to the formation of a series of sand bodies oriented along the depositional slope in a matrix of floodplain deposits of low permeability.

Consequences on the hydrogeological system:

Groundwater circulation within these rocks is mainly characterised by two systems:

1. Groundwater circulation through the fractures
2. Groundwater flow within the sands of the channel-fills.

The frequency distribution and the interconnection of fractures is not well understood. The extent of valid extrapolation can hardly be determined. Major faults or fractured zones could be even more efficient as migration paths than the small scale fractures themselves. Indeed, in the field it is very difficult to recognize the extent of such features by means of traditional investigation methods (e.g. drilling).

The frequency distribution and the interconnection of preserved channel fragments mainly depends upon the knowledge of how the ancient river-system worked. A reliable facies-model could help predicting sand body geometry, lateral facies relationship and potential hydrogeological characteristics from a limited data set. Allen (1978) suggests that sandstone bodies are likely not to be connected in meandering stream deposits with less than 50% sandstone in a vertical sequence.

The next step in our discussion will be to describe the relations of the molasse sediments with the quarternary valley fills and the underlying Mesozoic rocks. In the valleys, molasse rocks are covered with Quaternary deposits. The interconnections of these two units are important for possible hydrogeological processes. The heterogenities described above form the pathways through which groundwater flows from the molasse formations and results in surface springs. These springs yield a $Ca-Mg-HCO_3$-type of water with a mineral content of about 5 10 meq./l (Schmassman et al., 1984). The residence time of the water flowing out of these springs is usually short, in the range of only a few years. As indicated by stable isotope analysis, the recharge areas are located close to the springs. This groundwater is quite different from that found in the deeper formations of the molasse, below the morphological base line of the valleys (e.g. Balderer 1985 or Table 1).

Quarternary valley fills:

The recent glacial valleys are filled with gravels, sands, moraines, glacial tills and lake sediments. The gravel deposits make up most of the groundwater reservoirs in Switzerland. The importance of water resources for agricultural use and public consumption is bound to grow in the future (Bachmann 1987).

The lack of information about the transport of particles and chemical compounds in such subsurface systems led to active research on the influence of sedimentary

Fig. 6a   Sedimentary-structures within fluvio-glacial gravel deposits of the Northern
Swiss Midlands. (Pit Hüntwangen, Profile E-W).
A.) Well sorted gravel beds without fine-fraction (open framework zones)
B.) Sand layers

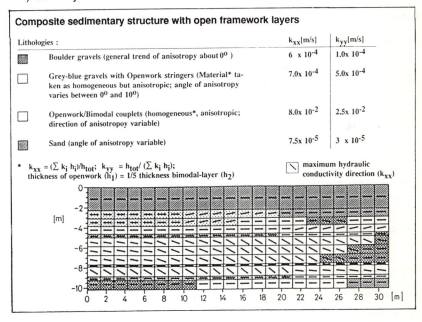

Fig. 6b   Idealized representation of the outcropping sedimentary structures of the
E-W profile (fig. 6a) as a finite element grid.

structures on groundwater travel times. In the beginning, investigations concentrated on gravel deposits in the eastern part of Switzerland (Huggenberger et al. 1987). Figure 6a gives an idea of how complex such systems can be, even without taking into consideration of their interfingering with periglacial sediments, e.g. moraines, eskers or lake sediments.

The typical depositional environment of these outwash gravels is the so- called braided river system. This environment is still poorly understood and appears to be the result of interactions among several depositional processes. Various facies-models have been developed (e.g. Beschta et al. 1978/1980, Walker 1979, Ashmore 1982, Steel & Thomson 1983, Smith 1983, Miall 1985). Most of them are derived from specific locations and cannot be easily applied to other regions. An actualistic approach -comparisons with modern systems- is limited by the uncertainties in the estimation of the preservation potential of certain structures. Another recently developed model (Siegentaler & Huggenberger, in prep.) tries to explain the gravel sequences in terms of only a few of the main facies elements (scour pools, horizontal gravelbeds & diffuse gravel sheet, flood deposits).

Investigations of depositional environments in periglacial systems have been made by Pugin (in prep.), in the western part of Switzerland. This particular environment is typical of areas located behind the frontal moraines of the last glaciations. His results clearly illustrate the complexity of their subsurface composition (Fig. 7). In terms of material properties, moraines and glacial tills are heterogeneous systems which cannot be fully described by scattered data such as those derived from borehole measurements.

Effects of sedimentary structures:

At the outcrop scale, gravel deposits display a complex pattern of sedimentary structures including channel fills, bars, silty beds etc.. The structures include deposits with variable hydraulic conductivities. The directions of the hydraulic conductivities are influenced by the horizontal layerings and cross-beddings. A minor part of the deposits comprises stringer gravels without any fine fraction (layers "A" on Fig. 6a). The hydraulic conductivity of such zones is several orders of magnitude higher than in the surrounding sediments. The degree of interconnection and the lateral extent of these layers determine the easiest flow path. Deterministic numerical models help to visualize the influence of such sedimentary structures on groundwater circulation on different scales. Fig. 6b shows an

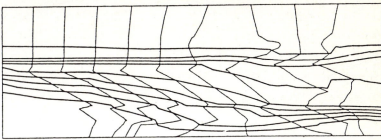

| Potential
— Streamfunction

Bulk hydraulic conductivity
Specific discharge

Gradient : 1 %oo
Porosity : 20 %o
$K_b = 2.33 \cdot 10^{-2}$ [m/s]
$q = 0.233 \cdot 10^{-4}$ [m/s]

Fig. 6c   Streamfunction and potential contour lines.
Boundary conditions:
A. Potential calculation : potential left side: $\varnothing = \varnothing_0$, on right side: $\varnothing = \varnothing_1$,
B. Streamfunction calculation:
upper boundary $¥_1 = 0.23$ (L^2/T)
lower boundary $¥_0 = 0$ (L^2/T)
Dense streamline pattern means faster groundwater velocities assuming uniform porosities.

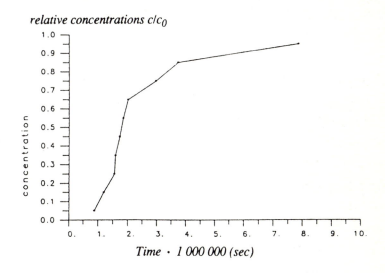

Fig. 6d   Breakthrough curve of an input step function.

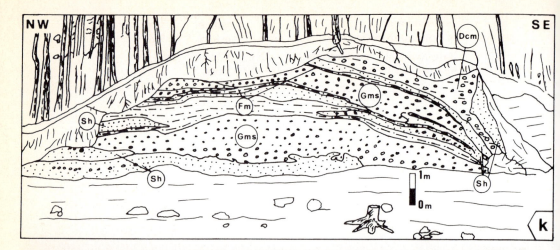

Fig. 7   Subaqueous outwash deposits. View from the paleoglacier on a gravely outwash deposit. Massive mud has been injected as dykes into the gravel, Locality: Pelevuit, Bulle area.
Sh: sand, variable grain size (horizontal lamminated)
Fm: silt, mud
Gms: massive matrix supported gravel (debris flow deposits)
Dcm: diamicts

(Photograph and sketch with permission of A. Pugin)

idealized representation of the outcrop situation (Fig. 6a) reduced to finite elements characterized by appropriate material properties (hydraulic conductivity, angle of anisotropy). The hydraulic conductivities have been estimated on the basis of the grain-size distributions of the different sediments. For the calculations, it is assumed that the potential gradient as well as the specific discharge throughout the block section profiles remain constant. The top and bottom of the section are no-flow boundaries. Fig. 6c illustrates potential and streamfunction contour lines for this simulation. Potential and streamfunction contour lines are not perpendicular due to the anisotropy of the hydraulic conductivities. A dense line pattern means higher groundwater velocities under conditions of constant porosity. Sand bodies -depending on their composition- can delay the groundwater flow. Travel times for an input step-function (an abstraction of a tracer experiment) have been determined from a streamfunction density pattern. Only advection (no dispersion or molecular diffusion) has been considered in the calculation. The breakthrough curve (Fig. 6d) shows a rapid increase in the beginning and a long tail caused by the slower flow paths. Differences in groundwater velocity, caused by the sediment structures and textures, are of importance for problems dealing with the transport and mixing of pollutants in the groundwater. Horizontal to subhorizontal highly-permeable zones -depending on their lateral extent and their linkage with associated structures- lead to significant differences in travel times because they act as high velocity conducts on a one hundred meter scale. In such cases, the amount of water flowing through less than a few percent of the aquifer cross-sectional area is likely to be equal to the flow through the rest of the same area (Huggenberger et al., 1988).

3.3 Mesozoic and Permo-Carboniferous rocks:

The major Mesozoic units underlying the molasse sediments in NE Switzerland comprise, from top to bottom: Malm limestones, Dogger and Liassic formations including limestones, marls and shales, and Triassic sediments deposited mainly in an evaporitic-carbonate-detritical environment. The Triassic sediments overlie continental deposits of the Carboniferous - Permian age. These continental sediments were deposited within graben-like structures faulted into the crystalline basement, and include conglomerates, sandstones and coal etc. (Nagra, NTB 84-25, 1984; Laubscher, 1986).

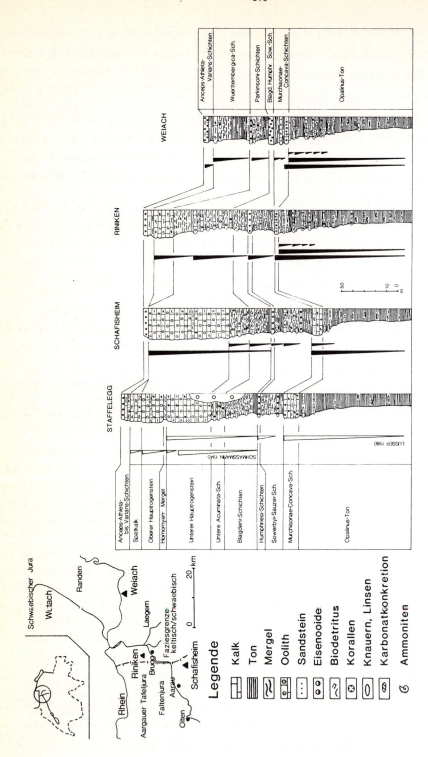

Fig. 8   Dogger profiles from Bläsi, 1987. Opalinus-clay at the base of the sections. Schematically indicated are sand or carbonate lenses and layers . (copy with permission of the author).

An interaction between deep groundwaters and groundwater in molasse strata was determined in the Nagra's deep borehole in Weiach and in the southern part of Germany, north of the Swiss border and of Lake Constance (region of Singen-Beuren Welschingen) [Table 1]. Hydrochemical and isotopic data as well as geological observations made in oil exploration wells indicate that the groundwater from Malm aquifers is hydraulically connected with the main aquifers of the molasse deposits, especially with the upper marine or the lower fresh-water molasse (as observed in the Weiach borehole) [Balderer 1985]. Other systems have been discribed which are characterized by local recharge-discharge or by deep water circulation (see Table 1) . The interactions among the different systems mainly follow the major tectonical lines.

For the Mesozoic sedimentary units, the principal lithology which is now considerd as the primary host-rock candidate for the final storage of non-radioactive waste is the Opalinus-clay. Clay stone is thought to have favorable characteristics i.e. low permeability, plastic behaviour and excellent sorptive properties based on handspecimen studies. In general, the very low permeability of clay lithologies is demonstrated by their ability to act as cap rocks for oil and gas accumulations and to maintain the abnormally high fluid pressures encountered in many oil fields, as well as by their role as aquiclude or aquitard in groundwater flow systems (Milnes ,1985). The current state of geological knowledge on the characteristics of the Opalinus Clay Formation was compiled by Bläsi, 1987. It is based on the Nagra's deep boreholes in Schafisheim, Riniken, Weiach and on the geological profiles of the Staffelegg and Wutach (Germany) regions (Fig. 8). The very lowest part of the section consists mainly of silty clays with a small amount of quartz (Weiach: carbonate content in the lower part: 5-9%; quartz content: 10%) [Bläsi,1987]. The depositional processes which have imparted small-scale and large-scale heterogeneities are not well understood. They are the result of changes in the sediment supply, the climate or the water currents.

The plastic behaviour of shales is well known from their reaction to folding and thrusting in orogenic zones and is reflected in the sparsity of joints and fractures within many shale formations (Milnes, 1985). This "plasticity" may create some geotechnical problems during the digging of cavities or open land disposal sites.

A relatively strong alteration of marly limestones due to repeated wetting and drying processes was recorded by Overbeck (1981).

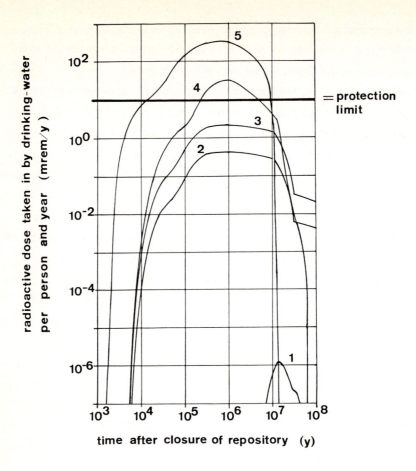

Fig. 9   Influence of circulation system on calculated dose release rates for radioactive waste repository in basement rocks (HSK, 1987).
Curve (1): Kakirite model (similar results to NAGRA's).
Curve (2) & (3) As Kakirite does not seem to be the dominant flow system in the Böttstein borehole, the data of which have been used for the safety performance analysis by NAGRA, HSK took the more appropriate dyke model. (HSK = Federal division for security of nuclear plants)
Curve (4) & (5) by comparison with (2) & (3) assumption
of higher solubility and larger amount of waters were used. Notice in the case of (5) the limits would just been reached after 10'000 years.

3.4 Basement rocks

The main results of the Nagra research project have shown the surprising extent of groundwater circulation within the crystalline basement. Groundwater circulation has been observed not only on the weathered surface of the basement but also in unaltered zones at greater depths. The major pathways for migrating fluids are:

-   Fracture zones with fine-grained wallrock fragments (Kakirite) and quarz veins
-   "Kakirite" associated with dykes
-   Fractured dykes

The mineralogical composition and the morphology of the different fracture zones determines the retardation factors and the release rates for migrating radionucleides. The most important conclusion drawn from these observations, -which can also be relevant for non-radioactive waste disposal- is that the choice of the type of flow mechanism in the system can lead to a miscalculation of the individual dose rates of several orders of magnitude (Fig.9, HSK 1987).

3.5 The composition of groundwater in the different circulation systems

The chemical composition of groundwater and the age estimates based on isotopes (Table 1) derive from Balderer (1985) and Schmassmann et al. (1984). The data shows the existence of various local and regional circulation systems with different flow directions and consequently different recharge zones and conditions. But the origin of most of the water and the main mechanisms driving these systems remain unknown (Balderer 1985).

4. Natural analogues for final storage quality

The disposal concept proposed by the Swiss Environmental Protection Agency (BUS), 1987, includes the segregation of inorganic wastes into non-hazardous and hazardous,

NEAR SURFACE GROUND WATER SYSTEMS

Geological Formation (near surf. system.)	Characterisation of ground water systems	Chemical composition	Age (estimated from isotopes)	Interactions with other systems
Molasse (near surf. system.)	(recharge areas located near the springs)	Ca-Mg-HCO3 (5 -10 meq./l)	few years	interaction with gw in the unconsolidated quarternary sediments
Malm, Dogger, Hauptrogen- stein	local flow system, springs in folded or tabular Jura	Ca - Mg - HCO3 low mineralisation	recent to several years to about 30 y	often mixtures with deeper circulation systems
Dogger, Hauptrogen- stein, Lias	local flow systems	Ca - Mg - HCO3 - SO4	ditto	ditto
Dogger, Hauptrogen- stein	local flow systems......	Ca - HCO3	ditto	ditto
Lias	Spring in Weissenstein railroad -tunnel ev. regional outflow point	Na -Ca -(Mg)-HCO3 - SO4 (16.5 meq./l)		
Keuper	local systems	Ca-Mg-SO4- HCO3 (20-35meq./l)	several -20y	
Muschel- kalk	near surface waters are undersaturated with respect to gypsum	Ca-Mg-HCO3- SO4 or Ca-Mg-SO4-HCO3	2 -20 y	In the folded Jura a mixture with waters of several flow systems seems to occur
Buntsand- stein	different circulation systems with different flow directions, different recharge zones	chemistry depending on interrelation with other aquifers	broad range of residence times (more than 30 y to 10'000 y	direct contact with underlying crystalline basement or permo-carboniferous rocks
Crystalline -basement	Saeckingen, mixtures with poorly mineralized and recent waters	Na-SO4-HCO3-Cl type, Na-(Ca)-Cl type, and a highly mineralized Na-Ca-HCO3-SO4-C type	Younger and older ages possible depending on recharge system	Some seem to be in contact with Buntsandstein or Permian waters.

DEEP GROUND WATER SYSTEMS

Geological Formation	characterisation of ground water systems	Chemical composition	Age (estimated from isotopes)	Interactions with other systems
Upper marine molasse	Depths 300-600m (Zurich, Konstanz, Mainau)	Na-HCO3- (Cl)-SO4, (5 -15 meq./l) high Na content (200mg/l)	1000 - (10'000) y	
Lower fresh-water molasse	Deep systems	high mineralisation: 40 - 150 meq./l	100 -10'000 y	
Malm	Deep gw systems (known in the NE part of CH and southern part of Germany			hydraulic connections with main aquifers of the molasse
Dogger	gw known in the Rhine-Graben Zone north of Basle	Dogger gw of Neuwiller (F) similar chemistry as upper marine molasse		
Keuper	deep systems above crystalline bedrock	Riniken: mineralisat.: 15700mg/l Na-Mg-Ca-SO4-Cl -HCO3 type	mean residence time several thousands - 10'000 y	
Muschel- kalk	deep boreholes and natural hot springs connected to tectonic zones	Losdorf, Weiach Ca-Mg-SO4- HCO3, hot springs of Baden, Beznau, Böttstein und Riniken: Na-Ca-Cl-SO4 type	several 100 to thousands of years for the first type; 1000-10'000 y for second type gw	
Permian- deposits	Deep borehole Riniken	highly mineralized (18g/l) contains H2S and CO2	1000 - 5000	
Crystalline -basement	Böttstein: highly mineralized gw about 1000 m below the top of the crystalline rock (origin has not yet been solved)	Na-SO4-HCO3-Cl type, Na-(Ca)-Cl type, and a highly mineralized Na-Ca-HCO3-SO4-Cl type	Younger and older ages possible depending on recharge system	Some seem to be in contact with Buntsandstein or Permian waters.

Table 1 Chemical composition and age of water in groundwater systems in the NE part of Switzerland (compiled after Balderer 1985)

315

reactive or soluble materials. It suggests that the latter should be transformed into a "mineralized" form (Oggier; 1987, BUS; 1986) and built into a geotechnically stable matrix (Leitbild für die schweizerische Abfallwirtschaft; 1986). This concept, however, needs to be discussed in more detail, taking into account the variable composition of the waste its "geotechnical matrix", and the changing physical and chemical conditions in a final storage site over a geologic time period.

Natural equivalents of stabilized wastes are thought to exist in the form of mineral deposits or earth-crust-like materials (Baccini, 1987). Mineral deposits are local accumulations or concentrations of rocks and minerals that can be recovered at a profit. Almost all ore-forming processes involve a slow movement of fluids at geological rates (liquids & gases), as in the case of meteoric water percolation or fluids associated with metamorphic or igneous activity.

The main differences between unmined and mined ore mineral conditions are to be found in their state of oxidation, surface and porosity, pH, pressure, mineral associations, and finally in the amount and chemical composition of percolating waters. The geotechnical form of the waste strongly determines the long-term behaviour of the inorganic constituents. A simple mechanical stabilization of the wastes as that outlined in the concept of the BUS (1986) resembles the mined stage more closely than the unmined. The behaviour of mineralized wastes could therefore be studied by analysing mill tailings.

Groundwater contamination is particularly created by minerals on spoil heaps. Such accessory minerals are genetically associated with the main ore. They may create potential hazards such as iron sulfides or arsenic sulfides associated with metallic ores or coal. Exposure to air and rain water causes them to oxidize into complex hydrated iron ion oxides or ochres and sulphuric acid, and to release arsenic into percolating meteoric water. These three products alone can seriously affect downslope water-pathways and vegetation.When sulphuric acid dissolves other elements from the spoil, the impact is much greater (Cloud, 1969). Mining within the Coeur d'Alene District of northern Idaho has been continuous for over 85 years. The data demonstrates that solution of metal sulfides from early day mining and milling wastes strongly influences the quality of present day groundwater (Mink et al. 1972). Other sources of contamination are known to be related to uranium mill tailings (Milnes, 1985).

As in laboratory experiments with hazardous radioactive wastes, a transformation of inorganic wastes into mineral analogues or ceramic waste forms could be considered. Ceramic waste forms are essentially synthetic crystalline rocks in which the chemical constituents are bound in the crystal lattices of the various mineral-ike phases. They are manufactured by crystallizing melts or recrystallizing powdered aggregates at high temperatures. Under experimental hydrothermal conditions they have proved to have a higher short-term stability than glass (e.g. Milnes, 1985). However, considerations on the volume produced and the energy necessary for such a solution require a careful evaluation of the waste form for the different chemical compounds to be deposited.

In Switzerland maximum contamination levels for inorganic constituents are based on the drinking-water guidelines published by the European Community (Richtlinien des Rates, 1980) and on the recommendations of the World Health Organization (WHO Guidelines for Drinking-Water Quality, 1984). In the U.S. except for NO_3^- and Cr, these constituents are rarely reported as a cause of significant deterioration of groundwater quality (Cherry et al., 1984). However, we do not yet know enough about the effect of trace metals on human health and ecology to discard all doubts.

The two main approaches used in predicting the chemical behaviour of inorganic contaminants in groundwater require (1) the use of distribution coefficients (which are incorporated in a simple retardation term of the advection-dispersion equation) and (2) the use of chemical equilibrium models based on thermodynamics. Recently, Cederberg et al. (1985) developed a model taking into account transport and chemical reactions, focusing more specifically on the formation of complexes on oxide surfaces. They first formulate the equations for the chemical interactions of a multicomponent-system. This leads to a certain number of differential equations which are related to the differential equations describing the transport in the system.

The chemical behaviour of most toxic inorganic contaminants mainly depends on the redox and pH conditions. When the pH is very low or very high, some heavy metals or metalloids are usually mobile. A very low redox condition generally leads to immobility. Field investigations have established that when the pH is neutral, metals and metalloids are not usually mobile, except for Cr(+VI) and Se, which occur in anionic forms and tend to be mobile when oxidizing conditions prevail. The most critical task in the prediction of the chemical behaviour of most toxic contaminants is to define the pH and redox conditions (Cherry et al 1984)

Attenuation capacity of geological materials

It is generally accepted that segregation of wastes and deposition of wastes in mono-disposal sites helps predict the attenuation characteristics of the geological materials. It also prevents reaction between incompatible wastes and reduces the number of parameters to be estimated for the safety analysis specific to a particular site.

However, this strategy implies a considerable concentration of inorganic materials in a small number of sites. The attenuation capacity of any geological formation has a limit, which, if exceeded by the volume of leachate that enters it, will allow contaminants to pass almost unretarded (Cartwright, 1981). Therefore, the attenuation capacity of each site's geological materials must be the limiting factor for the amounts of wastes to be disposed of there.

5. Critical comments on enclosing media (artificial barriers)

In the current discussion about disposal of hazardous wastes, the use of artificial seals (e.g. clay backfill, clay liners or geo-textils) in cavern repositories is proposed to reduce the effect of direct interaction between formation groundwaters and the stored wastes. The geotechnical and chemical characteristics of these materials are expected to delay the access of groundwaters to the repository and to act as a chemical buffer. But Siegenthaler (1987) has demonstrated that such clays are susceptible to "hydraulic fracturing", which is the formation of cracks within the rock or soil mass due to excessive fluid pressure. The precise sequence of hydraulic and mechanical events after the deposit of the wastes and the definitive sealing and abandonment of a repository is difficult to determine precisely. Depending on how fast the original hydrostatic pressures in the host rock build up and on how fast the clay saturates, a situation may arise in which high normal hydrostatic fluid pressures are created in the host rock when put in contact with the clay; while much lower pore pressures (air or water) and stresses continue to prevail in the inner layers of the clay which are not yet affected by the waters seeping in. These low pressures are local residues of the pressure during disposal operations (1atm). They can be expected to persist after the sealing of the repository because of the low permeability of the clays. Such a situation with large fluid pressure gradients is susceptible to lead to hydraulic fracturing (Siegenthaler, 1987).

Experiments simulating an abandoned underground repository were made with bentonite (clay composed of swelling minerals of the smectite group). They showed that fractures opened rapidly (within less than a second) and reached widths of 1.5 and 0.5 mm. Similar processes have been described in flysch-type sediments (shales and sandstones) (Dzulinski, 1956) or in dams in which abnormally high leakages have been attributed to hydraulic fracturing in the clay core (e.g. Lun, 1985). Such processes increase hydraulic conductivities and should be considered when evaluating possible causes of contaminant migration.

In shallow repositories, the use of trench liners may prove to be a critical factor in the mechanical compression and consolidation of clays on hill-slopes or cliff-walls in specific sites (Märki et al. 1974). The hydraulic behaviour in the real field system doesn't always correspond to that observed in a laboratory under controlled conditions.

6. Monitoring vs final storage

A monitoring system actually contradicts the definition of a final storage which implies that the repository should not have to be monitored (see final discussion). But because of lack of experience, monitoring is still essential for final storage repositories. It should provide sufficient warning of potential pollution problems. Not only the site itself but also the geological/hydrogeological system should be integrated into the concept. Such a program should not be reduced to a routine sampling program, designed by politicians in charge to reassure the public, and which does nothing but create a pile of data which often can not be interpreted. [According to Gibb et al. (1981), collecting "representative" water samples from monitoring wells is not an easy task to perform, see comments about measurement methods.] Our present knowledge of such systems does not exclude wrong selection of monitoring parameters. Therefore a monitoring system should be planned in such a way as to allow a continuous upgrading of our knowledge on the processes associated with final storage systems.

Advantages and disadvantages of deep burial storage

A deep burial of toxic wastes has some advantages over a shallow burial or surface storage systems. One of the most important advantages is that contaminants that may become dissolved in groundwater cannot migrate directly to the surface. The increased length of the groundwater flow path allows time for sorption, reaction, decomposition and dilution.

Deep burial within hard-rock hostmaterials (e.g. caverns or deep boreholes) can create great problems though, if not properly planned and constructed . They are difficult to monitor. If errors are made or unexpected flaws in the repository are uncovered, the removal of the waste in order to place it in a better location will be costly (Davis, 1984). Remedial actions as proposed for hazardous shallow land-disposal sites are not feasible.

7. Final comments and conclusions

At one stage practically non-permeable shales were recommended as possible bed-rock materials (e.g. In 1974 Jäckli proposed chalk, homogeneous glacial tills, molasse-marls, opalinus-clay and keuper-marls).

The complexity of natural subsurface systems and the number of significant transport and retardation processes in different geological formations make it impossible to develop "ready to cook" recipes which would satisfy safety requirements.

The topics discussed in this paper show that the proposed strategy for final storage repositories raises questions which cannot be answered immediately. The first question concerns the method of "final storage" repositories -which by definition should not have to be controlled- because the safety analysis would have to prove that safety limits would never be exceeded. I generally agree with the wish to find a final solution for a civilisation problem instead of keeping it in store for future generations. But do we really have enough experience and knowledge to guarantee that the stipulated safety requirements are fulfilled ? If not, this strategy had better be replaced by controlled accessible repositories.

The development of waste management is characterized by repeated changes of concepts, because of an increasing number of waste types, accidents and encounters of unpredicted processes; in other words, it is a trial-and-error method. One of the dangers of these concepts is the generalizations of certain statements such as "marls and clays are impermeable", "dry basement rocks" which are misleading because they are not always true for the repository as a whole. Statements such as "marls are impermeable", (CSD, 1978) strongly influenced site selection processes (e.g. choice of ancient clay quarries for hazardous waste disposal). They turned out to be oversimplified and led to erroneous and dangerous conclusions.

As a result of the basic premise outlined in chapter 2.2, the following requirements must be fullfilled to make a proposed repository site acceptable for waste disposal: (1) a good knowledge of the regional hydrogeological system, (2) a detailed understanding of the geological, geochemical and hydrogeological conditions of the site, (3) a capacity for accurate predictions on the migration of contaminated water through the hydrogeological system to the biosphere, taking different scenarios of leachate migration into consideration, (4) an understanding of microbiological processes and of possible biological accumulations.

Discussion of the different aspects which could affect the behaviour of a final storage system should illustrate that the processes we have to deal with in natural systems are manifold; and that many of them are still not well understood. The use of geological data (facies, dimensions & degree of heterogeneity, mineralogy, composition of groundwaters, isotopic composition etc.) in hydrogeological studies (distribution of hydraulic conductivities, porosities, retardation factors) is a major difficulty in estimating release rates from a final repository (see Fig 9). It is however a very important process, because these rates have to be compared with the limits defined by governmental regulations. Because of these difficulties groundwater circulation and transport models have mostly had a conceptual character and are based on simplifications such as uniform flow parameters. They should not be taken for granted, because they probably overlook processes of which we are not yet aware.

As a consequence, the site selection process becomes very important. A good example of the main points to be considered within such a process is given by Milnes (1985). He distinguishes four phases of investigation, alternating with steps involving political, economical and technical decisions, collectively refered to as social evaluation. This process involves: (1) reconnaissance: definition of boundary conditions, development of a strategy giving the following guidelines: forecasts of waste type, volume, isolation time, and engineering concepts based on experience in other fields, (2) problem-oriented investigation: detailed waste management concepts, generic models, the aim of which it is to explore the possible effects of different processes, data collection. This phase should end up with recommendations, possibly within the framework of a statement about generic environmental impact, intended as a contribution to the public debate, (3) site characterization (including drilling) , (4) construction (prototype testing). "Performance assessment for disposal of radwastes is essentially site-specific; it forms part of the safety analysis which should prove that a particular site will at all times in the future satisfy the stipulated requirements" (Milnes 1985). I think that the same condition should also apply to the site selection process of hazardous nonradioactive wastes.

Implications for geological / hydrogeological research:

The problems hydrogeologists usually have to deal with has changed in recent years from a quantitative to a more qualitative aspect. Questions concerning groundwater quality requires much more information on subsurface geology and chemical processes. It has become a research field with a strong orientation towards chemistry and hydraulic-engineering. Knowledge about subsurface composition will always be a key for success in the research of subsurface processes. Rapid development of stochastic theory now enables treatment of the spacial variability of multiple parameters (e.g. Gelhar and Axness, (1983); Hufschmied, (1986). These methods require fundamental assumptions about subsurface geology. In the future, sedimentologists, structural geologists and mineralogists may provide important new contributions in the process of flow-system characterisation.

A major effort should be made in the development of measurement techniques (e.g. drilling and sampling devices, packer-systems, geophysical measurement techniques for subsurface coverage), a very costly and time-consuming process which might have a considerable influence on the quality of all the parameters mentioned above.

Work initiated at the EAWAG, Dübendorf CH 8600 (Switzerland) and completed at the University of Geneva, Dept. of Geology and Paleontology, 13 rue des Maraîchers, 1211 Genève 4 (Switzerland)

Acknowledgements:
The author greatfully acknowledges the helpful suggestions and careful reading of E. Kupfer, Prof. G. Gorin and K.Kelts . I am also grateful to H.Weissert, A. Zingg, A. Losher, J.Davis, A. Lagerkvist and Prof. W.Wildi for their criticism and helpful comments.

References

Allen, J.R.L., Studies in fluviatile sedimentation: An exploratory quantitative model for the architecture of avulsion controlled alluvial suites: Sed. Geology, v. 21, p. 129-147, 1978.

Allen, J.R.L., A review of the origin and characteristics of recent alluvial sediments: Sedimentology, v.5, p. 89-191,1965c.

Ashmore, P.E., Laboratory Modelling of Gravel Braided Stream Morphology, Earth Surf. Processes, 7, 201-225, 1982.

Aziz Al-Azawi et al., Prognosen über Wasserhaushalt und Schadstofftransport im Untergrund einer geplanten Sondermülldeponie, Z. dt. geol. Ges. 137, 195-209, 1986.

Baccini, P., Die Schweiz ist gut versorgt - wie wird sie entsorgt?, Chimia 41, Nr. 7-8, p. 195, 1987.

Bachmann, M.R., Nahrung für die 5. und 6. Miliarde, Versuch einer Prognose zur Welternährungslage, NZZ, Forschung und Technik, Nr.226, p.77, 1987.

Balderer, W., The NAGRA investigation project for the assessment of repositories for high-level radioactive waste in geological formations, Mineralogical Magazine, Vol. 49, 281-288, 1985.

Beschta, R.L., and W.L. Jackson, The intrusion of fine sediments into a stable gravel bed, Jour. Fish. Res.Bd. Canada, 36, 204-210, 1979.

Bläsi, H. R., Lithostratigraphie und Korrelation der Doggersedimente in den Bohrungen Weiach, Riniken und Schafisheim, Eclogae geol. Helv., Vol. 80, Nr.2, 415-430, 1987.

Cant, D.,J., Fluvial Facies Models and Their Application, in Sandstone Depositional Environments, Editor, P. A. Scholle,the American Assoc. of Petr. Geol, Tulsa 1982.

Cartwright, K., Shallow Land Burial of Municipal Wastes, in Studies in Geophysics,Groundwater Contamination, National Academic Press, 67-78, 1984.

Cartwright, K., Gilkeson, R.H., Griffin, R.A. Johnson,T.M. Lindorff, D.E., and DuMontelle, P.B., Hydrogeologic considerations in hazardous wastes disposal in Illinois, Illinois State Geol. Surv., Environ. Geol. Note 94, 20pp.,1981.

Cederberg, G.A., and Street, R.L., and Leckie, J.O., A Groundwater Mass Transport and Equilibrium Chemistry Model for Multicomponent Systems, Water resour. Res., 21, (8), 1095-1104, 1984.

Cherry, J.A., and Gale, J.G., The Canadian Program for a high-level radioactive waste repositary: A hydrogeological perspective, Geological Survey of Canada, paper 79-1C, 35-47, 1979.

Cherry, J.A., Gillham, R.W., Barker, J.F., in Studies in Geophysics:Contaminants in Groundwater, Chemical Processes, Groundwater Contamination, National Academic Press, 46-67, 1984.

Cloud, P, Resources an Man, W.H. Freeman & Co, 1969.

CSD, Ing. Büro, Sondermüll: Ein weiterer Schritt zur umweltgerechten Beseitigung, Umweltschutz/Gesundheitstechnik, 9, 1978.

Davis, S.N., Deep Burial of Toxic Wastes, in Studies in Geophysics: Groundwater Contamination, National Academic Press, 78-93. 1984.

Dzulinski, S. & Radomski, A., Ann. Soc. Geol. Pologne, v. 26/3: 225-264, 1956.

EKA (Eidg. Komm. für Abfallwirtschaft), Leitbild für die Schweizerische Abfallwirtschaft, Schriftenreihe Umweltschutz (BUS) Nr.51, 1986.

Farlane, D.S., and Cherry, J.A., and Gillham, R.W., and Sudicky, E.A., Migration of Contaminants in Groundwater at a landfill: A case study, 1, Groundwater Flow and Plume Delineation, J. Hydrol., 63 (1/2), pp. 1-29, 1983.

Farquhar, G. J., Rovers, F.A., Gas production during refuse decomposition, Air, Water and Soil Pollution 2, 483-495, 1973.

Farquhar, G.J. and Rovers, F.A. Leachate attenuation in undisturbed and remoulded soils, Proc, Res. Symp. Gas and Leachate from Landfills: Formation, Collection , and Treatment, New Brunswick, N.J., U.S. Environmental Protection Agency, National Environmental Research Center, Cincinnati, Ohio, 1975.

Fisk, H. N., Fine Grained Alluvial Deposits and their Effects on Mississippi River Activity, 82pp., Mississippi River Commission, Vicksburg, Miss.,1947.

Freeze, R.A., A stochastic conceptual analysis of one-dimensional Groundwater Flow in non Uniform Homogeneous Media, Water Resour. Res., ll, 725-741, l975.

Freeze, R.A., Cherry, J.A., Groundwater, Prentice-Hall, 1979.

Gautschi, A. et al., Die geologische Charakterisierung der Wasserfliessysteme im kristallinen Grundgebirge der Nordschweiz, Nagra informiert, Nr 2, 15-21, 1986.

Gillham R.W. and Cherry J.A., Contaminant Migration in Satured Unconsolidated Geologic Deposits, Geol. Soc. Am. / Sp. Paper 189, 1982.

Gibb, J.P., Schnuller, R.M, Griffin, R.A., Procedures for the collection of representative water quality data from monitoring wells, Illinois State Water Survey/Illinois State Geol. Survey Cooperative, Groundwater Report 7, 61 pp., 1981.

Goodall, D.C., Quigley, R.M., Pollutant Migration from two sanitary Landfill Sites near Sarina, Ontario, Canadian Geotechnical Journal, 14, 223-236, 1977.

Griffin, R.A., et al., Attenuation of pollutants in municipal landfill leachate by clay minerals, Part 1, column leaching and field verification, Illinois State Geol. Surv.,Environ. Geol. Note 78, 34 pp. 1976.

Griffin, R.A., et al., Attenuation of pollutants in municipal landfill leachate by clay minerals, Part2, heavy metal adsorption, Illinois State Geol.Surv., Environ.Geol. Note 79, 47pp. 1977.

Guilette, B., Hydrogeology of the Taylor Marl Formation, Blackland Prairies of Central Texas, EOS, Vol.67, No.44, (Conference abstract) 1986.

Hoehn, E. and Santschi, P., Interpretation of Tritium Displacement during Infiltration of River water to Groundwater, Water Resour. Res. 1987.

HSK: Hauptabteilung für die Sicherheit der Kernanlagen, Technischer Bericht zum Gutachten über das Projekt Gewähr, 1987.

Huber, M., & Huber, A., Der Aufbau des kristallinen Grundgebirges - Kristallinkörper und Schollen, Nagra informiert, Nr.2, 1986.

Huggenberger, P., Siegentaler, Ch., Stauffer, F., Influence of Highly-Permeable Zones on Groundwater Velocity Distribution in Fluvio-Glacial Gravel Deposits, Water Resources Research, in press.

Huggenberger, P., Siegentaler, Ch., Stauffer, Grundwasserströmung in Schottern; Einfluss von Ablagerungsfromen auf die Verteilung der Grundwasserfliessgeschwindigkeit, Wasserwirtschaft, vol.78, Nr. 5,1988.

Jäckli, H., Die Beurteilung von Deponiestandorten im Hinblick auf den Gewässerschutz, Gewässerschutz/Lufthygiene, Heft 4, 1974.

Kiraly, L., FEM 301-D three dimensional Model for Groundwater Flow Simulation, NAGRA, Technical Report , 84-49, l985.

Lallemand-Barres and Peaudecerf, Recherche des relations entre la valeur de la dispersivité macroscopique d'un milieu aquifère, ses autres caractéristiques et les conditions de mesure, Bulletin du BRGM, Ze Serie, Section III, p.4, 1978.

Lasaga, A.C., Chemical Kinetics of Water-Rock Interactions, J. geophys.Res., 89(6), pp.4OO9-4O25, 1984.

Laubscher, H.P., Struktur des Grundgebirges und des Paläozoikums in der Nordschweiz, Expertenberichte HSK, 1986.

Long, C.S.,Gilmour,P. & Witherspoon,P.A., A Model for Steady Fluid Flow in Random Three-Dimensional Networks of Disc-Shaped Fractures Water Resour. Res., 10(21), 1105-1115, 1985.

Long, J.C.S and Wilson, C.R. and Remer, J.S. and Witherspoon, P.A., Porous Media Equivalents for Networks of Discontinuous Fractures, Water Resour. Res., 18(3), l982.

Lun, P.T.W., Critical review of hydraulic fracturing in dams, CSIRO Division of Geomechanics, Mt. Waverly, Australia, v. 138, 1985.

Märki, E., et al., Die Sondermülldeponie "Schlauen" in Oeschgen im Kanton Aargau,Hydrologische Grundlagen und technischer Aufbau der Deponie,Wasser- und Energiewirtschaft, Nr.10, 1974.

Miall, A.D., Glaciofluvial transport and deposition,In: Eyles, N. (ed.), Glacial Geology, Pergamon Press, New York, 168-183, 1984.

Miall, A.D., Architectural Element Analysis: A new Method of Facies Analysis applied to fluvial Deposits, Earth Sciences Reviews, 22, 1985

Milnes, A.G., Geology and Radwaste, Academic Press Geology Series, 1985.

Mink, L.; Williams, R. and Wallace, A., Effect of Early Day Mining Operations on Present Day Water Quality., Ground Water, Special Issue, Vol 2., pp. 17-26, 1972.

Nagra,Gewährsberichte (NGB), Endlager für hochaktive Abfälle:Sicherheitsbericht, NGB 85-5, 1985.

Nagra, Erläuterungen zur geologischen Karte der Nordschweiz, 1:100'000, NTB 84-25, 1984.

Narasiman, T.N., Numerical Modeling in Hydrogeology, Geol. Soc. Am. / Sp. Paper 189, 1982.

Narasimhan, T.N., Recent Trends in Hydrogeology, Geol. Soc. Am. / Sp. Paper 189, 1982.

Oggier, P., Neue Deponie-Typen als Ausgangspunkt Für Zukünftige Gesetzliche Regelungen, Umweltschutz in der Schweiz, Bulletin des Bundesamtes für Umweltschutz , Vol.2 12-18 1987.

Overbeck, R., Verwitterung von Mergelkalken durch Trocknungs - Befeuchtungs - Wechsel, Ber. 3. Nat. Tag. Ing.-Geol., Ansbach, 225-232, 1981.

Pugin, A., Deglaciation in an overdeepened perialpine valley (Bulle Area, Western switzerland), Facies Model and Paleogeography, Vortrag SNG, Luzern, 1987.

Pugin, A. Facies models for deglaciation in an overdeepened alpine valley (Bulle area, Western Switzerland), Palaeogeography, Palaeoclimatoloy, Palaeoecology, in preparation.

Richtlinien des Rates vom 15. Juli 1980 über die Qualität von Wasser für den menschlichen Gebrauch (80/778/EWG), Amtblatt der Europäischen Gemeinschaften Nr. L 229/11-29, 1980.

Schmassmann, H.J., Balderer, W., Kanz, W., and Pekdeger, A., Regionale Untersuchungen über die Beschaffenheit der Tiefengrundwasser, NAGRA Techn. Rept. NTB 84-21, 1984.

Siegentaler, Ch., Hydraulic Fracturing - A Potential Risk for the Safety of Clay-Sealed Underground Repositories for Hazardous Wastes, Hazardous Waste & Hazardous Materials, Vol 4, number 2, 1987.

Siegentaler, Ch., Huggenberger, P., Late Pleistocene Rhine gravel: a depositional braided river model, in preparation.

Smith, N.D., Proglacial Fluvial Environment, in: Glacial Environments, SEPM Short Course No.16, 1985.

Stauffer, F., and Job, D., Infiltration in Geschichteten Böden und Reaktion des Grundwasserspiegels, Experiment und Simulation, Inst. für Hydromechanik und Wasserwirtschaft ETH-Z, R 18-82, 1982.

Steel, R.J., and D.B. Thompson, Structures and textures in Triassic braided stream conglomerates ("Bunter" Pebble Beds) in the Sherwood Sandstone Group, North Staffordshire, England, Sedimentology, 30, 341-367, 1983.

Tsang, Y.W., Chanel Model of Flow Through Fractured Media, EOS, Vol.67, No.44, (Conference abstract) 1986.

URAN, Geschichte, Mineralogie, Erkundung und Ausbeutung, Lagerstätten, Schweizer Vorkommen, Vorräte, Gefahren, Sonderausstellung Geologisches Institut ETH, Zürich, 1980.

Walker, R.G. (ed.), Facies Models, 2nd ed., Geoscience Canada, Reprint Series 1, Geol. Assoc. Canada, 317 pp. 1984.Allen, J.R.L, Studies in fluviatile sedimentation: An exploratory quantitative model for the Architecture of avulsion controlled alluvial suites, Sed. Geology, Vol. 21, 129-147, 1978.

WHO, Guidelines for Drinking-Water Quality, Vol.1 Recommendations, Vol2 Health Criteria and Other Supporting Information. World Health Organization, 1211 Geneva 27 (1984).

TRANSPORT MODELS FOR LEACHATES FROM LANDFILLS

E. Bütow and H.-P. Lühr
Institut für wassergefährdende Stoffe
Technische Universität Berlin

Abstract

During the last ten years, transport models to describe the solute migration in the subsurface were more and more developed. The simulation of solute transport can include the main processes influencing the amount of spreading, as advection, dispersion/diffusion, sorption and degradation. To use a mathematical model with confidence, it is important to have an insight into the assumptions und limitations describing the movement of solutes. These restrictions depend on the type of hydrogeology, on the type of solution and on important processes influencing the migration. Generally, the best validation of simulation and field measurements is available for the advective flow in porous media. Strong differences are to state for simulations with a high dispersive migration of solutes. This is given for cases with a very heterogeneous aquifer and high contrasts in the composition of the aqueous solution. A proper simulation of sorption and degradation is possible for a simplification of the natural conditions only. To model the transport of leachates from a landfill, a better understanding of sorption and degradation processes is necessary. Geochemical models can provide important informations. Because of the high complexity of the aqueous solution leaching from a landfill, assistant laboratory and field studies are necessary to deliver the basics for the geochemical approach. The article gives an overview on the limitations and restrictions using a transport model to simulate the migration of leachates.

1. Introduction

Contaminant transport models can describe under specific conditions the movement of dissolved substances in groundwater. On the basis of partial differential equations models describe the relevant physical, chemical and biological processes for migration:

- advection (physical process)
- dispersion (physical processes)
- diffusion (physical process)
- sorption (chemical and biological processes)
- degradation/transformation (chemical and biological processes).

The different types of models include different types of processes. At the present state-of-the-art a model can not combine all the processes in one approach. With the exception of advection, the different solute transport processes are poorly under-stood. In general, a transport model takes into account advection and dispersion. The use of this type of model tends to underestimate the actual concentration level. This approach can include diffusion, sorption and degradation (decay), too. The mathematical description then is a simplified description of the processes occuring under natural conditons (van der Heijde 1985).

2. Basics

The International Groundwater Modeling Center (IGWMC) gives an overview on existing groundwater models (van der Heijde et al. 1985). From the 399 reviewed models 84 are transport models. Only 39 of these codes are commonly available and well enough docu-mented. To solve the basic equations different numerical methods are used.

Commonly, finite-difference and finite-element-methods are employed. The main differ-ences in both methods is shown by the mesh for discretization. The finite-difference mesh is rectangular and therefore the model area is a square or a rectangle. The fi-nite-element mesh is more flexible using triangular and rectangular elements. This advantage is usefull especially for areas with very irregular boundaries and complex geologic structures. A finite-element grid with irregular structures is difficult to handle for a 3-dimensional configuration.

One source of difficulty in the numerical solution of transport equations is the treatment of the advection term. For conventional finite-difference and finite-ele-ment solvers, either numerical dipersion or the overshoot-undershoot phenomena may be introduced. The sets of discretized equations are solved repeatedly using a direct or an iterative equation solver. Iterative techniques such as successive overrelaxation, Gauss-Seidel and iterative alternating direction implicit (ADI) are the most common. The numerical stability of different solvers is designed by the discretization in time and space. Some codes offer the opportunity to select different solution techni-ques to minimize numerical problems according to the treated problem.

The mathematical description of groundwater flow in water saturated porous media suppose the laminar flow and the validity of Darcy's law. The Darcy law implies that the flow is linear with respect to the head. Numerical models to compute the groundwater flow field are available and well developed. Most of them are tested by a validation or verification procedure. The groundwater flow field is identical with the advective part of migration for tracer concentrations. In the case of significant density effects the flow field depends on the dispersion (NEA 1988). Table 1 summarizes the status to model the advection.

In a fractured medium, the flow can be laminar and turbulent also. Therefore Darcy's law is not valid. The basic equations have to take into account in addition to the head the properties of fractures as fracture width and fracture frequency. Basic equations are developed for single fractures, but not useful to describe a field situation. For fracture systems different approaches are in research based on stochastic concepts for the networks of discontinuous fractures. Some of these approaches are leading to an equivalent porous medium permeability (Cacas et al. 1987).

A lack of basic informations is given for the flow in unsaturated media. Because of the two phase flow of water and air much more influencing parameters are detectable. The main problem is the validation procedure of this type of codes on an field situation.

Table 1: Overview on modeling of advection

	Saturated Flow		Unsaturated Flow	
	Porous Media	Fractured Media	Porous Media	Fractured Media
Assumptions	Laminar Flow	Laminar and Turbulent Flow	Two Phase Flow (Water and Air) Saturation Effects	Saturation Effects and Turbulent Flow
Basic Equations	Darcy Law	in Research (Statistical Approaches)	Fairly Known	Research Needed
Mathematical Modeling	Well Developed (for Tracer Concentrations) in Research (for Significant Density Effects)	in Research (Porous Media Equivalent Systems)	Developed (for Simple Cases only)	Not Available

The mechanical <u>dispersion</u> depends on the inhomogeneity of an aquifer which is passed through by pollutants. The spreading of dissolved substances is characterized by the deviation of pore velocities from the calculated mean value of advective flow. The commonly used dispersion theory assumes the Gaussian distribution of the spatial variability of hydrologic field parameters. For the mathematical description the migration is separated in a term parallel to the flow direction and a term accross of them. This leads to the longitudinal and transversal dispersion. To quantify these parameters laboratory or field measurements are necessary.

Observations of dispersion effects will depend on the measurement scale. At the laboratory the sample size is small and the measurements can determine microscopic scale effects only. Not comparable are results from field measurements with longer travel distances at macroscopic scale. Different mathematical formulations are developed to use the variability of field parameters for a description of macroscopic dispersion with stochastic approaches (Gelhar & Axness 1983).

The spreading of pollutants is caused by molecular diffusion also. Diffusion, expressed by Fick`s law, results from the migration of solute from higher to lower concentration regions. The term hydrodynamic dispersion (Bear & Verruijt 1987) describes the summary of mechanical dispersion and molecular diffusion. In the case of low flow velocities the solute transport by diffusion can be important.

The basics for the plume distribution in fractured media or for unsaturated flow are clearly different. For fractured media field experiments show complex migration processes described by the synomys "matrix diffusion" and "channeling" (Moreno et al. 1985). Table 2 summarizes the status to model the dispersion.

Most chemicals in solution or in suspension do not show the same mobility in groundwater as the water itself. If a substance is not effected by retardation, it could be considered as an "ideal" tracer. The retardation is controlled by different processes, like <u>sorption</u>, precipitation/dissolution, chemical substitution, isotopic exchange or by filtration. Sorption can be categorized in physical adsorption, electrostatic adsorption, e.g. ion exchange, and specific adsorption. All three types of sorption are dependent on the chemical form of the sorbate in solution. Parameters like Eh, pH and concentrations of complexing species influence adsorption by controlling speciation.

The simplest way to quantify the retardation is possible by a linear isotherme, respectively by the K_d-concept. The distribution coefficient K_d summarizes the different processes discussed before and is detectable in laboratory experiments. The linear isotherme is applicable for very low concentrations and supposed equilibrium conditions. Other sorption models take into account the solute concentration (Freund-

Table 2: Overview on modeling of dispersion

	Saturated Flow		Unsaturated Flow	
	Porous Media	Fractured Media	Porous Media	Fractured Media
Assumptions	Tracer Concentrations	Channeling, Tracer Concentrations	in Research	Unknown
Basic Equations	Dispersion Theory	in Research	in Research	-
Mathematical Modeling	Developed (Numerical Problems)	-	-	-

lich-isotherm) and solute concentration plus the sorption capacity (Langmuir-isotherm). Especially the K_d-concept is very useful for transport models (Kinzelbach 1987).

For a proper description of the different complex processes, the use of geochemical models is necessary. Commonly used geochemical models are useful for simple aqueous solutions and for stationary conditions only. A coupling procedure between transport models and geochemical models is required. Table 3 gives an overview on the modeling of sorption.

Table 3: Overview on modeling of sorption

Influenced by:	Modelled by:	Incorporated in:
* Adsorption * Ion Exchange * Complexation * Precipitation * Diffusion into the Matrix * Filtration * Microbiological Activity etc.	* Linear Isotherme (kd-Concept) ——> * Freundlich Isotherme ——> * Langmuir Isotherme ——> etc.	Transport Models
	for Simple Solutions ——> under Stationary Conditions	Geo-chemical Models

Table 4: Overview on degradation processes

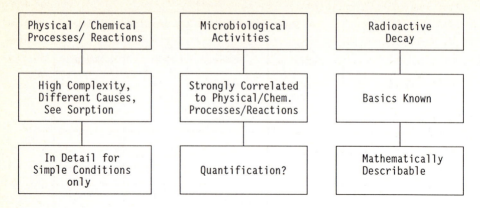

For a great number of substances e.g. organics, the <u>degradation</u> (Table 4) and formation of new compounds is very important. A geochemical model is usefull to handle
physical processes or chemical reactions only, which are leading to degradation. The
microbiological attack with a decomposition procedure is not to be described mathematically in the same way. The amount of biodegradation depends on the nutrient offer
for a microbial activity. It is possible to describe the degradation in a first approach over a half-life time in equivalence to the radioactive decay (Mattheß et al.
1985).

3. Concepts and methodical tools

A model serves as a tool designed to represent a simplified version of reality. The
application of a model to groundwater pollution requires the description of initial
and boundary conditions of the system, the characterization of its hydrogeology and
the definition of the relevant processes.

The use of a model includes laboratory and field studies of solute migration and the
analysis of the prediction that results from the simulation. Gaps in the data base
may be discovered. Such results may require some repetition in data collection efforts and may lead to the development of a new, improved model. The process of mathematical modeling includes the model testing. Model validation should test the capability of the model to simulate the actual physical processes (Naymic 1987).

A waste disposal site contains different types of solid and liquid waste. The leachate formed from these sources consists primarily of dissolved minerals, heavy metals and organic chemicals. For low contaminant concentrations, the advection process

can be regarded as decoupled of the dispersion. Strong differences of concentration, as often found for leachate, do affect the advection, though. The dispersion theory is valid for the assumption of miscible groundwater constituents. Special problems are associated with modeling the transport of chlorinated hydrocarbons which have a density greater than water and complex reactions induced by chemical and biological processes (Macalady 1986).

The modeling of leachate transport needs a separation into different problem areas. These problem areas have different types of scale. The essential processes for the transport are different in the near field, from the processes in the local or regional environment. Different types of models are necessary. Figure 1 shows possible model and problem areas: the site, the local environment and the regional environment (Ward et al. 1987). The figure shows a telescopic mesh refinement used for a finite - difference - code to model a hazardous waste site. This approach is also necessary to minmize the needed computer time for simulations. In addition, with increasing scale the available information generally decreases. A monitoring system is often available for the site and for the local scale only. These different scales are useful to identify different problem areas

The site scale (Table 5) is dominated by the source of leachates. In this area the infiltration of contaminants to the groundwater starts. The driving force is the fluid infiltration rate controlling the amount of leachates. To quantify the source term, knowledge on the solubilities, the two - phase flow in the unsaturated zone and the behaviour in high concentrated aqueous and non-aqueous solutions is necessary. In addition, a distinct temperature change and gas production can be expected. To model the complexity of processes at the site scale models are needed, which are not available at present. Only for single perseverant substances a conservative approach based on geochemical models is applicable.

At the local scale (Table 6) the transport of leachates occurs in the saturated aquifer and is mainly horizontal. The hydraulic head distribution forms the groundwater flow field. A mixing process between the different types of leachates starts. Because of changing Eh and pH conditions, enrichment of polluants in the subsurface can be observed. The various transport processes are still very complex and difficult to describe with just one model. For this area a coupling procedure between geochemical models and advection-dispersion type models is needed.

For the regional scale (Table 7) or the far field the commonly used transport models are applicable. The main assumption is that leachate occurs in tracer concentrations. At this scale the influence of geochemical reactions is decreasing.

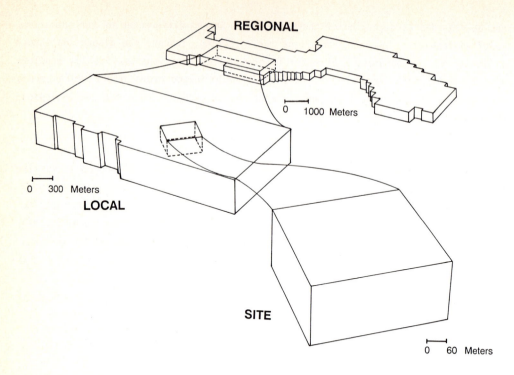

Figure 1: Conceptual diagram of modeling approach (Ward et al. 1987)

4. Practical application

In general, three different types of models are used in groundwater quality studies (van der Heijde 1985):

- . <u>flow models</u> to understand the flow field in the surrounding of the contamination,
- . <u>transport models</u> to predict the movement of solutions in groundwater and
- . <u>geochemical models</u> to interpret the chemical behaviour of an aqueous solution.

As discussed before, flow models are well developed and useful for all problems linked to advection dominated transport of contaminants. Flow models can simulate different system alterations which are influencing the hydraulic heads and the flow direction (e.g. groundwater pumping or reinjection, reduction of infiltration, subsurface drains). Transport models cannot be employed with the same grade of confidence because of problems caused by dispersive migration. Naymic 1987 gives an

Table 5: Site Scale (Near Field Area)

PROBLEM DEFINITION: - Source term for leachates - Mainly vertical directed transport - Driving force by fluid infiltration rate
ESSENTIAL PROCESSES: - Solubility controlled leaching rates - Two phase flow (gas, liquid) in the unsatured zone - Highly concentrated solutions - Strong interrelations between different soluted substances
TYPE OF MODEL NEEDED: - Model should include leaching processes, geochemical processes, the transport in the unsaturated zones, thermal effects
STATE-OF-THE-ART OF MODELING: - Flow in unsaturated zone fairly developed for porous media - Geochemical models are developed but for stationary conditions only
USEFUL APPROACH: - Conservative approach for single perseverant substances
RESEARCH NEEDED: - Use of geochemical models to understand the different processes in the waste - Because of high complexity start with simple examples necessary

overview on different practical applications of transport models for the U.S. and Canada.

A well documented case study for Germany is the Eppelheim site located in Baden-Württemberg (MELUF 1987). Groundwater contamination was caused by deposited municipal and industrial wastes, particularly by various organic substances. A plume of chlorinated hydrocarbons, as Trichloroethene, Tetrachloroethene and cis-1,2-Dichloroethene migrated in the upper alluvial aquifer to different drinking water wells.

Table 6: Local Scale (Direct Surrounding)

PROBLEM DEFINITION: - Strong effect of leachates on groundwater quality - Flow in a saturated aquifer, mainly horizontal directed - Driving forces given by hydraulic head distribution
ESSENTIAL PROCESSES: - Geochemical and biological processes are leading to enrichement of polluants in the subsurface - Mixing process of different types of leachates starts - Effects of highly concentrated solutions and of temperature are decreasing
TYPE OF MODEL NEEDED: - Model should include advection, dispersion, geochemical and biological processes
STATE-OF-THE-ART OF MODELING: - Transport models including advection and dispersion are developed for porous media - Geochemical models are developed for stationary conditions only
USEFUL APPROACH: - Calculations to understand the local flow field are possible - Check of effectiveness from monitoring systems on hydraulics influencing activities - Transport calculations for single substances only
RESEARCH NEEDED: - Coupling of geochemical models with advection-dispersion type models - Better understandig of dispersive processes

Research teams have studied the situation at the Eppelheim site in detail, especially the hydrogeology, the groundwater hydraulics, the physico-chemical and microbiological behaviour of the groundwater and the migration of leachates in the subsurface. A transport model was used to interpret the various field measurements and to understand the transport processes. It was possible to detect the flow field and the migration behaviour on the base of parameter variations.The following characteristics of the plume behavior were found by modeling:

average groundwater velocity (supposing no sorption),

Table 7: Regional Scale (Far Field)

PROBLEM DEFINITION: - Flow in saturated aquifer system - Small effects of leachates on groundwater quality - Driving forces given by hydraulic head distribution
ESSENTIAL PROCESSES: - Leachates in trace concentrations - Advection and dispersion are dominant - Geochemical reactions with decreasing influence
TYPE OF MODEL NEEDED: - Model should include advection, dispersion, linear   sorption

STATE-OF-THE-ART OF MODELING:
- Solute transport models are available
- Advective dominant influences are to model with
  confidence, dispersive dominant influences are to handle
  carefully

USEFUL APPROACH:
- Regional scale calculations are possible for the advec-
  tion, for dispersion with limitations
- Linear sorption is useful for simple ions
  (k_d-values necessary)

RESEARCH NEEDED: - Better understanding of dispersion at large scale

- two different locations of sources for leachates,
- different time dependent source rates and
- longitudinal and the transversal dispersivity

The various simulations were made with different parameters in a way of a curve fitting between simulated and measured data. The used model is based on a random - walk - procedure (Kinzelbach 1986).

Figure 2 shows as an example an intercomparison between the average yearly concentrations (designed in bars) and the simulated concentration distribution for four different observation wells (A, B, C and D). These observation wells are located in different distances downstream of the sources. The observation well A is nearest to the first source of lechates. The well B is located on the stream line from A in a distance from about 5 km and slightly influenced by dispersive spreading of the second source of leachates. The concentration level of the other both wells is the result of dispersive flux of the two different sources. Because of the different

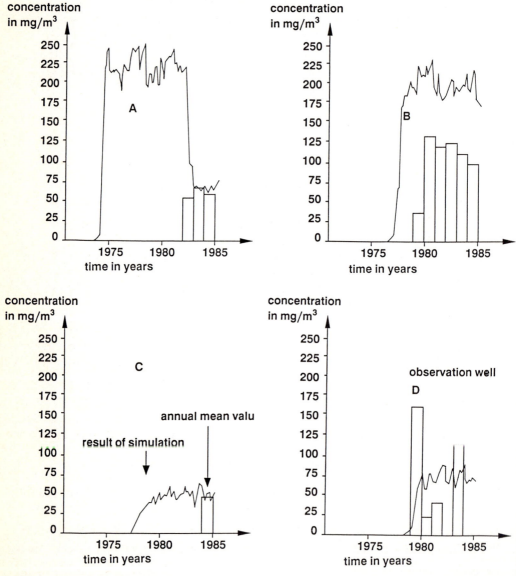

Figure 2: Comparison between measured and calculated concentrations for Trichloroethene at the Eppelheim site (MELUF 1987)

source rates the quantity of superposition of spreading depends on the location of the well and on the particular time of observation.The authors discuss the opportunity that further parameter variations could lead to a better curve fitting. It is important to state that this procedure needs a detailed and well documented data base. The information level for the four observation wells shown in figure 2 is different. Field values are available for well C at 1985 only.

Because of the used random-walk model the simulated results are showing a discontinuous time dependent concentration. The discontinuity is specific for the used approach and is to interpret as variations of results around a mean value. In the case of application of a finite-element or finite-difference code the results are smoothed. But then problems are increasing by numerical dispersion with an overestimation of leachate spreading.

5. Conclusion

A mathematical model should be a part of a safety analysis for a landfill or waste disposal site. The model can serve as a tool to help to understand the groundwater flow field and it can help to optimize the monitoring system. Also, it can provide information on different clean-up strategies. A useful approach seems to be a step by step analysis of the problem. Starting point should be a regional model. With increasing scale, generally the complexity of influences on the leachate transport is decreasing. In addition, the groundwater quality is less effected in the far field than in the near field.

The presently available transport codes are useful to understand the flow conditions in the regional and in the local scale. Generally, the spreading of leachates is overestimated by solute transport codes. This leads to an underestimation of the maximum concentration level. The advective part is fairly useful for porous media aquifers, whereas fissured rock aquifers are at present not to be modelled with confidence. For high contrasts in the hydraulic conductivities numerical problems arise for porous media taking the Darcy law as a basis.

At last, the successful implementation of models requires a modeling group which is familiar with the used code, a properly integrated mathematical methodology in the code and a continuous information exchange between the modeling group and the staff exploring the field situation.

340

6. References

Bear, J. 1972, Dynamics of fluids in porous media. - Elsevier,New York.

Bear, J. & Verruijt, A. 1987, Modeling Groundwater Flow and Pollution. -
Reidel Publishing Comp, Dordrecht.

Cacas, M.C.; Ledoux,E.; de Marsily, G. & Tillie, B. 1987, The use of stochastic
concepts in modeling fracture flow. - Presentation at Conference Groundwater Contamination: Use of Models in Decision Making; Amsterdam, Oct. 26-29,1987.

Gelhar,L.W. & Axness, C.L.1983, Three dimensional stochastic analysis macro-
dispersion in aquifers. - Water Resources Research, Vol. 19, p. 161.

van der Heijde, P.K.M 1985, Modeling Contaminant Transport in Groundwater.
- Presentation at Washington Conference on Groundwater Protection and Cleanup,
Arlington, Nov. 12 - 13, 1985.

van der Heijde, P.K.M.; Bachmat, Y.; Bredehoeft, J.; Andrews, B.; Holtz, D.;
& Sebastian, S. 1985, Groundwater Management: The Use of Numerical Models. -
Water Resources Monograph No. 5, Second Edition, American Geophysical
Union, Washington.

Kinzelbach, W. 1986, Groundwater Modeling - An Introduction with Sample
Programs in Basic. - Elsevier.

Kinzelbach, W. 1987, Numerische Methoden zur Modellierung des Transports
von Schadstoffen im Grundwasser. - Schriftenreihe gwf Wasser · Abwasser,
Band 21,Oldenbourgh Verlag München/Wien.

Macalady, P.L. 1986, Transport and Transformations of Organic Contaminants. -
Journal of Contaminant Hydrology, Vol. 1 ,Special Issue.

Mattheß, G.; Isenbeck, M.; Perdeger, A.; Schenk, D. & Schröter, J. 1985, Der
Stofftransport im Grundwasser und die Wasserschutzgebietsrichtlinie W 101. -
Umweltbundesamt Bericht 7/85

NEA 1988, HYDROCOIN, Level 1, Code Verification. - Nuclear Energy Agency/OECD,
Paris.

MELUF 1987, Ministerium für Ernährung, Landwirtschaft, Umwelt und Forsten Baden
Württemberg (Editor) Grundwassergefährdung durch Altablagerungen am Beispiel
Eppelheim. - Wasserwirtschaftsverwaltung, Heft 17.

Moreno, L.; Neretnieks, I. & Eriksen, T. 1985, Analysis of some
laboratory tracer runs in natural fissures. - Water Ressource Research,
Vol. 21, p. 951

Naymic, T.G. 1987, Mathematical Modeling of Solute Transport in the Subsurface. - Critical Reviews in Environmental Control, Vo. 17, Issue 3.

Ward, D.S.; Buss, D.R.; Mercer, J.W.; & Hughes, S.S. 1987, Evaluation of
a Groundwater Corrective Action at the Chem-Dyne Hazardous Waste Site Using
a Telescopic Mesh Refinement Modeling Approach. - Water Resources Research, Vol. 23,
No. 4, p. 603.

ECOTOXICOLOGICAL CRITERIA FOR FINAL STORAGE QUALITY
Possibilities and Limits

Josef Zeyer and Joseph Meyer
Swiss Federal Institute for Water Resources and Water Pollution Control

1. Introduction

Traditionally, hazard and risk assessments have been made primarily with respect to humans. Within the last decade, however, it became a generally accepted view that not only man, but ecosystems as a whole, have to be protected from hazardous impacts. The Swiss Federal Law relating to the Protection of the Environment (1), which became effective in 1985, begins with the statement: "The purpose of this law is to protect persons, animals and plants, their biological communities and habitats against harmful effects or nuisances and to maintain the fertility of the soil. Early preventive measures shall be taken in order to limit effects which could become harmful or a nuisance." In order to match the new requirements, classical toxicology had to be supplemented with ecotoxicology. Ecotoxicology encompasses not only the impact of a selected chemical compound on a gene, a cell or an organism, but also the effect on a population or a whole ecosystem. Moreover, as we will see below, ecotoxicology also has to consider the effects of the ecosystem on a chemical compound and to evaluate the fate of the compound in the environment. A large number of ecotoxicological tests (e.g., biodegradability, bioaccumulation, Daphnia, Microtox test) to evaluate the behavior of chemicals have been developed and reported in detail (2). Therefore, this article will not focus on test methods to assess the hazard and risk of landfill emissions, but rather on the conceptual problems associated with these evaluations. The paper will also not address the acute hazards generated by a landfill (noxious gases, fire and explosion hazards, emissions with acute toxicity) but discuss the long term problems.

Landfills containing municipal or hazardous wastes are usually characterized by a high chemical and biological activity and the emissions often require specific treatments to meet air- and water-quality standards (Fig. 1). These types of landfills are designated "Reactor Type Landfills" by the Swiss Federal Office for Environmental Protection (3). As a future alternative, this agency proposes the concept of a "Final Storage Landfill", whose chemical and biological activities are low and whose emissions are predictable and don´t require a specific treatment. Although this new concept is widely supported, fundamental political and technical problems prevent its rapid practical implementation (4). We will only be able to move slowly towards the "Final Storage Landfill" concept by testing and adopting methods like sorting the waste based on physical, chemical and biological characteristics, recycling valuable components, incinerating organic compounds, pretreating and immobilizing inorganic waste and carefully selecting appropriate geological sites for final storage. Basic mass balances and flow diagrams illustrating this type of waste management have recently been published (5). In addition, reducing the quantity of waste at the source whenever possible is of paramount importance. Presently in Switzerland, 500 kg of sewage sludge (95% water content), 400 kg of municipal waste and 30 kg of hazardous waste are produced per capita per year, and any attempt to turn these quantities into final storage quality appears to be hopeless. Consequently, within the next few years, a landfill will hardly ever be entirely designable according to the "Final Storage Landfill" concept, and its emissions will be of considerable ecotoxicological relevance, rather than being negligible.

An ecotoxicological evaluation of a "Reactor Type Landfill" and a "Final Storage Landfill" should not only focus on the emissions but should also consider the physical, chemical and biological processes within the landfill (Fig. 1). A chemical and ecotoxicological examination of the emissions alone may be adequate for a preliminary regulatory decision, but it will not have much predictive power. A careful analysis of the physical (e.g., transport), chemical (e.g., hydrolysis) and biological (e.g., biotransformation, gas evolution) processes within the landfill,

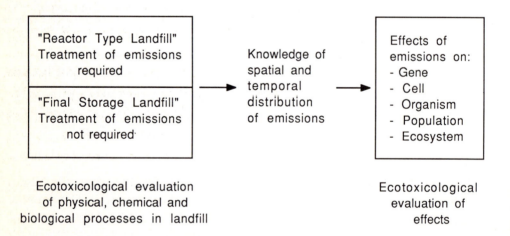

Fig. 1: Ecotoxicological aspects of landfills.

however, may allow the establishment of a model able to roughly predict the spatial and temporal behavior of the components within the landfill and of the emissions. Only a good knowledge of these processes will allow the establishment of a meaningful monitoring program and prevent regulatory agences from being taken by surprise by changing emissions.

2. Processes in landfills relevant for an ecotoxicological evaluation

2.1. Physical-chemical processes

Mass flow within and in the near-field of a landfill is not only determined by the hydrogeology but also by a variety of processes including evaporation, dissolution, advective and diffusive transport in gas and liquid (e.g. aqueous, organic solvent) phases, sorption/desorption processes, and chemical and biological transformation reactions. Many of these processes are interrelated and, therefore, have to be considered together. Most of these processes have been treated in earlier sessions of this workshop. This paper will only summarize a few aspects of biological transformations, since these reactions are of particular interest with respect to an ecotoxicological evaluation of a landfill.

2.2. Biological processes

Physiological and xenobiotic compounds in a landfill are subject to various types of biotransformations, such as redox reactions (oxidation and reduction), hydrolytic cleavage (e.g., cleavage of esters) and coupling reactions (e.g., conversions of anilines to azocompounds). From a microbiological point of view, redox reactions are crucial transformations since they yield energy and often produce physiological building blocks. Therefore, they support microbial growth in a landfill. Ecotoxicologically, redox reactions are also very important since xenobiotic compounds can be converted to highly toxic products (e.g., tetrachloroethylene to vinylchloride) or mineralized to harmless endproducts (e.g., aromatic hydrocarbons to CO_2 and H_2O). Conceptually, xenobiotic compounds can either be oxidized (serve as electron donors, Fig. 2A) or reduced (serve as electron acceptors, Fig. 2B). The oxidation of organic xenobiotic compounds is coupled with the reduction of molecular oxygen to water, nitrate to molecular nitrogen, sulfate to hydrogen sulfide and, eventually, carbon dioxide to methane (Fig. 2A). This sequence is based on thermodynamic considerations (6), and the emission of nitrogen, hydrogen sulfide, methane and hydrogen (a precursor of sulfate reduction and methanogenesis) from a landfill is therefore roughly predictable (7). Due to the stripping effect, these gases often contain high concentrations of hazardous volatile compounds like dichloromethane and vinyl

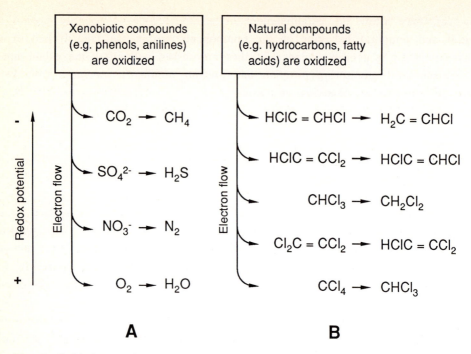

Fig. 2: Oxidation and reduction of xenobiotic compounds. A: Xenobiotic compounds (e.g., phenols, anilines) serve as electron donors and molecular oxygen, nitrate, sulfate or carbon dioxide serve as electron acceptors. B: Natural compounds (e.g. hydrocarbons, fatty acids) serve as electron donors and xenobiotic compounds serve as electron acceptors.

chloride (8). Certain groups of xenobiotic compounds can serve as electron acceptors in a landfill (Fig. 2B). For example, both the transformation of nitrobenzene to aniline and the reductive dechlorination of chlorinated aromatic and aliphatic compounds are electron-consuming reactions. In particular, the reduction of halogenated aliphatic compounds has been intensively studied over the last few years (9). It was found that the sequence according to which these xenobiotics are dechlorinated is dominated by the redox potential of the landfill and therefore, to a certain extent, is predictable. It is noteworthy that tetrachloroethane can rather easily be dechlorinated to vinylchloride at low redox potentials. However, vinylchloride is likely to be stable even under methanogenic conditions.

In summary, a careful examination of the content, the microbial activity and, in particular, the redox potential of a landfill will allow a qualitative prognosis of the type of emissions to be expected. A major problem associated with microbial transformations is the quantification of the turnover rates. In contrast to physical and chemical processes, biological processes depend partially on a number of undefinable factors such as adaption, evolution and predation which prevent an accurate quantification of the microbial activity. Biotransformation rates are very susceptible to changes in the environment (pH, humidity, temperature, availability of surfaces, etc.), and extrapolations of kinetic data measured in the laboratory to field conditions are therefore questionable.

3. Ecotoxicological evaluation of emissions from a landfill

3.1. Ecotoxicological evaluation of landfill emissions based on compound properties

An analysis of the compounds evolved from a landfill and a careful consideration of their specific characteristics (e.g., hydrophobicity, biodegradability, toxicity) may allow a preliminary hazard assessment. Additional data with respect to the temporal and spatial distribution of the compounds may even allow a rough risk assessment.

Several ways to conduct a risk assessment have been described, and two proposals will now be discussed in detail. A very common approach, mainly adopted by Germany and France, is based on a system in which various compound-specific properties are scored and weighted (Tab. 1, ref. 10). Higher scoring represents increasing risk; properties reducing the risk (e.g., ready biodegradability) are included in the system as negative scores. This approach is very feasible but can be misleading, since it integrates many diverse factors into a single value. The selection of the criteria (e.g., which toxicity tests are

Tab. 1: Simplified scheme of an ecotoxicological evaluation of a compound according to a scoring system. Modified from ref. 10.

Criteria	Score	Weights		
		Air	Water	Soil
Biodegradability	-2 to 0	0	1/2	1/2
Bioaccumulation	0 to 2	0	1/4	1/4
Skin irritation	0 to 1	1/4	0	1/4
Oral toxicity	0 to 4	1/2	1/4	1/1
Total score		-2 to 8.5		

included?), the valuation of the properties (is high bioaccumulation potential or high persistence more worrysome?) and the consideration of the distribution pattern (is a compound in the water more harmful than in the air?) is very arbitrary and hardly ever adequate to a specific situation.

A more sophisticated stepwise approach to evaluate emissions has been proposed by a research group of the Swiss Federal Office for Environmental Protection (Fig. 3, ref. 11). In this procedure, first a hazard assessment based on a few ecologically relevant data is made. In the second step, it is decided whether contaminations and adverse biological effects are likely to occur and a risk assessment is established based on a decision tree.

3.2. Ecotoxicological evaluation of landfill emissions based on biotests

Both the scoring approach (Tab. 1) and the decision tree (Fig. 3) rely entirely on a precise qualitative and quantitative analysis of the emissions, a prerequisite which is often not fulfilled. In many cases, the chemical composition of a landfill emission is largely unknown and bioassays are employed to evaluate its ecotoxicological risk. An analogous situation occurs with many municipal and

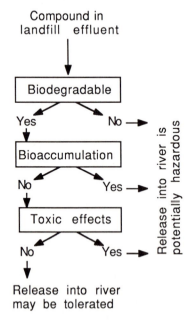

Fig. 3: Simplified scheme of a decision tree to evaluate whether a compound in a landfill effluent can be released into a river. Modified from ref. 11.

industrial effluents, and a biomonitoring (water quality-based) approach is currently being implemented by the United States Environmental Protection Agency to evaluate potential effects of such discharges on aquatic biota (12).

Biotests range from simple, standardized in vitro and in vivo tests (e.g., Ames test) to sophisticated studies in model ecosystems (Fig. 4, ref. 13,14). No test system, however, can meet all requirements. Simple tests are cheap, easy to employ and adequate for monitoring programs. They do not, however, consider the complexity of a whole ecosystem, and the data can rarely be extrapolated to field situations without providing large margins of safety in the interpretations (15). In contrast, model ecosystems can represent natural structures and processes to a certain degree, but their maintenance is time consuming and expensive and they are hardly usable for monitoring programs. In addition, the

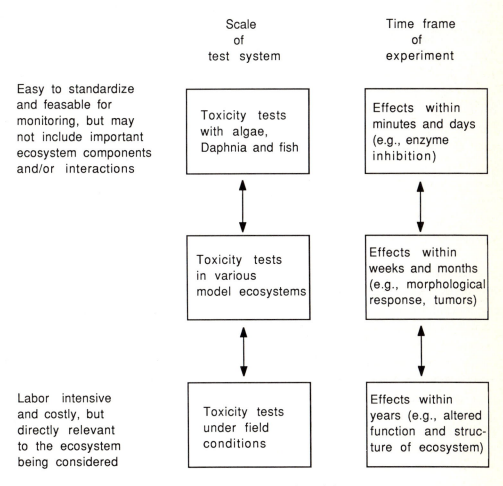

Fig. 4: Potential and limits of toxicity tests.

response time of biotests should be short in order to allow prompt regulatory decisions. Unfortunately, biotests designed to meet this requirement cannot indicate chronic effects that are only expressed within months or years (13,14). At best, an attempt can be made to use short-term tests whose results correlate well with chronic effects. Even then, problem compounds cannot be identified without complementary chemical analyses and/or fractionation studies of the emission.

3.3. Scientific and political considerations in evaluating landfill emissions

The hazard and risk assessment of a landfill relies to a large extent on known toxicity data (chemical compounds identified in emissions, Tab. 1, Fig. 3) or on results from bioassays (emissions mainly unknown, Fig. 4). Interpretation of these data in turn is either directly or indirectly based on a dose-effect relationship (Fig. 5, ref. 14). Each test system has an inherent statistical sensitivity, below which an applied dose does not generate a detectable response (no effect level, NOEL). Toxicity tests are usually performed at doses high enough to generate a statistically relevant effect within an appropriate exposure time. However, concentrations of hazardous chemicals found in landfill emissions are often very low and may have subacute or chronic effects that routine tests cannot detect (16). The relative sensitivity of toxicity tests needed to evaluate potential hazards may depend on whether effects on humans constitute a major part of the risk assessment.

Risk analyses for human exposure are usually concerned with low-level effects and are based on results from mammalian toxicity tests. However, mammalian assays for chronic effects can be insensitive and highly variable, and the extrapolation of data obtained at high doses to effects in the chronic range is associated with great uncertainities. Mathematical extrapolation models are often not justified by the quality of the data; and the model, rather than the data, may dominate the final conclusions. Assuming that a typical 2-year rodent carcinogenicity study would be performed with several hundred animals, then a cancer occurring at a frequency of 1 in 1000 would almost certainly not be detectable (16). In practice, however, a 0.1%-frequency translates into 6000 cases of cancer in the current Swiss population. Moreover, extrapolations from animal experiments to man, and in particular to special groups within an exposed population (babies, adults, individuals with kidney or liver insuffiencies), are questionable since differences in metabolism may bias the conclusions. Synergistic and antagonistic effects of the mixture of chemicals usually found in landfill emissions can further complicate the prediction of effects.

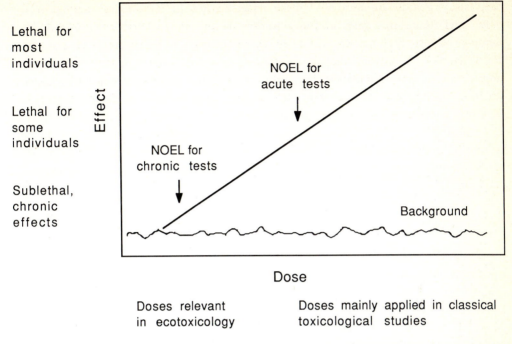

Fig. 5: Dose-Effect diagram. Modified from ref. 14.

NOEL: No effect level

On the other hand, risk analyses for aquatic and terrestrial (non-human) populations often are not directly concerned with low-level effects such as carcinogenesis and mutagenesis. Instead, the survival of a viable population at a specified location or within a specified region may be the management/regulatory goal. Unless gross deformities, tumors or illness are observable and objectionable to the general public, a 0.1% incidence of any lethal or sublethal effect can be ignored. Even 10% changes in population size usually go unnoticed and cannot be detected as significantly different from normal variability. Hence, chronic toxicity tests in which growth or reproduction of aquatic and terrestrial organisms are monitored generally require less extrapolation to the regulatory endpoint. Moreover, they are relatively sensitive and have relatively low variability. Unfortunately, test methods for terrestrial organisms are not as well developed as they are for aquatic organisms. And no matter how well characterized the test systems are, there are always uncertainties associated with extrapolations of toxicity tests from the laboratory to field situations and from single-species tests to population-, community- and ecosystem-level responses.

Regulatory boards are aware of these uncertainities in interpreting ecological test data, and therefore only a concentration below the NOEL (determined in a number of test systems) is considered as reasonably safe and taken as a criterion for legislation. However, whether a safety factor of 10^{-1}, 10^{-2} or 10^{-3} is considered appropriate is not a scientific but rather a political question which often depends on the economic and technical feasibility. Criteria also depend partially on whether a large population (population risk) or a limited number of individuals (individual risk) is exposed to the emission (17). Most landfills pose a risk to only a limited geographic area, and the population risk may be small. To what extent these considerations are taken into account is again mainly a political question.

4. Conclusions

Landfills are complex chemical and biological reactors whose internal processes are often beyond the immediate control of process engineers. Therefore, the concept of a "Final Storage Landfill" may be deceptive. Furthermore, traditional approaches to establishing discharge criteria and treatment requirements for industrial effluents may not work well for landfill emissions. Factories can often be treated as steady-state processes whose inputs and outputs are predictable; however, landfills are batch reactors whose contents and emissions may be unknown and will vary temporally and spatially. If the contents of a landfill are known, the sequence of chemical reactions can be predicted qualitatively. Even if that sequence is predictable, though, quantitative ecotoxicological criteria will be difficult to establish, and risk assessments based on chemical "laundry lists" will be questionable.

The situation is not hopeless, though. New approaches can be developed to monitor and predict landfill emissions. We believe these will include (1) testing (biological and chemical) of internal components of landfills as well as emissions; (2) development of laboratory and/or field methods in which the chemical and biological evolution of landfills can be studied at accelerated rates, thus allowing better prediction of future emissions; and (3) flexible ecotoxicological criteria that are adaptable to the evolving nature of landfill emissions. These criteria should be based on complementary chemical analyses and biological tests that fit into a hierarchical (decision-tree) hazard assessment strategy.

5. References

1. Federal law relating to the protection of the environment. 1983. Swiss Federal Office for Environmental Protection, Berne.

2. Sheehan P.J., F. Korte, W. Klein and P. Bourdeau. 1985. Appraisal of tests to predict the environmental behaviour of chemicals, Scope 25, John Wiley and Sons, Inc., New York.

3. Oggier P. 1987. Neue Deponie-Typen als Ausgangsprodukt für zukünftige gesetzliche Regelungen. Bulletin Bundesamt für Umweltschutz: 2, 12-18.

4. Heim T. 1986. Sonderabfälle: ein Diskussionsansatz. Bund für Naturschutz Baselland, Liestal, Schweiz.

5. Baccini P. und P.H. Brunner. 1987. Die Umweltverträglichkeitsprüfung von Entsorgungsanlagen. Eidg. Anstalt für Wasserversorgung und Gewässerschutz (EAWAG), Dübendorf, Schweiz.

6. Stumm W. and J. Morgan. 1981. Aquatic Chemistry. John Wiley & Sons, Inc., New York.

7. Rettenberger G. 1987. Gashaushalt von Deponien, p. 292-310, in Karl J. Thomé-Kozmiensky (Hrsg.), Deponie: Ablagerung von Abfällen, EF-Verlag für Energie- und Umwelttechnik, Berlin.

8. Wiemer K. und G. Widder. 1987. Emissionsminimierung und -kon-trolle bei der thermischen Deponiegasbehandlung, p. 639, in Karl J. Thomé-Kozmiensky (Hrsg.), Deponie: Ablagerung von Abfällen, EF-Verlag für Energie- und Umwelttechnik, Berlin.

9. Vogel T.M., C.S. Criddle and P.L. McCarty. 1987. Transformations of halogenated aliphatic compounds. Environ. Sci. Technol.: 21, 722-736.

10. Klein W., W. Kördel, D. Kuhnen-Clausen und Ch. Boose. 1986. Improvement of a procedure for the environmental hazard classi-fication of new chemical substances. Project 106 04 022, Fraunhofer Institut für Umweltchemie und Ökotoxikologie, Schmallenberg-Grafschaft, Deutschland.

11. Weber A. und H. Barben. 1984. Beurteilung ökologisch relevanter Daten neuer organischer Chemikalien. Chimia: 12, 443-456.

12. U.S. Environmental Protection Agency. 1985. Technical support document for water quality-based toxics control. Publication No. EPA 440/485-032, Office of Water Enforcement and Permits and Office of Water Regulations and Standards, Washington, DC.

13. Sheehan P.J. 1984. Effects on community and ecosystem structure and dynamics, p. 51-99, in P.J. Sheehan, D.R. Miller, G.C. Butler and P. Bourdeau (Eds.), Effects of Pollutants at the Ecosystem Level, Scope 22, John Wiley and Sons, Inc., New York.

14. Sheehan P.J. 1984. Effects on individuals and populations, p. 23-50, in P.J. Sheehan, D.R. Miller, G.C. Butler and P. Bourdeau (Eds.), Effects of Pollutants at the Ecosystem Level, Scope 22, John Wiley and Sons, Inc., New York.

15. Giesy J.P. and P.M. Allred. 1985. Replicability of aquatic multispecies test systems, p. 187-247, in J. Cairns (Ed.), Multispecies Toxicity Testing, Pergamon Press, New York.

352

16. Wilkinsen C.F. 1987. Being more realistic about chemical carcinogenesis. Environ. Sci. Technol.: 21, 843-847.
17. Travis C.C., S.A. Richter, E.A.C. Crouch, R. Wilson and E.D. Klema. 1987. Cancer risk management. Environ. Sci. Technol.: 21, 415-420.

Preamble

Everything we dispose of in landfills today may reappear in the biosphere. Ecotoxicological tests only are able to show whether a substance is harmful, but they cannot guarantee whether it is harmless.
Therefore the total amount of waste, and in particular the use of hazardous chemicals must be minimized.
The following conclusions are based on recent experience and should be
considered in future concepts of final storage.

Waste quality

How is the pollution potential of a landfill defined, and what type of
waste is unacceptable for final storage?

The pollution impact can be determined by performing a hazard assessment and a risk assessment. The hazard of a compound is given solely by
its chemical and physical properties, whereas the risk assessment also
includes the probability of a compound to be released into the environment around the waste body. These probabilities are mainly determined
by the barrier qualities. All toxic effects are concentration dependent
and we have very limited knowledge of the chronic effect of chemicals
at low concentration and extended exposure time. Therefore an increase
of pollutants over ambient concentrations has to be evaluated with
great care and possibly avoided. This goal requires that certain
chemicals must be excluded from the landfill input. It is essential to
have a precise knowledge of the landfill's content and its chemical,
physical and biological characteristics.
Even pretreated waste (e.g. after incineration) contains residues of
organic compounds. Some of these organic compounds are undesirable in
final storage landfills, because they can initiate redox reactions
(serving as electron acceptors or donors) or they can be transformed
into highly toxic compounds. An indication of the TOC or DOC content of
the waste is necessary but not sufficient, since it does not provide

any prediction of transformation products. Safety criterions for organic compounds should not only be based on their toxicity but should also consider their potential to turn into highly hazardous products.

The chemical composition of waste is very variable, as is the composition of natural rocks. In any case, wastes represent a geochemical anomaly compared to natural environments in one or more aspects. Particularly important is the concentration of heavy metals, which are known to pose a hazard in the biosphere. We envisage that waste with high heavy metal concentrations should be treated separately from waste with a composition similar to the earth's crust. The waste should be in thermodynamic equilibrium with its environment, in particular as insoluble in water as possible. Highly soluble salts such as sodiumchlorides should be removed before final storage.
For many harmful elements such as Cd, Hg, Pb, Ni, Zi, Cu, the lowest solubility is attained in a reducing environment, and therefore, this redox state is desirable. Such a redox state can possibly be guaranteed in a marine environment. In other geological environments, such as typically encountered in Switzerland, it might be difficult to maintain reduced conditions over extended time periods. Therefore we must also consider final storage of harmful elements in their oxidized forms. This poses a problem with regard to the reactor landfill, which must be expected to reach final storage quality under anaerobic conditions. We therefore conclude that waste emanating from reactor landfills and pretreated waste must be dealt with separately.

Site quality

What does a landfill final storage facility encompass?

The landfill final storage facility encompasses the waste body, the engineered barrier systems and the surrounding geosphere.

What are the geotechnical requirements on the landfill body?

The principal geotechnical requirements include adequate mass-stability, low settling (or differential compaction), low erodibility and high compacted density.

What is the purpose of the engineered barrier system?

The *primary* purpose of the liner is to collect leachate in order to check on the performance and quality of the waste body which itself represents the first barrier. A *secondary* role occurs solely in the concept of a multi-barrier system. As such this barrier has to limit release of pollutant leachate to the geosphere in the event of failure of landfill body quality. The role of the cap is to limit infiltration of water into the waste body, to maintain a specified physico-chemical environment (milieu) within the waste body, and to prevent uncontrolled gas emissions.

What is the role of the geosphere?

The geosphere should enhance the performance of the geotechnical barrier system (i.e. provide a good foundation, and be free from geological hazards). In the longterm, it should provide an additional margin of safety as a third barrier in the multi-barrier concept. As such its role is in dilution and attenuation of eventually released pollutants.

What should be the performance of the final storage, and how can it be evaluated?

The performance of final waste storage should be evaluated by measuring the leachate quality at the interface between waste-body and liner (i.e. in the primary leachate collection system). This leachate should be compared with local or regional groundwater quality standards and not exceed established limits. These limits should not be significantly higher than the local or regional groundwater quality standards and should be established considering ecotoxicological aspects. In any case we foresee the need for monitoring the emissions. This monitoring should not solely be designed to show whether the emissions meet quality standards, but should also allow to expand the knowledge of the ongoing processes.

Unanswered questions and future research

A number of questions relating the various aspects of final storage quality could not be discussed by our group because of existing knowledge gaps. These gaps are at least in part due to lack of available data, basic understanding of processes and to the short duration of the workshop.

We feel that the following points have to be clarified by future research:

1) Development of new pretreatment techniques for waste which would allow significant reductions in the TOC-level in incinerated waste.

2) Diagenesis (i.e. transformation of waste into a consolidated "rock-like" form by physical and chemical processes in the landfill)

3)Ecotoxicological tests on chronic toxicity at low concentrations and extended exposure, which encompass organism networks and not only single organisms.

4) Risk assessment studies for the longterm evolution of landfills. They should include the evaluation of pH, Eh, deterioration of caps and liners. These scenario studies must include variations in global climate and its consequences on groundwater quality and flow as well as erosion. Furthermore, they must take into consideration changes in society.

5) Techniques of monitoring the waste after disposal aimed at understanding the processes responsible for the observed changes.

6) Advanced hydrogeologic modeling on complex systems such as fluvial and glacial sedimentary environments. This should include the analyses of hydraulic conductivities and pathways in complex, anisotropic sedimentary strata.

METHODOLOGY FOR THE EVALUATION OF THE FINAL STORAGE QUALITY

Chairman:	Paul H. Brunner (CH)
Lecturers:	Hans-Peter Fahrni (CH)
	Chester W. Francis (USA)
	Gerald Milde (FRG)
Troublemaker:	Hans-Peter Müller (CH)
Rapporteur:	Jürg Krebs (CH)
Discussion participants:	Hervé Billard (F)
	Francino Filippini (CH)
	Peter Lechner (A)
	Jörg Messmer (CH)
	Dominique Stämpfli (CH)

SPECIFIC QUESTIONS:

1) What is "final storage quality"?

2) What are the main problems in choosing tests for waste characterization?

3) What are the relevant parameters in emissions from final storage deposits which are to be measured to meet the required safety in the ambient water (ground and surface water), air and soil (soil/dust)?

4) How are environmental impacts of landfills evaluated and reduced by remedial actions in order to attain final storage quality?

5) On which legal and organizational concepts waste fluxes can be oriented towards final storage quality?

METHODOLOGY FOR THE EVALUATION OF THE FINAL STORAGE QUALITY

Paul H. Brunner

Swiss Federal Institute for Water Resources and Water Pollution Control

One of the important and yet unresolved problems in waste management is the lack of a reliable methodology to qualify waste materials as suited for landfilling. Landfill operators as well as local authorities and waste producers ask for a simple, inexpensive and universal leaching test, which is to be applied uniformly to all waste materials and landfill sites. Based on the experience of the past (cf. the comprehensive review of existing leaching tests by Löwenbach, 1978), it will most likely never be possible to develop such an omnipotent test method. The variations in the properties of waste materials, landfills and landfill environments are so large, that individual, tailor made test methods seem more appropriate to qualify wastes as suitable for landfilling.

In the case of "final storage landfills", the problem of developing a test method seems less exorbitant because such a landfill will receive only selected, inert materials of similar properties. Indeed, in the future it may be one of the main advantages of the "final storage" concept, that wastes, which are treated to achieve "final storage quality" can be evaluated by reliable, appropriate and cost effective test methods, while it is doubtful that the fate of the same wastes, untreated and disposed of in a mixed reactor landfill, may ever be predicted for long time periods by any method.

What are the requirements for the development of a "final storage quality" test?

1. The "conditio sine qua non" is, that the term "final storage quality" is defined and preferably quantified. Since the difficult process of defining "final storage quality" opens up a new field in waste management and earth science and is still going on, it was necessary for

group B2 to agree quickly on a hypothesis for "final storage quality" and to proceed to address the main questions about the testing methodology. The working hypothesis for a final storage landfill (leachate concentrations in the order of magnitude of groundwater quality, cf. conclusions B2) proved to be useful for further discussions but should be taken as a working hypothesis only.

The definition of "final storage" has to include the practice of landfilling as well. Mixed waste landfills or mono waste landfills require different investigations of the leaching behavior of a material. Also the time period to be observed for "final" storage has to be specified in order to design leaching tests properly (kinetic investigations).

2. In future waste management, it will be necessary to have more information on the composition of the waste materials. It is a unique feature of waste processing that in this business biochemical, chemical and physical reactions are often applied to unknown substrates. As long as the substrates (wastes) are unknown, the quality of the products (residues from waste treatment) can not be guaranteed either. The concept of "final storage" is based on the knowledge of the composition of the input into the final storage. The first step of a testing method therefore should be the determination of the composition of the waste, be it by a laboratory analysis, or by a material flux analysis of the process which produced the waste material. It is important that the need for this information is regulated as well as the leaching assessment methods; the proposed waste ordinance of Switzerland (Fahrni) requires for the first time specific information on the composition of the matrix as well as the trace substances of a waste, and even limits certain substances (water soluble materials, organic carbon etc.) in final storage wastes.

3. Elements and compounds of concern in terms of leachate quality have to be selected. Again, it is important how "final storage" is defined: If the definition allows only wastes of known composition in final storage landfills, the list of potential leachate constituents will be orders of magnitude smaller than for unknown waste materials, where a priori every element or compound may be present.

There are several possibilities to select the "parameters of concern". One approach is to select certain metals and nonmetals which are representative for the behavior of others too, to analyze these substances in the lab leach tests, and to extrapolate to other elements and compounds not measured. An attractive approach has been taken by Milde and

Arneth: Their investigations into the leachate composition of many existing landfills allows them to identify those substances which are most abundant. And the combination of the abundancy of a substance with the physico-chemical parameters of this substance makes it possible to generalize the fate of many additional compounds with similar properties.

For legislative purposes, it might be necessary to have a uniform list of selected parameters (e.g. cf. Fahrni). Even if this list is quite long it cannot be sufficiently comprehensive for all possible waste materials; it will be necessary to add some kind of toxicity test as a safeguard.

In any case, the elements and compounds to be used for "final storage quality" control have to be selected carefully according to the individual situation (waste material, landfill type and site). It is expected, that in the near future parameters will be selected and published for many waste classes (filter ash from MSW incineration, bottom ash from MSW incineration, fly ash from coal power stations, electroplating sludges etc.). It must be pointed out that parameters may be selected much more efficiently and quicker if the wastes have been mineralized before disposal (in which case only traces of organic constituents are present).

4. When the composition of the waste is known and the (in terms of landfilling) relevant parameters have been selected, a test concept has to be developed. Many concepts for the leaching of waste materials exist (Löwenbach, 1978), but up to now no single method has been established as the best method. That it may never be a single test method becomes clear when Francis critical review of current methods is observed: It seems much more appropriate to apply chemodynamical principles to the interaction of the known composition of the waste material with the landfill environment (precipitation, atmosphere etc.), and to combine such considerations with additional laboratory experiments. In any case the kinetics of the reactions have to be taken into account since even very slow processes can become quite important for final storages with very long residence times. The concept for the "final storage quality" test should also take into account that leaching fluxes are needed to model the transport of substances from the landfill to the groundwater. Such transport models are being developed (e.g. Francis).

5. Investigations, which allow to compare field data from actual landfills (leachate concentrations, groundwater concentrations, gas flows

etc.) with the results of the chemodynamic considerations and the laboratory experiments are necessary to evaluate the chosen "final storage quality" test procedure. Eventually the testing procedure has to be improved.

6. Figures such as groundwater quality standards or background concentrations are needed in order to evaluate the figures found in the testing procedure and in the field test.

The six topics listed above are considered the main prerequisites when attempting to design a "final storage quality" test. It is out of the possibilities of a pure desk workshop to elaborate these topics and to develop the testing procedure. The main priority of group B2 therefore had to focus on a few questions only. Such questions concerned the definition of the "final storage quality", the selection of the relevant parameters, and the chief problems encountered when developing a leaching test. In addition, it was discussed how to determine most appropriately the impact of a final storage landfill on the environment, emphasizing the groundwater.

Reference

Löwenbach W. (1978), Compilation and Evaluation of Leaching Test methods, EPA/600/2-78-095, U.S. Environmental Protection Agency, Cincinnaty, USA.

METHODICAL GUIDELINES IN FEDERAL ORDINANCES TO ASSIGN WASTES
TO TREATMENT AND FINAL STORAGE

H.P. Fahrni
Bundesamt für Umweltschutz
Bern (Switzerland)

SUMMARY

The federal Office for Environmental Protection is actually preparing
an ordinance containing stringent criteria for the choice of disposal
methods and requirements for materials suitable for final storage. This
requirement shall guarantee that even untreated leachate of a final
storage respects the existing limits for waste water effluents and may
be discharged directly in surface waters.

In a final storage site only substances are allowed which do not react
with itself, with ambient air or water. To avoid slow biological fermentation organic chemicals are not suitable for final storage. Inorganic compounds shall not be soluble in water. For that reason the results of a leaching test are crucial.

1 The need for an improved standard in treatment and disposal of wastes

In Switzerland, the realization of urgently needed facilities for the
treatment and disposal of wastes is actually very difficult and is partially blocked up by the reigning mistrust of people. This is due to
the formerly practised disposal which focused on the lowest possible
expenses. Unfortunately the experience of some of our old hazardous
waste disposal sites shows that the cocktail of organo-chemical substances mixed up with fly ash and soluble salts leads to a heavily polluted leachate. This leachate has to be treated during the next centuries.

In 1986 the Federal Commission for Waste Management published principles for a waste disposal compatible with the environment [1]. These
guidelines demand that disposal systems should produce only substances
which can be recycled or materials that can be deposited in a final
storage.

Final storage means a disposal in a landfill that does not pollute the
environment, water, soil and air to an intolerable extend neither now
nor in the near or remote future.

2 Fundamental ideas of assigning wastes to treatment

The fundamental idea of treatment is to transform wastes into substances existing naturally. For this purpose inorganic matter, especially compounds containing heavy metals, have to be treated until
their properties are similar to those of a mineral or an ore. This
guarantees a low toxicity and a low solubility of these pollutants.

Salts of the metals sodium, potassium, calcium, magnesium with the anions chloride, sulfate, nitrate can be reused or may be discharged directly into surface water. As a matter of fact the flux of these salts
in waste water is ten to twenty times higher than the flux in solid
wastes [2].

Organic material, synthetic compounds derived from mineral oil, has to
be transformed in carbon-dioxide, water and alkaline salts of halogens.
This mineralization allows to transform organic material back to the
compounds which served as substrate in the formation of organic material in biological processes. In philosophical terms also solvents and
waste-oil can be recycled to natural substances. The mineralization can
be done in an incineration plant. Hazardous or toxic substances should
be burnt in high temperature incineration, e.g. a rotary kiln.

Well-developed and very efficient systems for the flue-gas treatment
allow to retain hydrochloride acid and dust containing heavy metals.
The emissions of sulphur dioxide and nitrogen oxides of an incineration
plant are comparable with those of a medium-sized industrial boiler.

3 Final storage quality

A final storage site shall not pollute ground water, ambient air and surface water even without any treatment of effluents and gas.

This condition can only be fulfilled if the future reactions of the disposed materials are known with great probability. Neither chemical processes due to changing redox-potential or due to changing pH-values nor biological fermentation should occur. So, reactive materials, biologically unstable materials and wastes which react either with water, air or other wastes, are not suitable for final storage.

Municipal waste should be incinerated and may only be disposed directly in landfills if there isn't sufficient capacity in incineration plants.

To examine wastes the prepared ordinance provides an extensive chemical analysis and leaching tests for special wastes.

The criteria for final storage of special wastes are the following:

1. The chemical composition has to be known up to the degree of at least 95 % of the waste.

2. Wastes must not contain more than 5 % of organic carbon.

3. Wastes must not contain more than 5 % of material soluble in water.

4. Wastes must not contain reactive materials (alkaline-metals, corrosive substances).

5. Wastes must not contain more than 10 mg per kg of organo-chlorine.

As the reaction with water will determine the loss of pollutants to natural environment, the results of a leaching test are most important.

There exists a big variety of other leaching tests. As we design a final storage site without contact to biological active materials we do not intend the use of tests with liquids with the character of percolates of a landfill for municipal waste.

There are wellknown leaching tests like the ULP (Uniform Leaching Procedure) proposed by the United States Environmental Protection Agency
[3]. This procedure which consists of 12 successive tests is for a frequent use too complicated and too lengthy.

Nevertheless the leaching test should have a certain time-accelerating
effect. For this purpose we intend to use as leaching agent either an
acidified solution or water, which is continuously saturated with carbon dioxide. The sum of the concentrations leachate of two successive
tests with a slightly acidified leaching agent must not exceed the
quality standards of the Ordinance for Waste Water Discharge [4].

Aluminium	1	mg/l
Arsenic	0,1	mg/l
Barium	5	mg/l
Lead	0,5	mg/l
Cadmium	0,1	mg/l
Chromium III	2	mg/l
Chromium IV	0,1	mg/l
Cobalt	0,5	mg/l
Copper	0,5	mg/l
Nickel	2	mg/l
Mercury	0,01	mg/l
Silver	0,1	mg/l
Zink	1	mg/l
Tin	2	mg/l
Cyanides	0,1	mg/l
Fluorides	10	mg/l
Nitrites	1	mg/l
Sulfites	1	mg/l
Chlorinated solvents	0,1	mg Cl/l
Ammonia	5	mg N/l
Lipophilic, non volatile chlorinated compounds	0,05	mg Cl/l
Total organochloro-pesticides	0,005	mg/l
Total organic carbon	200	mg/l
Hydrocarbons	5	mg/l
Phosphate	10	mg/l
pH-value (without acidified leaching agent)	6 - 11	

Table 1   First draft of proposed quality standards for leachate of
          wastes suitable for final storage

If waste water with quality of leachate demanded in this test is discharged into a river, the dilution by factor 10 to 50 is responsible for preventing damage in rivers and lakes. But the leachate of landfills should not be discharged directly into the groundwater bearing strata, because it is very difficult to quantify the tolerable pollution of ground water and to assess the dilution at a water point. In a final storage site for inorganic residues, the leachate has to be collected and measures have to be taken for the treatment of the leachate in case this should be necessary. But all the possible should be done to assure the quality standards mentioned above, such as careful evaluation of the site, impervious liners and efficient (speaking in long-terms) leachate collection systems.

The flux of metals in gases can be ignored, as mostly inorganic compounds are deposited.

4 The long term impacts of a final storage

We can conclude that there is no local problem if there is one final storage site in a big watershed area which respects the quality criteria mentioned above. But we have to take in account the annual disposed amounts of pollutants, specially of heavy metals.

Applying the best possible technique for the treatment of the wastes and for the pretreatment of the inorganic residues is no guarantee for completely stopping the flux of pollutants out of a final storage. It is essential to know whether the small remaining flux is essentially influencing the ecosystem in the near future as well as in the long run. If e.g. the flux of metals in the municipal wastes continues without changing over the next 200 to 400 years, we will have 10 to 40 times more metals in our landfills and thus we have to reckon with an environmental impact which is approximately 10 to 20 times higher than now. As preventive measure, a considerable reduction of the mobility of hazardous substances in landfills is therefore needed, even if the former and current practice have not yet resulted in serious damages. The guidelines for the waste management in Switzerland are based on these ideas. From a final storage (Endlager) only environmentally acceptable fluxes of substances should leach out.

Let us look at a landfill receiving all the filter ashes that are produced in Switzerland over 10 years.

Filter ashes from incineration plants for municipal solid wastes represent a concentrate of highly volatile heavy metals, such as cadmium and
lead. But main constituents are aluminium-oxides, silicates and
calcium-oxides [5], [6]. Organic hazardous substances are not essential
for the assessment of the ecological impact as they can be destroyed by
heating up to temperatures around 600 ^{O}C.

Our supposed landfill receives 400'000 tons of filter ashes that are
pretreated, together with 100'000 tons of cement. All this is deposited
as solidified material. Taking into account a density of 1.75 tons per
m^3, we calculate a volume of 285'000 m^3 which means a surface of 33'000
m^2, covered with wastes, 8.5 m high.

In our country we have to reckon with an annual precipitation of 1'000
mm, totalling 33'000 m^3 of rain. Let us assume that about 30% or 10'000
m^3 of the precipitation is percolating through the landfill. Under the
assumption that all this water is polluted up to the limit values mentioned above, only 5 kg of lead and 1 kg of cadmium is washed out annually. It is obvious that technical precautions, such as improved cover
(only about 10% percolation) and vertical drainage systems, help to reduce the calculated maximum load considerably.

Already a dilution by a small river with a minimum of 10 liters per
second is enough to reach acceptable concentrations. Therefore there
won't be any problems locally.

Regionally, one has to compare these fluxes with those of our rivers
that have not caused serious damages up to now.

Rhein and Rhône export annually more than 60'000 kg of lead and 6'000
kg of cadmium - the extra 5 and 1 kg respectively can really be ne-
glected [7]. Thus, even in larger regions we will not have problems
caused by the leachate of several final storage sites. This situation
allows us to maintain the fluxes of metals unchanged over the coming
centuries. But it is anyway very probable that in the long run efforts
to reduce and to recover substances will be successful, and thus the
flux of hazardous substances will gradually decline.

References

[1] Bundesamt für Umweltschutz: Leitbild für die Schweizerische
     Abfallwirtschaft, Schriftenreihe Umweltschutz, Heft Nr. 51,
     Bern, Juni 1986.

[2] H.-P. Fahrni: Stoffflüsse im Siedlungsabfall, Bulletin des
     Bundesamtes für Umweltschutz Nr. 2, 1986.

[3] United States Environmental Protection Agency (EPA): Guide
     to the Disposal of Chemically Stabilized and Solidified
     Waste. September 1982.

[4] Ordinance for Waste Water Discharge, December 1975.

[5] Bundesamt für Umweltschutz: Behandlung und Verfestigung von
     Rückständen aus Kehrichtverbrennungsanlagen, Schriftenreihe
     Umweltschutz, Heft Nr. 62, 1987.

[6] P. Baccini, P.H. Brunner: Behandlung und Endlagerung von
     Reststoffen aus Kehrichtverbrennungsanlagen, Gas-Wasser-
     Abwasser, 7, 1985.

[7] J. Zobrist: Die Belastung der Gewässer mit Schadstoffen aus
     Abwässern und Niederschlägen, Gas-Wasser-Abwasser, 3, 1983.

PHYSICAL AND CHEMICAL METHODS
FOR THE
CHARACTERIZATION OF HAZARDOUS WASTES

C. W. Francis[1], M. P. Maskarinec[2], and D. W. Lee[3]

OAK RIDGE NATIONAL LABORATORY[4]

Oak Ridge, Tennessee 37831

ABSTRACT

Numerous test methods have been proposed and developed to evaluate the hazards associated with handling and disposal of wastes in landfills. The major concern is the leaching of toxic constituents from the wastes. The fate of hazardous constituents in landfilled wastes is highly dependent on the physical and chemical characteristics of the waste. Thus, the primary objective in the selection of waste characterization procedures should be focused on those methods that gauge the fate of the waste's hazardous constituents in a specific landfill environment. Waste characterization in the United States has centered around the characteristics of ignitability, corrosivity, reactivity, and toxicity. The strategy employed in the development of most regulatory waste characterization procedures has been a pass or fail approach, usually tied to some form of a mismanagement scenario for that waste. For example, USEPA has chosen the disposal of a waste in a municipal waste landfill as a mismanagement scenario for the development of the waste leaching tests to determine the toxicity characteristic. Many wastes, such as large-volume utility wastes or mining wastes, are not disposed of in municipal waste landfills. As a consequence, more effort is needed in the development of waste leaching tests that determine the long-term leaching characteristics of that waste in the landfill environment in which the waste is to be disposed. Waste leaching models also need to be developed and tested as to their ability to simulate actual disposal environments. These models need to be compared with laboratory leaching tests, and, if practical, coupled with groundwater transport models.

---

[1] Environmental Sciences Division
[2] Analytical Chemistry Division
[3] Energy Division
[4] Operated by MARTIN MARIETTA ENERGY SYSTEMS, INC. for the U.S. DEPARTMENT OF ENERGY, INC. under Contract No. DE-AC05-84OR21400. Publication No. 3066, Environmental Sciences Division, ORNL.

INTRODUCTION

Reason for Physical and Chemical Characterization of Wastes

Acceptable disposal of hazardous wastes in landfills should be determined from the physical and chemical characteristics of the wastes themselves. The principal objectives in using the physical and chemical characteristics of wastes as suitability criteria for landfill disposal are to ensure that (1) the health and safety of the personnel who handle and dispose of the waste are not adversely affected, (2) the air and water that come in contact with the wastes are not polluted, (3) no significant impacts are made on the ecological balance at the site, and (4) the disposal does not seriously affect future reuse of the land over the long-term.

The importance of obtaining background information on the wastes under investigation cannot be overemphasized. This process includes determining the industrial process or processes likely to have generated the waste. First, such information is valuable in identifying potential contaminants. Second, the information aids in the characterization approach and, most importantly, alerts the investigator to potential safety concerns. Initial analyses should include hazard assessment checks by measuring pH, flash point, and reactivity. Spot tests should be conducted to determine the presence of cyanides, sulfides, and other potentially harmful constituents that may be emitted as gases or liquids during handling and analyses. Radioactivity checks should be made routinely as well. Initial analyses should also include total elemental analysis using a nondestructive technique such as X-ray fluorescence, inductively coupled plasma, or atomic absorption spectroscopic methods following sample dissolution by acid digestion or fusion procedures. The presence or absence of volatile and extractable organic compounds needs to be verified using gas chromatography and/or gas chromatography/mass spectrometry analyses. While these initial analyses identify potential contaminants, in many cases they are of limited value in determining whether a waste can be landfilled in a safe and environmentally acceptable manner. In certain instances, knowledge of simple physical characteristics such as low-percent solids and high concentrations of soluble contaminants known to be highly mobile in subsurface environments (e.g., nitrate) makes disposal in landfills a poor waste management practice. Recent criteria of acceptance for disposal of a waste in landfills have been based predominantly on the waste's leaching characteristics. This is especially the case in the United States.

Historical Aspects of Leaching Tests

One of the most comprehensive reviews of hazardous waste leaching tests was performed by Lowenbach (1978). This review was completed approximately two years after enactment of the Resource Conservation and Recovery Act (RCRA) of 1976, which regulated the management of hazardous waste. Consequently, the review revealed almost as many plausible leaching tests as there were people to run them. The methods employed and descriptions of approximately 31 leaching tests, including nine tests from individual states, were complied. Lowenbach evaluated each leaching test to determine both its relevance to those factors that control leaching in actual landfills and its potential with respect to regulatory interpretation. Most tests used distilled water as the leaching medium; however, several used either acetic acid or CO_2-saturated distilled water in an attempt to simulate municipal waste leachate generated in a sanitary landfill. All were "shake/batch" tests assumed to represent chemical equilibrium between the liquid and solid phases. None were dynamic column tests. The parameters found to be most significant in the leaching of constituents from waste were (1) the pH of the extraction medium, (2) the ratio of the extraction medium to the waste (liquid-to-solid ratio), and (3) the redox environment of the leach test. His first paragraph of the Summary and Conclusions section was most revealing of the task involved in the design of a waste leaching test. After ten years, his comments are still pertinent.

> The selection or design of any leachate test will ultimately be decided by a number of practical, rather than theoretical, considerations. It must be recognized at the outset, however, that a single test will not be optimal for all wastes. Nevertheless, from a regulatory point of view, developing different tests for each different waste is clearly impractical and probably unworkable.

One of the early leading waste leach tests was that developed at the University of Wisconsin under funding from the U. S. Environmental Protection Agency (USEPA), see Ham et al. (1978, 1979a, 1979b). Compared to the other leaching tests, its leaching medium, made from acetic acid, sodium acetate, ferrous sulfate, and pyrogallol, most accurately represented municipal waste leachate characteristics with respect to pH, redox potential, and buffering capacity. Its disadvantages were that the leaching medium was relatively unstable (i.e., the medium could not be stored for a week or more) and its organic complexing capacity was several orders of magnitude greater than that of natural leachates. Its most important detriment was the leaching medium's inherent toxicity in many of the biotesting protocols. This toxicity led to the test's nonacceptance by USEPA because one of the criteria in the development of a waste leaching test was its compatibility with biotesting protocol.

In 1980, USEPA promulgated the Extraction Procedure (EP) as a method to determine the toxicity characteristic of hazardous wastes required under subtitle C of the Solid Waste Disposal Act as amended by the RCRA of 1976 (USEPA 1980). The EP is a 24-h batch-type laboratory extraction procedure that uses acetic acid to acidify the liquid-to-solid suspension (20:1) to a pH of 5. Maximum concentration of acetic acid is 2 meq per gram of waste. Regulatory limits for EP toxicity characteristics were based on the concentrations of eight inorganic elements (arsenic, barium, cadmium, chromium, lead, mercury, selenium, and silver), four pesticides (Endrin, Lindane, Methoxychlor, and Toxaphene), and two herbicides (2,4-D and 2,4,5-TP Silvex) for which National Interim Primary Drinking Water Standards (NIPDWS) have been established.

The intent of the EP was to simulate the leaching action that an industrial waste would undergo if the waste were mismanaged by disposal in a municipal waste landfill. Acetic acid was used as a first-order approximation of the leaching by the low-molecular- weight carboxylic acids generated in an actively decomposing sanitary landfill. Acetic acid has been chosen as a simple and relatively accurate model of municipal waste leachate by other investigators. For example, acetic acid at a concentration of 5000 mg/L and adjusted to pH 5 with sodium hydroxide had been shown to simulate the leaching of metals from a synthetic metal hydroxide sludge by a 2-year-old municipal waste leachate (Wilson and Young 1983). The EP is further described in the next section, and complete details concerning the equipment and procedures required to conduct the test are provided in USEPA (1980).

The implementation of the EP as a regulatory waste leaching test was severely criticized by the industrial and scientific communities for several reasons. The most important ones were that the ability of the EP to simulate a real-world disposal environment had not been verified (Larson et al. 1981) and that solid waste leachates containing toxic constituents other than those listed above were not included in the toxicity criteria (i.e., the EP was optimized to evaluate the leaching of elemental inorganics rather than the leaching of toxic organic compounds). In 1984, the U.S. Congress amended RCRA, directing USEPA to make changes in the EP to ensure that it accurately predicted leaching potential and to identify additional characteristics of hazardous waste, including measures or indicators of toxicity.

In June 1986 (USEPA 1986a), USEPA proposed to replace the EP with another waste leaching test called the Toxicity Characteristic Leaching Procedure (TCLP). The proposed replacement of the EP with the TCLP was based on research conducted at Oak Ridge National Laboratory (ORNL) over a 3-year period. This research compared concentrations of selected inorganic and organic waste constituents in laboratory extracts with their respective concentrations in field-generated leachates that used actual municipal waste leachate as the leaching medium (Francis et al. 1984, 1986; and Francis and Maskarinec 1986).

The leaching data obtained to support the TCLP were obtained by filling large-scale "lysimeters" (1.8 m in diameter and 3.6 m in height) with approximately 1.5 Mg of municipal waste and adding distilled water to form an anoxic leachate. After a 6-month incubation period, this leachate was in turn used to leach potentially hazardous constituents from industrial wastes. The leaching of industrial wastes was conducted in two phases. Phase I, conducted in 1982, consisted of leaching four wastes (an incinerator ash, a solvent-production distillation bottom, a paint sludge, and an electroplating waste) for 79 days in the field. During Phase I, laboratory methods of waste extraction were compared; for example, rotary-batch and upflow-column methods at four liquid-to-solid ratios (2.5, 5, 10, and 20:1) were compared. Also compared were four extraction media: (1) pH 5 sodium acetate (0.1 $\underline{M}$ with respect to acetate), (2) a weak carbonic acid solution (CO_2-saturated deionized distilled water), (3) deionized distilled water, and (4) fresh municipal waste leachate from the lysimeters. Phase II, conducted in 1983, consisted of leaching seven wastes (a settling sludge, an ammonia lime still bottom, mixtures of an organic solvent waste with settling sludge and vermiculite, mixtures of an organic waste solvent with settling sludge and vermiculite, and fly ash) for 103 days.

Any constituent that exhibited a distinct peak in effluent concentration from the waste relative to its influent concentration in control municipal waste leachate was defined as a target constituent. The method of specifying concentrations for target constituents in the field leachates was determined by Kimmell and Friedman (1986). These field concentrations, called AMC20 concentrations, were the average maximum concentrations measured within a 20:1 liquid-to-solid leaching volume centered around the maximum leachate concentration.

Concentrations measured in laboratory extracts were compared with the field-measured AMC20 concentrations to evaluate which of the laboratory extraction methods best simulated field leaching conditions. Phase I results indicated that the best laboratory extraction method for simulating field leaching conditions was a rotary-batch extraction at a liquid-to-solid ratio of 20:1. The rotary-batch extraction method better simulated AMC20 field concentrations and exhibited much fewer operational difficulties than the upflow-column extraction method. Jackson et al. (1984) also concluded that the rotary-batch extraction method was superior to the upflow-column method because it offered greater reproducibility and a simpler design. A statistical comparison among the extraction media, using a multivariate analysis approach, revealed that for all wastes tested, the pH 5 sodium acetate buffer best simulated field AMC20 target concentrations. However, it is important to point out that there were significant interactions between wastes and laboratory extraction media, verifying Lowenbach's earlier conclusions that a single test is not optimal for all wastes (Lowenbach 1978).

The EP and TCLP are regulatory tests intended to give a pass/fail mark with respect to disposal of a waste in a nonhazardous landfill. Chemical equilibrium between the liquid phase and solid phase is assumed. Further characterization of leaching requires additional methods of investigation to identify important leaching parameters. Standardized methods for such characterization have not been developed. Developing such methods requires significant commitments in scientific personnel and funding. To date, priorities for research efforts in waste characterization have been established by federal and state authorities in an effort to increase the regulation of waste management practices.

In an effort to standardize physical and chemical characterization methods for solid wastes, members of industry, federal and state governments, and academia have worked together through the American Society for Testing and Materials (ASTM). One of the subcommittees (D-34.02 - Physical and Chemical Characterization) developed and issued a standardized leach test for solid wastes (D3987 - Shake Extraction of Solid Wastes with Water). Briefly, the test involves leaching 700 g of waste with distilled water at a liquid-to-solid ratio of 4:1 over 48 h (ASTM 1987). A column extraction procedure has also been proposed and is being evaluated in interlaboratory testing. Variations in the D3987 batch leaching procedure are also being investigated (e.g., operating the test in a sequential mode and adding soil to evaluate soil attenuation characteristics). Other tests being addressed are permeability of wastes and biodegradation potential of wastes.

The D-34.02 subcommittee has shown considerable interest in the development of test methods for solidified wastes. For example, what effects do wet/dry and freeze/thaw cycles have on the stability of these waste materials? Little research has been directed at assessing the characteristics of solidified hazardous wastes in landfill environments. It appears that considerable information could be procured by evaluating existing techniques to characterize the leaching of solidified radioactive wastes (Strachan et al. 1982). For example, the American Nuclear Society (ANS) has recommended a standardized leaching test (ANS-16.1) for solidified radioactive wastes (ANS 1986). The Nuclear Regulatory Commission (NRC) has promulgated regulations specifying minimum waste characteristics for nuclear wastes (NRC 1985). Tests for compressive strength, biodegradation, leachability, thermal degradation, and freestanding liquids are recommended. A recent conference (ASTM, in press) has addressed many of these issues, and, for this reason, they are not addressed in this report.

CURRENT REGULATORY TEST METHODS IN THE UNITED STATES

To assist generators of waste and operators of waste treatment and disposal facilities, USEPA has published a manual entitled "Test Methods for Evaluating Solid Wastes" (USEPA 1986b). Methods in this manual are those approved for obtaining data under the RCRA of 1976. The manual contains procedures used for collecting representative samples of solid waste and for determining the physical and chemical characteristics of wastes necessary to comply with RCRA regulations relative to handling and disposal. The manual, in its third edition, is composed of two volumes and is supplemented on a continuing basis. Volume I, which includes three sections, pertains to laboratory methods of analyses, and Volume II pertains to field methods. Some of the more important procedures, those used to identify the hazardous characteristics of the waste, are addressed in the following paragraphs.

Paint Filter Liquids Test (PFLT)

This test is used to determine the presence of free liquids in a sample of waste. Approximately 100 g of waste is placed on a commercially available conical paint filter (mesh No. 60). If any portion of the material passes through and drops from the filter within a 5-min test period, the material is considered to contain free liquids. Free liquids are those that readily separate from the solid portion of the waste under ambient temperature and pressure. USEPA procedures are intended to minimize the presence of free liquids in containerized waste to be disposed of in landfills. A common practice has been disposal of liquid waste in metal drums. Over time the drums corrode, leak, and then collapse because of overburden of the landfill. This often results in subsidence and enhanced movement of surface water into the landfill, thus accelerating groundwater contamination. The presence of hazardous liquids can dramatically affect the integrity of the landfill as well as increase the mobility of hazardous constituents leaving the landfill. For example, free liquids in waste can cause liner failure through a variety of chemical and physical mechanisms.

Liquids Release Test (LRT)

This test is designed to prevent the improper disposal of wastes that have failed the previously discussed PFLT. Also, in certain instances, the LRT is used to analyze wastes that have met the criteria for PFLT-generated freestanding liquids (a distinct liquid layer above the solid layer in the container) after transportation to the disposal site. Regulations passed in 1982 make it illegal to dispose of freestanding liquids in containers. Some owners or operators who landfilled containers exhibiting freestanding liquids began to treat the liquids in the containers by adding adsorbents of various kinds to pass the PFLT. Many of the adsorbents were biodegradable (e.g., ground corncobs, wood fibers, etc.) and were not structurally stable (e.g., free liquids were generated as the adsorbents were compressed). As a consequence, USEPA proposed (Federal Register, Vol. 51, No. 247, pp. 46824-46844, Dec. 24, 1986) that sorbents containing more than 1% organic carbon could not be used to solidify liquid wastes and that any admixture generated by adding such adsorbents would have to pass the LRT. Under this proposed regulation, the owner or operator of a landfill is required to determine if the generator of the waste has added a nonbiodegradable adsorbent (one whose organic carbon content is <1%) to a containerized liquid hazardous waste. If this is the case, the LRT is then used to test the waste.

The LRT is conducted using a modified version of the Zero-Headspace Extractor (ZHE) developed in conjunction with the TCLP to test the presence of volatile organic compounds in wastes. The ZHE uses gas pressure to force a piston against the sample to squeeze any free liquid from the sample onto adsorptive filter paper mounted against a perforated disk at the open end of the ZHE. Although the TCLP and the LRT use the same waste extraction device, the two tests are essentially unrelated. (The properties of the TCLP are described later in the text.) Single laboratory testing for ruggedness and precision and a multilaboratory collaborative study have been completed for the LRT. However, final protocol has not been published because the multilaboratory study has revealed the need for further development and for changes in procedure (Hoffman et al. 1987). Initial parameters include the pressurization of 100 g of waste at 50 psi for 30 min. Visible detection of any liquid squeezed from the waste represents failure to pass the test. Some suggested changes are using increased pressure to shorten the test times and using colored filter paper to improve visual detection of a liquid release. A second collaborative study is planned for 1988.

Ignitability

USEPA's objective in defining the ignitability characteristic was to identify both wastes either present fire hazards under routine storage, disposal, and transportation and those that can severely exacerbate a fire that has started. Regulatory definitions of wastes (including liquids, nonliquids, compressed gases, and oxidizers) are listed in 40 CFR 261.21. The test methods for determining ignitability are still in the formative stages.

Corrosivity

Under USEPA's definition, a waste exhibits a corrosivity characteristic when its disposal might pose a hazard to human health or to the environment because of its ability to (1) mobilize toxic metals if discharged into a landfill environment; (2) corrode handling, storage, transportation, and management equipment; and (3) destroy human or animal tissue in the event of inadvertent contact. These characteristics, defined in 40 CFR 261.22, are based on the pH of the liquid phase of the waste (pH <2 or >12.5 constitutes corrosivity) and the rate at which the liquid corrodes type SAE 1020 steel. (A corrosion rate in excess of 6.35 mm/year at 55°C constitutes corrosivity.)

Reactivity

Defining the reactivity characteristic was intended to identify those wastes that might be a hazard to human health or the environment because of their extreme instability and tendency to react violently or explode during handling or disposal. The regulatory definition of the reactivity characteristic has been established in 40 CFR 261.23 to identify wastes that (1) readily undergo violent chemical change; (2) react violently or form potentially explosive mixtures with water; (3) generate toxic fumes when mixed with water, or in the case of cyanide- or sulfide-bearing wastes, when exposed to mild acidic or basic conditions; (4) explode when subjected to a strong initiating force; (5) explode at normal temperatures and pressures; or (6) fits within the Department of Transportation's forbidden explosives; Class A explosives, or Class B explosives classifications.

Extraction Procedure Toxicity Characteristic

The Extraction Procedure (EP) is a waste leach test designed to identify the toxicity characteristic of a waste's leachate if the waste were disposed of in a municipal waste landfill. The EP was designed to simulate the leaching that an industrial waste would undergo if the waste were mismanaged by disposal in a municipal waste landfill. The liquid extract of the EP is analyzed for eight metals, two herbicides, and four pesticides. Regulatory levels of these 14 constituents in the extract were determined by USEPA using a generic dilution/attenuation factor of 100 times the National Drinking Water Standards to account for the movement of the leachate out of the landfill to groundwater.

The EP is used extensively to determine the toxicity characteristic of wastes. If the wastes contain a solid phase that is <0.5% of the waste, the solid phase is discarded and the solution phase is analyzed for the 14 regulatory constituents. The solid phase must pass through a 9.5-mm-diam standard sieve. If not, the solid phase has to be subjected to the Structural Integrity Procedure (SIP). The SIP tests the ability of the waste to retain its monolithic form after disposal. The test involves repeated dropping (14 times) of a 0.33-kg weight from a height of 15.25 cm onto a 3.3-cm-diam by 7.1-cm core of waste. If the waste does not remain intact, the waste must be ground to pass the 9.5 mm sieve prior to the EP.

At least 100 g of waste is extracted for 24 h with an aqueous medium at a 16:1 liquid-to-solid ratio whose pH is maintained at or below 5 with 0.5 $\underline{N}$ acetic acid. Maximum acetic acid that can be used is 4 mL of 0.5 $\underline{N}$ acetic acid per gram of waste. After extraction, the final waste suspension is diluted to a 20:1 liquid-to-solid ratio with distilled water. The suspension is then filtered through a 0.45-um (effective pore size) membrane filter prior to the analyses of the filtrate for the 14 regulatory constituents.

Toxicity Characteristic Leaching Procedure (TCLP)

The primary objective in the development of the TCLP was to address the leaching of organic compounds from landfilled wastes. In 1984, Congress amended RCRA directing the USEPA to make changes in the existing EP to predict more accurately the leaching potentials of hazardous wastes and expand its application to a greater number of toxic constituents. The TCLP was first published as part of the proposed Land Disposal Restriction (Federal Register, Vol. 51, No. 9, Jan. 14, 1986). Here, the TCLP was used to specifically evaluate the leaching of solvent- and dioxin-containing wastes destined for disposal in landfills. The final land disposal restrictions rule became effective July 8, 1987. On June 13, 1986, the TCLP was proposed to replace the EP by expanding the toxicity characteristic to include 38 additional compounds (Table 1) and establishing concentration limits by applying compound-specific dilution/attenuation factors generated from a groundwater transport model (Federal Register, Vol 51, No. 114). The final ruling is expected to be promulgated in May 1988.

The philosophical approach used in screening a waste prior to its leaching is significantly different for the TCLP than for the EP. For example, under the EP protocol it is not permissible to use total analysis of a waste as a prescreening test to determine if the waste contains levels of constituents that exceed regulatory levels. Such a prescreening test is allowable under the TCLP protocol. Also, the EP protocol allows the use of the SIP to evaluate the structural stability of monolithic wastes. The TCLP protocol, as currently written, does not include such a SIP; thus all wastes are required to be of particle sizes small enough to pass through a 9.5-mm-diam sieve (or a surface area > 3.1 cm^2/g) prior to extraction. Rationale for the omission of the SIP is based on the premise that physical stability of a waste in itself was not of sufficient merit. The strategy taken by USEPA was that parameters that occur in the natural weathering of monolithic wastes, such as the effect of wet/dry and freeze/thaw cycles, needed to be characterized as to their effect on the chemical stability of such wastes rather than simply relying on a structural integrity test.

The TCLP uses one of two acetic acid leaching media, depending on the alkalinity of the waste. Acetic acid was chosen as a candidate leaching medium because it is one of the more dominant carboxylic acids present in municipal waste leachate. A 2-year study involving the leaching of approximately 100 target constituents from 11 wastes under simulated

Table 1. Toxicity characteristics contaminants
and regulatory levels for the TCLP

Contaminant	Regulatory level	Contaminant	Regulatory level
	mg/L		mg/L
Acrylonitrile	5	Isobutanol	36
Arsenic	5	Lead	5
Barium	100	Lindane	0.06
Benzene	0.07	Mercury	0.2
Bis(2-chloroethyl) ether	0.05	Methoxychlor	1.4
Cadmium	1	Methylene chloride	8.6
Carbon disulfide	14.4	Methyl ethyl ketone	7.2
Carbon tetrachloride	0.07	Nitrobenzene	0.13
Chlordane	0.03	Pentachlorophenol	3.6
Chlorobenzene	1.4	Phenol	14.4
Chloroform	0.07	Pyridine	5
Chromium	5	Selenium	1
o-Cresol[a]	10	Silver	5
m-Cresol[a]	10	1,1,1,2-Tetrachloroethane	10
p-Cresol[a]	10	1,1,2,2-Tetrachloroethane	1.3
2,4-D	1.4	Tetrachloroethylene	0.1
1,2-Dichlorobenzene	4.3	2,3,4,6-Tetrachlorophenol	1.5
1,4-Dichlorobenzene	10.8	Toluene	14.0
1,2-Dichloroethane	0.4	Toxaphene	0.07
1,1-Dichloroethylene	0.1	1,1,1-Trichloroethane	30
2,4-Dinitrotoluene	0.13	1,1,2-Trichloroethane	1.2
Endrin	0.003	Trichloroethylene	0.07
Heptachlor	0.001	2,4,5-Trichlorophenol	5.8
Hexachlorobenzene	0.13	2,4,6-Trichlorophenol	0.3
Hexachlorobutadiene	0.72	2,4,5-TP (Silvex)	0.14
Hexachloroethane	4.3	Vinyl chloride	0.05

[a]o-,m-, and p-Cresol concentrations are added together and compared
with a treshold of 10 mg/L.

landfill conditions showed that an 18-h laboratory extraction procedure
using an acetate buffer (0.1 $\underline{N}$ with respect to acetate and a pH of 4.93)
as its leaching medium at a 20:1 liquid-to-solid ratio best duplicated
landfill leachate concentrations (Francis et al. 1984, Francis and Maskarinec
1986). This acetate buffer system (called the TCLP No. 1 extraction
medium) supplies 0.7 meq of acetic acid for each gram of waste extracted.
For those wastes whose alkalinity exceeds 0.7 meq/g, USEPA chose to use a
more highly acidic leaching medium because highly alkaline wastes were
not adequately represented·in the original 11 wastes tested. Also, other
leaching data on highly alkaline wastes indicated that leaching rates

were not significantly reduced after continued leaching with municipal waste leachate at liquid-to-solid ratios in excess of 20:1. Thus for wastes whose alkalinity exceeds 0.7 meq/g, the extraction medium is made by diluting 5.7 mL of glacial acetic acid with distilled water to a volume of 1 L (final pH of 2.88, called the TCLP No. 2 extraction medium). The milliequivalents of acetic acid used in this extraction medium are identical to the maximum allowed under the EP protocol (i.e., 2 meq of acetic acid for each gram of waste extracted).

One of the major benefits cited in the proposed rule for replacement of the EP with the TCLP was the improved operational aspects of the TCLP. For example, the EP involves continual pH adjustment with 0.5 $\underline{N}$ acetic acid to a pH of 5.0 $\pm$ 0.2 (in essence, a titration of the waste). For many wastes this requires considerable operational attention and has been a major source of variability between laboratories and operators, especially for the extraction of lead. Another source of variability is the long times required in the EP for filtering waste suspensions of highly dispersed, finely divided particulates through 0.45-um membrane filters. A 0.6- to 0.8-um glass fiber filter is specified in the TCLP. Some investigators have observed filtration times with the EP to be on the order of 5 h with as many as ten changes in filters as compared with filtration times of less than 1 h with no filter changes for the TCLP (Dhaliwal et al. 1987).

The TCLP also differs from the EP in specification of extraction vessels, method of extraction, and time of extraction. The EP was based on a 24-h extraction period as compared with the proposed 18-h period for the TCLP. The type of extraction vessel or method of extraction was never specified in the EP regulations. Two general extraction methods were considered to be acceptable under EP regulations: one was a stirrer-type method that used a small fanlike blade to mix the extraction fluid with the waste, and the other was a rotary-type method that involved tumbling closed bottles of the waste/extraction fluid in an end-over-end fashion. The TCLP specifically requires the rotary-type extraction method (a fixed agitation rate of 30 $\pm$ 2 rpm). The temperature during extraction is also specified to be maintained at 22 $\pm$ 3 °C.

The most significant change involved with the TCLP was the development of the ZHE vessel for the extraction of volatile contaminants from wastes. The ZHE allows the liquid/solid separation and extraction of wastes in the absence of headspace. The units are made of 316 stainless steel, with a 90-mm-diam filter mounted on one end. Up to 25 g of solids can be extracted in these units using the No. 1 TCLP extraction medium (volume capacity between 500 and 600 mL to accommodate a 20:1 liquid-to-solid ratio). The ZHEs, with minor modifications, are also to be used in the proposed LRT.

Multiple Extraction Procedure (MEP) and the Oily Waste EP

Both of these extraction procedures have been used by USEPA to delist individual wastes from specific generators. USEPA has recognized that generic listing of certain process streams may inaccurately classify individual wastes because of differences in processes, raw materials, and cleanup steps of individual generators. Thus, USEPA has established a delisting program where generators can demonstrate that their particular waste does not constitute a hazardous waste. In general, the delisting approach has used the EP protocol for metal contaminants with the application of a more conservative dilution/attenuation factor and/or a change in extraction medium.

The MEP is designed to simulate the leaching that a waste would be subjected to from repetitive precipitation of acid rain on an improperly designed sanitary landfill. It involves at least nine sequential extractions of the waste at a liquid-to-solid ratio of 20:1 with a pH 3 synthetic rainwater made from sulfuric and nitric acids. If, after completing the ninth sequential extraction, the tester finds the concentration of any of the constituents of concern is higher than that found in the seventh and eighth extractions, the waste is continually extracted until the concentration in the extract ceases to increase.

The EP for oily wastes is used to determine the mobile metal concentration (MMC) in these wastes. This process involves sequential extraction of the solid-phase sample with tetrahydrofuran and toluene in a Soxhlet extractor. The EP is then run on the dry residue. The original liquid, combined tetrahydrofuran and toluene, and the EP extracts are analyzed for the EP metals.

California Waste Extraction Test (WET)

Prior to the promulgation of the EP leaching test by USEPA, many states were using a variety of waste leaching tests to classify wastes. Most states now use the EP. California is an exception in that it has developed and is using its own Waste Extraction Test (California WET 1985). Operationally, the test differs significantly from the EP. First, the waste must pass through a 2-mm-diam standard sieve (as compared with a 9.5-mm-diam sieve for the EP). Second, the liquid-to-solid ratio is much lower than that used in the EP (10:1 as compared with 20:1 for the EP). The WET extraction solution consists of 2 meq of citric acid/citrate per gram of waste (0.2 $\underline{M}$ sodium citrate at pH 5) as compared with a maximum of 2 meq of acetic acid/acetate (pH 5) per gram of waste for the EP. In the WET procedure, the waste-extract suspension is purged with nitrogen gas to remove or exclude atmospheric oxygen from the extraction medium. The sealed suspension is then extracted for 48 h as compared with 24 h for the EP. Soluble Threshold Limit Concentrations (STLCs) of inorganic and organic persistent and bioaccumulative toxic substances are used to determine if the waste is classified as hazardous. For toxic metals, the WET is usually a much more aggressive waste leaching test than the EP or TCLP extraction medium (Table 2). Some investigators (Dhaliwal et al. 1987) have faulted the WET because it has some key operational problems and does not use an extraction medium that is meaningful with respect to simulating the leaching of a waste in a municipal waste landfill.

FUTURE NEEDS IN CHARACTERIZATION OF HAZARDOUS WASTES

Methods of Sampling and Sample Preservation

Extensive testing and characterization of wastes have become an integral part of the strategy for determining disposal options. Because of the costs involved in such testing, it is important to ensure that the sample collected is representative of the waste. A key issue is the use of composite samples. If many individual samples can be composited in the field, the work load of the testing laboratory and the associated cost are greatly reduced. Composite sampling is well understood from a statistical standpoint, but several analytical problems exist. If one contaminated sample is composited with a large number of uncontaminated samples, the net effect is to dilute the importance of that one sample. Thus, the analytical method used must be sufficiently sensitive so that the dilution does not result in a false negative. Attention must be given to the use of standard sample preparation techniques for ensuring homogeneity, a practice which should increase the sensitivity of the testing procedure by providing larger concentration factors. The major drawbacks to this approach that must be studied are the loss of information concerning single samples and the possibility that analytical background will prevent the further concentration required.

Table 2. Concentrations of toxic metals in waste leach extracts of resource recovery ashes[a]

Metal	Extraction method	Resource recovery ash (mg/L)				
		Blank	Chicago	Sumner	Hampton	Auburn
				mg/L		
Arsenic	WET	0.001	0.061	0.185	0.950	0.017
	EP	<0.001	0.003	0.009	0.017	0.002
	TCLP	<0.001	0.005	0.001	0.002	0.003
Cadmium	WET	0.0005	1.6	0.81	1.52	0.18
	EP	<0.0005	0.71	0.24	0.50	0.02
	TCLP	<0.0005	0.19	0.52	0.33	0.03
Chromium	WET	0.024	1.0	1.02	1.18	1.72
	EP	<0.003	0.03	0.042	0.035	0.011
	TCLP	<0.003	0.01	0.006	0.005	0.016
Copper	WET	0.022	1.90	0.041	0.05	211
	EP	0.006	1.67	6.04	0.642	4.3
	TCLP	0.003	1.06	0.15	0.212	4.2
Lead	WET	0.070	29	35	46	29
	EP	0.007	5.8	6.4	11	3.1
	TCLP	<0.003	0.50	0.28	1.6	4.2
Selenium	WET	<0.002	0.006	<0.002	<0.002	<0.002
	EP	<0.002	0.003	<0.002	0.002	<0.002
	TCLP	<0.002	0.004	0.005	0.007	0.004

[1] Data taken from Francis and White (1987). The resource recovery ashes were milled to <2 mm diameter as outlined by WET procedure. The TCLP extraction was conducted using No. 1 extraction medium (0.1 M NaOAc pH 5 buffer).

The second key issue is the maintenance of the sample integrity during sampling. Integrity is expected to be problematic for those classes of analytes subject to rapid change in concentration during sampling. A case in point is the class of volatile organic compounds. Information to date suggests that the concentration of these analytes in the sample changes almost immediately during sampling, containerization, and storage. Two possible solutions to this problem are proposed. The sampling system itself could be redesigned to eliminate these changes, taking into account variables such as temperature, encapsulation, and stabilization. Alternatively, the sampling could include the initial sample preparation--collection on a sorbent trap, for example--and therefore provide assurance that the sample is not compromised further during shipment and analysis. Since this class of analytes has potential for contaminating both air and groundwater, it is clear that sampling and testing procedures must be developed specifically for these analytes. As mentioned earlier, the ZHE portion of the TCLP is an attempt to address this issue in leaching tests. However, it is not clear that the sample can be collected and presented to this test in an accurate and unperturbed manner.

For many common contaminants, preservation techniques are recommended prior to analysis. This is usually appropriate for analytical tests but almost never appropriate for leaching and physical tests. Preservation of a waste sample for metals by acidification is clearly untenable if the sample is to be subjected to a leaching test. Little or no data are usually available on the stability of a particular waste sample. It is clear that major efforts are required in this area in the future if meaningful test results are to be obtained.

A final area in which much remains to be done is the development of reference materials for use in the evaluation of new testing procedures. The lack of such materials hinders the comparison of new test methods with those already established. Because of the diversity of waste forms, it may be impractical to expect that reference materials cover the universe of samples to be tested, but the availability of at least several well-characterized reference waste samples would greatly facilitate the evolution of new testing procedures.

Need for a Monofill Leaching Test

The strategy used in the development of the EP and TCLP leaching tests is based on a mismanagement scenario (i.e., codisposal of an industrial hazardous waste in a municipal waste landfill). USEPA has maintained that the codisposal scenario still represents a reasonable worst-case mismanagement scenario in spite of a large number of wastes being disposed of in monofill landfills (landfills in which only one waste of known physical and chemical characteristics is disposed). In the background document used in promulgation of the TCLP (USEPA 1986c), USEPA acknowledged that "industrial facilities dedicated to the management of only one waste, or the waste form from only one generator, are likely to pose less of a hazard than would general sanitary landfills, since the design and operation problems are simpler and the operator has much more information on the properties of the wastes before and while the facility is in operation." However, from the standpoint of defining the toxicity characteristic of hazardous waste, USEPA considered that the mismanagement scenario for the disposal of an industrial waste in a municipal waste landfill was most suitable. Thus, an acid leaching medium was selected. Also, in the case of the California WET, an acid medium (pH 5) and a strong chelator for metallic cations, citric acid, were selected as a leaching medium to model a worst-case scenario. These leaching media, however, are not relevant with respect to the modeling of leaching that might take place in a monodisposal scenario.

Research and development of a leaching test designed to characterize the applicability of disposing large-volume wastes in monofill landfills are needed. There are a number of reasons for the development of such a test (or battery of tests). In the first place, wastes, particularly alkaline wastes that fail the EP, TCLP, or WET, in many instances will fail to produce a toxic leachate if disposed of in a landfill environment that is nonacid generating. It is simply more economical to place these wastes in landfill environments whose characteristics (geochemistry and flow of water into the landfill) do not promote the leaching of the waste's toxic contaminants than to dispose of them in expensive high-maintenance hazardous waste landfills. Second, more information is needed about the long-term leaching characteristics of those wastes that are not classified as "hazardous" by such short-term leaching tests as the EP, TCLP, or WET. For example, because of their physical and chemical properties and the disposal method or characteristics of the site in which they are disposed some wastes not classified as hazardous may generate a leachate that can potentially contaminate surface water and groundwater proximate to the landfill. The primary objective of leaching tests should be the modeling or simulation of the leaching environment in which the waste is disposed.

Factors that need to be considered in developing a monofill leach test are (1) acid-generation capacity of the waste; (2) initial leaching characteristics in distilled water; (3) long-term leaching characteristic in distilled water (successive leaching); (4) influence of pH, Eh, and groundwater characteristics; (5) effect of leachate macroconstituents (those cations, anions, and organic compounds most prevalent in the waste leachate) on speciation and leachability of trace toxic constituents; (6) rates and mechanisms of dissolution (e.g., diffusion or solubility limited), and (7) the development of a general model descriptive of leaching of that waste in the selected disposal environment.

Development of Waste Leach Models

Output from most leach tests, especially those designed for regulatory compliance purposes, is usually limited to a single value (e.g., a concentration in a leach solution or concentration based on quantity of waste). These values are in turn used in environmental transport models to determine dose or risk of a particular constituent of a population. Little effort is made to assess the leaching characteristics of a waste over time or to determine the influence of this rate on dose or risk to the population in question. For many wastes, concentrations of contaminants in the waste leach solution are highly dependent on the liquid-to-solid ratio used in the leach test. For example, higher concentrations of leachable constituents are usually observed at low liquid-to-solid ratios than at high ratios.

To model the leaching of soluble constituents from the surfaces of a waste, a general model modified after that proposed by Chapelle (1980) is presented here. Proposed refinements in the model include the determination of total leachable constituents on the waste (the dependent variable) and the expression of the independent variable in terms of an effective liquid-to-solid ratio (instead of time as proposed by Chapelle). The model takes the following form:

$$A/A_o = e^{-kr},$$ (1)

where

A = amount of leachable constituent remaining on the waste (mg/kg),

A_o = total leachable constituent on the waste (mg/kg),

B = a partition coefficient,

r = liquid-to-solid ratio (i.e., the ratio of the volume of leachate solution to the mass of waste in units of L/kg).

To estimate the quantity of A_o, the waste is leached over a range of liquid-to-solid ratios so that the quantity of leachable material can be determined as a function of liquid-to-solid ratio (see Fig. 1). This can be described mathematically by

$$A_o - A = B_o(1 - e^{-Br}).$$ (2)

The coefficients B and B_o can be estimated by nonlinear regression analysis using the leaching data over a range of liquid-to-solid ratios (see Fig. 1 where the smooth curve is fitted to the leaching data). Thus, at high liquid-to-solid ratios, the quantity of leachable constituent remaining on the waste (A) approaches zero, and A_o approaches B_o ($A_o = B_o$).

Experimentally, A_o and B can be determined by extracting the waste at liquid-to-solid ratios of 5, 10, 20, and 50. In this manner, the total quantity of leachable constituent (A_o) can be determined as well as the distribution of the constituent between the liquid and solid phases (expressed in terms of the partition coefficient, B, whose units are kg/L).

The usefulness of the model can best be appreciated by a consideration of its application to actual experimental data. For example, the data illustrated in Fig. 1 represent nickel concentrations in the column leaching of fly ash from a coal-fired power plant using distilled water as the leaching medium (Turner et al. 1983). The high linear regression coefficient (R^2 = 0.964, P > 0.001) indicates a good fit between the model [Eq. (1)] and the experimental data. To further test its usefulness, the model was used to determine values of A_o and B for other elements and wastes (Table 3).

The experimental determination of values for A_o and B is much easier to accomplish by batch rather than column experiments. Many wastes have low permeabilities making it time consuming and difficult to conduct column extraction tests. Experiments have indicated that batch and column leach tests are generally equivalent (Van de Sloot et al. 1982, and Jackson et al. 1984). Values of B and R^2 determined from batch extractions for sulfate (B = 0.0049 and R^2 = 0.999) and boron (B = 0.0183 and R^2 = 0.966) indicate that the model is equally applicable (Van der Sloot et al. 1982, Elseewi et al. 1980).

The model assumes that leachable constituents desorb from the wastes in proportion to the concentrations on the surface of the waste. This assumption probably explains why the simple leaching by water of soluble constituents from residuals of coal combustion and coal gasification, as illustrated in Table 3, fits the model so well.

This type of leaching model appears to be applicable to the leaching of wastes in a monofill disposal scenario. However, when leaching rates are dependent on the diffusion of constituents from monolithic waste forms or when dissolution from the waste form is dependent on pH and the

ORNL-DWG 87-17370

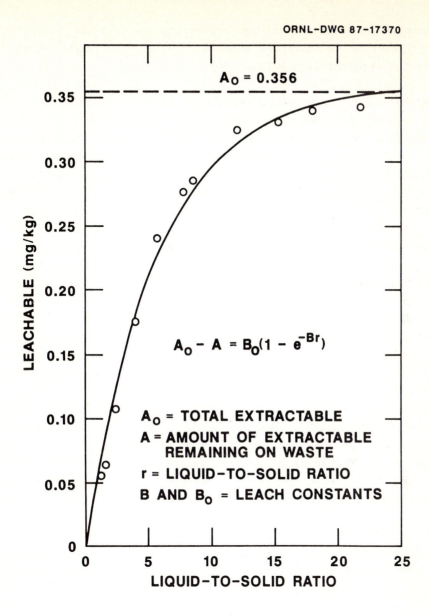

Fig. 1. Monofill leach model.

pH of the leaching medium changes significantly over time (e.g., that observed in a codisposal scenario of industrial wastes in a municipal waste landfill) then the model is not likely to be applicable. Also, constituents whose aqueous concentrations are regulated by the solubility limit of a solid phase would not expected to conform to the model. Other factors, such as biodegradation of the leaching species or changes in redox potentials, which affect the solubility or rate of dissolution and/or precipitation of chemical species from or onto waste surfaces, also negate the applicability of the model.

In certain cases, namely the aqueous leaching of wastes disposed of in monofill landfills, output from the model could be a useful source term to describe the quantity and rate of a contaminant available for leaching. For example, in an analytical sense, if the volumetric rate of water movement through a landfill is known, use of the B value at $A/A_O = 0.5$ can be used to estimate the time at which one-half of the leachable constituent is leached from the waste. Relative differences in the time required to leach specific fractions of the leachable component of various constituents can also be estimated [e.g., it takes approximately 40 times longer to leach the leachable component of aluminum than that of boron from a gasifier ash (see Table 3)].

Coupling of Leaching Models with Groundwater Transport Models

Physical and chemical characterization of hazardous wastes provides an understanding of the leachates that can be generated after disposal. The understanding gained has obvious applications to the regulation of waste disposal, development of improved waste forms and design of disposal facilities. These applications benefit from the capability of modeling the generation and transport of contamination in the groundwater regime with some confidence. Developing this confidence is a major task facing the physical and chemical characterization of wastes. Many issues, ranging over several disciplines, must be resolved before the ultimate fate of hazardous contaminants in the environment can be modeled accurately. A prominent issue is the coupling of the model for the leaching of hazardous wastes with the model for the transport of contamination in the groundwater regime.

The modeling of the transport of contamination in groundwater is based on a thorough understanding of the hydrodynamics of groundwater. Groundwater motion is generally modeled with Darcy's Law, an empirical rule that has been successful in establishing the gross behavior of groundwater. The geologic properties needed for applying Darcy's Law are often hard to determine with accuracy and precision because of the difficulty and costs of obtaining data. Field data, which generally are viewed to be more accurate, can vary from laboratory results by as much as two orders of magnitude. With limited sampling and large uncertainties in the collected

Table 3. Parameter estimates and fit to leaching model
for constituents leached from wastes

Constituent	Model parmeters		Regression coeffecient
	A_O	B	R^2
	mg/kg	kg/L	
Coal utility ash (Turner et al. 1983)			
Barium	6.9	0.05	0.999
Calcium	6700	0.12	0.966
Nickel	0.36	0.15	0.964
Coal gasifier ash (Turner et al. 1983)			
Barium	2.9	0.08	0.973
Calcium	23	0.47	0.981
Cadmium	0.05	0.05	0.975
Iron	8.3	0.52	0.928
Nickel	9.6	0.37	0.976
Coal gasifer ash (Francis et al. 1982)			
Alumimum	14	0.009	0.989
Boron	30	0.38	0.971
Calcium	690	0.13	0.983
Potassium	240	0.10	0.978
Magnesium	84	0.10	0.977
Nickel	1.5	0.20	0.977
Silica	250	0.07	0.943
Sulfate	2000	0.07	0.943

data, the modeling of groundwater hydrodynamics is a substantial undertaking. Since contaminant transport is largely dependent on groundwater motion, the modeling of contaminant transport is even more ambitious. Establishing the predictive skill of contaminant transport modeling in groundwater regimes with known source terms is an area of active research. The results of this research are anticipated to provide the necessary techniques for determining the transport of contamination in groundwater with a high degree of confidence.

The development of the modeling skill needed for predicting the transport of contamination from known sources is made more difficult when the generation of leachate from solid wastes is considered. For the generation of leachate by solid wastes to become a known source for the purposes of transport modeling, the concentration of the leachate generated by the solid wastes needs to be known as a function of time and space in the disposal environment. Additionally, the flux of the leachate from the disposal unit into the groundwater regime needs to be established as a function of time and space in the environment used for disposal. Once the flux and concentration have been established, the geochemical behavior of the leachate as a function of time, space, and concentration in the disposal environment needs to be established. This level of understanding is well beyond the current level incorporated into existing test methods for defining the physical and chemical characteristics of hazardous solid wastes. In spite of the difficulty, initial attempts at coupling leaching models with groundwater transport models have been performed.

Historically, the coupling of leaching processes with groundwater transport has focused on specifying the retardation of contaminants in soils with the source term determined by the solubility limit of the contaminant. Improvements in the understanding of the thermodynamics of geochemical environments have allowed for a more comprehensive understanding of the equilibrium geochemistry of contaminants in soil. Source-term models have typically considered the generation of leachate by measuring the release fraction of contamination as a function of time for various leaching configurations. Recent attempts to fit the formation of leachate to a first-order decay in concentration as a function of the liquid-to-solid ratio show some promise in further resolving the mechanisms responsible for the formation of leachate in specific environmental situations. While progress is being made beyond merely providing a simple pass/fail leaching test for characterizing solid waste, much remains to be done before a comprehensive leaching model can be proposed that meets the needs of a source term for a groundwater transport model.

As part of improving the present understanding of leaching processes and their relation to groundwater transport of contamination, some future needs can be identified to be addressed as the understanding of leaching processes improves. Most important, from the viewpoint of modeling, is separation of the effects of time and space. Some leaching models have attempted to address the variation of leachate concentration in time. The use of the liquid-to-solid ratio as the independent variable is an initial attempt to resolve the spatial distribution of waste and water within a disposal unit. In the actual disposal environment, both the effects of the spatial distribution of contamination and the kinetics of leachate generation occur. Attempting to quantify these variations is a formidable but necessary task for characterizing the physical and chemical nature of hazardous wastes. In addition to more rigorously determining the mechanisms of leachate generation, the demonstration of predictive skill in the transport of contamination from solid wastes is necessary. Even for simple geologic settings, predictive skill in leachate generation and transport is largely unknown.

The existing physical and chemical methods for the characterization of hazardous wastes provide some insight into leachate formation and useful techniques for classifying hazardous wastes. In spite of the advances made, formidable challenges need to be overcome before reliable models of leachate generation and transport are available. In the dynamic, heterogeneous world of a hazardous waste disposal unit, the hydrodynamics have not been resolved.

SUMMARY AND CONCLUSIONS

Health and environmental impacts resulting from final disposal of wastes in landfills depend on the physical and chemical characteristics of the wastes in the landfill environment. The dominant concern is leaching of toxic constituents from the wastes, but losses via the gaseous phase should not be underestimated. Numerous test methods have been proposed and developed to evaluate the hazards associated with handling and disposal of wastes in landfills. Waste characterization in the United States has centered around the characteristics of ignitability, corrosivity, reactivity, and toxicity as defined by the RCRA of 1976. Consequently, nearly all research has been associated with the development of pass/fail test methods meant to be used in assessing regulatory compliance.

Perhaps the most controversial characterization effort has been the development of waste leaching tests to determine the toxicity characteristic of a leachate that would be generated on disposal of a waste in a landfill. The evolution of three very important criteria is essential for the acceptance of such a test method. First, the disposal scenario needs to be defined. Second, the methods and materials used for conducting the test need to be developed based on that scenario, and, third, the criteria used for determining regulatory limits also need to be developed. For regulatory characterization, the disposal scenarios adopted are often mismanagement oriented. For example, USEPA in the development of the EP and TCLP leaching tests has chosen disposal in a municipal waste landfill. Many wastes, such as large-volume utility wastes or mining wastes, are not disposed of in municipal waste landfills.

Research concerning and development of a leaching test to characterize the applicability of disposing of large-volume wastes in monofill landfills are needed. More effort is needed to evaluate long-term leaching characteristics of a waste under its disposal environment. The significance of pH, Eh, and characteristics of the groundwater on the leaching of waste contaminants should be more fully understood. More emphasis should be placed on understanding the kinetics and mechanisms responsible for the dissolution of the waste than do the regulatory test methods now available. Research characterizing leaching mechanisms and rates at actual landfills needs to be conducted. Waste leaching models need to be developed and tested as to their ability to simulate actual disposal environments. These models need to be compared with laboratory leaching tests and, if practical, coupled with groundwater transport models.

REFERENCES CITED

American Nuclear Society (ANS). 1986. American national standard measurement of the leachability of solidified low-level radioactive wastes by a short-term test procedure. ANSI/ANS-16.1. American Nuclear Society, LaGange Park, Illinois.

American Society for Testing and Materials (ASTM). 1987. Shake extraction of solid wastes with water. Method D3987. Section 11. Water and environmental technology. Vol. 11.04. Pesticides; Resource Recovery; Hazardous Substances and Oil Spill Responses; Waste Disposal; Biological Effects. American Society for Testing and Materials, Philadelphia.

American Society for Testing and Materials (ASTM). Proc., Fourth International Hazardous Waste Symposium on Environmental Aspects of Stabilization/Solidification of Hazardous and Radioactive Wastes. American Society for Testing and Materials, Philadelphia (in press).

California Waste Extraction Test (WET). 1985. State of California Administration Code. Title 22. Social Security Division. 4. Environmental Health, Section 66700.

Chapelle, F. H. 1980. A proposed model for predicting trace metal composition of fly-ash leachates. Environ. Geol 3:117-123.

Dhaliwal, B., J. P. Snyder, and P. Mendoza. 1987. Comparison of the Environmental Protection Agency's Toxicity Characteristics Leaching Procedure and California Waste Extraction (WET). Paper presented at a Specialty Conference: Analytical Techniques and Residuals Management in Water Pollution Control, Los Angeles, June 1-2, 1987.

Elseewi, A. A., A. L. Page, and S. R. Grimm. 1980. Chemical characterization of fly ash aqueous systems. J. Environ. Qual. 9(3):424-428.

Francis, C. W., and M. P. Maskarinec. 1986. Field and laboratory studies in support of a hazardous waste extraction test. ORNL-6247. Oak Ridge National Laboratory, Oak Ridge, Tennessee.

Francis, C. W., and G. H. White. 1987. Leaching of toxic metals from incinerator ashes. J. Water Pollut. Control. Fed. 59(11):979-986.

Francis, C. W., W. J. Boegly, R. R. Turner, and E. C. Davis. 1982. Disposal of coal conversion solid wastes. J. Environ. Eng. Div. ASCE. 108:1301-1331.

Francis, C. W., M. P. Maskarinec, and J. C. Goyert. 1984. Mobility of toxic compounds from hazardous wastes. ORNL-6044. Oak Ridge National Laboratory, Oak Ridge, Tennessee. National Technical Information Service (NTIS). 1985. PB 85-117-034. National Technical Information Service, Springfield, Virginia.

Francis, C. W., M. P. Maskarinec, and J. C. Goyert. 1986. A laboratory extraction method to simulate codisposal of solid wastes in municipal waste landfills. pp. 15-35. In J. K. Petros, Jr., W. J. Lacy, R. A. Conway (Eds.), Hazardous and Industrial Solid Waste Testing: Fourth Symposium. ASTM Spec. Tech. Pub. 886. American Society for Testing and Materials, Philadelphia.

Ham, R. K., M. A. Anderson, R. Stanforth, and R. Stegmann. 1978. The Development of a leaching test for industrial wastes. pp. 33-46. In Land Disposal of Hazardous Wastes. Proc., Fourth Annual Research Symposium. EPA-600/9-78-016. Municipal Environmental Research Laboratory, U.S. Environmental Protection Agency, Cincinnati.

Ham, R. K., M. A. Anderson, R. Stegmann, and R. Stanforth. 1979a. Background study on the development of a standard leaching test. EPA-600/2-79-109. Municipal Environmental Research Laboratory, U.S. Environmental Protection Agency, Cincinnati.

Ham, R. K., M. A. Anderson, R. Stegmann, and R. Stanforth. 1979b. Comparison of three waste leaching tests. EPA-600/2-79-071. Municipal Environmental Research Laboratory, U.S. Environmental Protection Agency, Cincinnati.

Hoffman, P. A., R. S. Truesdale, P. F. Overby, and M. B. Meyers. 1987. Further development of the liquid release test. pp. 4-7. In Proc., Third Annual Symposium on Solid Waste Testing and Quality Assurance. United States Environmental Protection Agency, Office of Solid Waste and Emergency Response, Washington, D.C. 20460.

Jackson, D. R., B. C. Garrett, and T. A. Bishop. 1984. Comparison of batch and column methods for assessing leachability of hazardous waste. Environ. Sci. Technol. 18:668-673.

Kimmell, T. A., and D. Friedman. 1986. Model assumptions and rationalebehind the development of EP-III. pp. 36-53. In J. K. Petros, Jr., W. J. Lacy, R. A. Conway (Eds.), Hazardous and Industrial Solid Waste Testing: Fourth Symposium. ASTM Spec. Tech. Pub. 886. American Society for Testing and Materials, Philadelphia.

Larson, R. J., P. G. Malone, T. E. Myers, and R. A Shafer. 1981. Evaluation of the extraction procedure testing of hazardous industrial wastes. pp. 139-150. In R. A. Conway and B. C. Malloy (eds.), Hazardous Solid Waste Testing: First Conference. ASTM Spec. Tech. Pub. 760, American Society for Testing and Materials, Philadelphia.

Lowenbach, W. 1978. Compilation and evaluation of leaching test methods. EPA-600/2-78-095. Municipal Environmental Research Laboratory, U.S. Environmental Protection Agency, Cincinnati.

Nuclear Regulatory Commission (NRC). 1985. Waste Classification and Wastes Characteristics. 10 CFR 61.55 and 10 CFR 61.56.

Strachan, D. M., R. P. Turcote, and B. O. Barnes. 1982. MCC-1: A Standard leach test for nuclear waste forms. Nucl. Technol. 56:306-312.

Turner, R. R. and C. W. Gehrs, and P. D. Lowry. 1983. Comparison of solid wastes from coal combustion and pilot coal gasification plants. EPRI EA-2867. Electric Power Research Institute, Palo Alto, California.

U.S. Environmental Protection Agency (USEPA). 1980. Identification and listing of hazardous waste. In Environmental Protection Agency Hazardous Waste Management System. 40 CFR 261.24.

U.S. Environmental Protection Agency (USEPA). 1986a. Identification and listing of hazardous waste. In Environmental Protection Agency Hazardous Waste Management System. Federal Register 51(114):21648-216893.

U.S. Environmental Protection Agency (USEPA). 1986b. Test methods for evaluating solid wastes. SW-846. Third ed. United States Environmental Protection Agency, Office of Solid Waste and Emergency Response, Washington, D.C.

U.S. Environmental Protection Agency (USEPA). 1986c. Toxicity Characteristic Leaching Procedure (TCLP), Resource Conservation and Recovery Act Subtitle C - Hazardous Waste Management System, Section 3001, Identification and Listing of Hazardous Waste. PB87-154886. United States Environmental Protection Agency, Office of Solid Waste and Emergency Response, Washington, D.C.

Van der Sloot, H. A., J. Wijkstra, A. Van Dalen, H. A. Das, J. Slanina, J. J. Dekkers, and G. D. Wals. 1982. Leaching of trace elements from coal solid wastes. ECN-120. Netherlands Energy Research Foundation, Petten, Netherlands.

Wilson, D. C., and P. J. Young. 1983. Testing methods for hazardouswastes prior to landfill disposal. pp. 69-81. In C. W. Francis and S. I. Auerbach (Eds.), Environment and Solid Wastes: Characterization, Treatment, and Disposal. Butterworths, Boston.

WASTE DEPOSIT INFLUENCES ON GROUNDWATER QUALITY
AS A TOOL FOR WASTE TYPE AND SITE SELECTION
FOR FINAL STORAGE QUALITY

Jan-Dirk Arneth, Gerald Milde, Helmut Kerndorff and Ruprecht Schleyer
Institut für Wasser-, Boden- und Lufthygiene des Bundesgesundheitsamtes
Berlin

ABSTRACT

Leachates from deposits of wastes may, in the long run, adversely influence groundwater quality. Since tipping still constitutes the most important form of waste disposal, strategies must be developed which are capable of protecting groundwater against contamination from leachates. In the first instance such protective measures must provide for a minimization of contamination by setting up optimal barriers. Since it would seem difficult to reach this goal in a forseeable future, the avoidance of substances with a high potential for groundwater hazards has to be attributed much importance.

In former times, little attention was given to impermeability or avoidance of substances with a high potential for groundwater hazards contained in wastes. Therefore, results of the investigation of groundwater near abandoned sites can be used to optimize groundwater protection on future tipping sites. In the present study, the results of chemical investigation of groundwater from the vicinity of 92 waste disposal sites in the Federal Republic of Germany are presented and the changes in groundwater quality owing to the penetration of leachates are discussed separately for inorganic and organic contaminants.

The level of the contamination factor may serve to estimate the potential of a substance to become mobilized from wastes. Also the detection frequency of substances in groundwater permits the recognition of this potential. Substances characterized by a high contamination factor and/or a high detection frequency are defined as main contaminants. A comparison with similar studies made in the USA shows a good coincidence of results and in particular a predominance of volatiles in groundwater.

The contamination factor and the detection frequency are closely related to mobility, persistence and accumulation potential of substances which can be described by their physico-chemical properties, and they describe the groundwater currency of a substance. The potential groundwater hazard is not known before an additional evaluation of the toxicity and hygienic relevance of the substance has been made. Substances involving a high potential groundwater hazard are defined as priority pollutants. In the future, priority should be given to reduce their presence in wastes to be disposed. According to present knowledge, this refers to arsenic, cadmium, lead, chromium and nickel from the number of inorganic substances, and to tetrachloroethene, trichloroethene, cis-1,2-dichloroethene, benzene, vinyl chloride, trichloromethane, 1,1,1-trichloroethane, xylenes, trans-1,2-dichloroethene, toluene, ethylbenzene, dichloromethane, dichlorobenzenes, chlorobenzene and tetrachloromethane from the organic ones.

Since, for a variety of reasons as e.g. heterogenity of material, control of tipping, it does not appear to be possible to avoid completely these substances in wastes, there should be an additional protection of groundwater by a suitable selection of future sites, taking into account present and future uses of the groundwater in this area, as well as by an optimal sealing of tips.

INTRODUCTION

The annual production of wastes in the Federal Republic of Germany amounts to ca. $250*10^6$t. Wastes are of heterogeneous nature and may include comparatively harmless types such as construction wastes as well as dangerous material such as chemical wastes with a high potential environmental hazard. While a minor portion of wastes is incinerated or recycled, the major share is disposed of by tipping. It should be the objective of a responsible final storage to ensure that hazardous substances cannot, on the long run, penetrate into the environment (water, soil and air).

In the past, this was hardly given any attention. Before the Federal Waste Disposal Act came into force in 1972, wastes of all types were deposited, sometimes even without any authorization and control, in former gravel, sand and clay pits, abandoned quarries and other artificial or natural land depressions. At present, some of these abandoned waste sites constitute one of the most severe environmental problems, in particular due to the penetration of leachates into groundwater. Thus, the prime objective of any final storage of wastes should be a general protection of groundwater against leachates, by way of a selection of optimal sites and a maximal impermeability of tips. Today, this is already accounted for prior to tipping operations by developing tips on naturally impermeable ground (clay, loam etc.) and also by artificial sealing of the tip bottom with clay, plastic or bituminous material etc.

There is only limited knowledge at present on the longterm behaviour of natural or artificial sealing of the bottom of sites with respect to final storage. A longterm (> 100 years) and complete barrier effect would, at the present state of art, seem to be rather improbable as adverse influences from "sealed" deposits have already been demonstrated in groundwaters.

For this reason, it is of great importance to minimize potentially hazardous substances in wastes. In order to do so, it will be necessary to know these substances and their behaviour within the path waste-soil-groundwater. Substances found in the groundwater downstream of abandoned deposits are following this path. It can be expected that in the case of improperly sealed future deposits, above all, these substances are likely to penetrate into groundwater. Thus, the presence of the toxicologically relevant ones among these substances will have to be preferentially minimized in wastes to be disposed of.

Since both a complete impermeability and a complete avoidance of substances hazardous to groundwater in refuses would principally seem to be improbable, areas should be selected as sites for future refuse tips where there is no contaminable groundwater or where a use of groundwater is not envisaged.

1. DATA BASE

For the purposes of detecting and listing cases of groundwater contamination from abandoned waste deposits, extensive studies were conducted by the Institute for Water, Soil and Air Hygiene of the Federal Health Office (Kerndorff et al., 1985; Arneth et al., 1986; Brill et al., 1986). As part of these studies, groundwater in the vicinity of 35 waste deposit sites was systematically examined for the presence of certain inorganic and organic groundwater constituents. Meanwhile, the authors have studied another 27 abandoned sites. Available results from several authorities in the Federal Republic of Germany have also been included in the statistical evaluation, to provide a representative picture of the situation. The ensuing evaluation of a total of 92 sites (Fig. 1) has not yet come to an end and gives evidence of the present state of knowledge.

In accordance with the rough geological structure of the territory of the Federal Republic, the sites may be classified as follows: Unconsolidated tertiary and pleistocene sediments - 72; Mesozoicum - 16; Pre-varisticum - 4. This grouping was taken into account when evaluating the inorganic groundwater constituents as only groundwaters from unconsolidated tertiary and pleistocene sediments have been examined. With respect to the organic substances all 92 sites were evaluated.

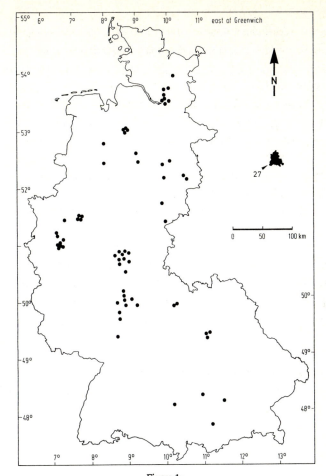

Figure 1:
Locations of 92 waste sites in the Federal Republic of Germany and Berlin (West) evaluated in this study

2. INORGANIC GROUNDWATER CONSTITUENTS

Leachates from waste deposit sites contain a number of inorganic substances which also may be present in groundwater for geogenic reasons. In order to recognize such substances as contaminants, their natural, i.e. non-anthropogenic concentrations (background values) must be known.

The left part of Table 1 shows the mean geogenic concentrations (i.e. those not being influenced by waste deposits) of inorganic groundwater constituents and their variation range upstream of 33 sites in unconsolidated tertiary and pleistocene sediments. They refer to the most important type of aquifer and the most important reservoir for drinking water in the Federal Republic of Germany. A comparison of mean and maximal concentrations with the maximal acceptable concentrations (MAC) stipulated by the Drinking Water Regulation of the Federal Republic of Germany (BMJFG, 1986) reveals that the largest portion of concentration values is clearly below the MACs and thus gives evidence of the hygienic

		uncontaminated groundwater					contaminated groundwater					CF
		min	mean	max	std.dev.	n	min	mean	max	std.dev.	n	
HCO$_3$	mg/l	203	214	227	12.1	3	145	439	2260	356	42	2.1
SO$_4$	mg/l	12.1	72.5	138.5	38.6	18	0.8	223	1630	303	90	3.1
Cl	mg/l	3.2	27.3	63.1	16.2	18	14.7	243	3680	504	90	8.9
NO$_3$	mg/l	<0.5	6.6	31.4	11.5	18	0.5	173	11500	1242	90	26.2
NO$_2$	mg/l	<0.2	<0.2	1.3	0.3	17	<0.5	0.4	7.4	1.2	72	>1.8
F	mg/l	<0.1	<0.1	0.12	0.07	3	<0.1	0.26	0.73	0.2	42	>2.6
Br	mg/l	<1	<1	<1	---	15	<1	<1	5	0.7	46	1.0
PO$_4$	mg/l	<0.5	<0.5	<0.5	---	17	<0.5	1.7	75	9.7	68	>3.4
CN	µg/l	<10	<10	<10	---	2	<10	<10	20	4.0	27	1.0
K	mg/l	0.6	1.0	1.2	0.3	3	0.7	27	330	65	40	27.0
Na	mg/l	<5	9.6	21	6.4	18	<5	140	1520	246	90	14.6
Ca	mg/l	<5	90	138	32	18	22	185	840	151	90	2.1
Mg	mg/l	<2	8.1	21	5.4	18	<2	36	252	22	90	4.4
NH$_4$	mg/l	0.12	0.19	0.25	0.09	2	<0.05	11.8	153	32	27	62.1
Fe	µg/l	546	6990	24200	7835	17	<10	10400	172000	23600	81	1.5
Mn	µg/l	19	498	1350	481	18	<10	2020	24700	3750	89	4.1
Sr	µg/l	<40	147	430	104	18	<40	465	6350	745	90	3.2
As	µg/l	<2	3.2	25	6.3	17	<2	109	1930	360	70	34.1
Pb	µg/l	<2	3.4	27	7.3	17	<2	17	336	55	76	5.0
Cd	µg/l	<0.2	<0.2	0.6	0.14	16	<0.2	1.3	19	3.6	74	>6.5
Ni	µg/l	<5	9.4	16	5.6	17	<5	28	330	49	89	3.0
Zn	µg/l	10	36	95	24	17	<10	170	2620	417	82	4.7
Cu	µg/l	<5	5.8	43	10.9	18	<5	27	373	64	90	4.7
Cr	µg/l	<5	<5	8.0	2.5	18	<5	25	1710	180	90	>5.0
B	µg/l	<20	<20	27	11.5	18	<20	1270	31200	4420	90	>63.5

Table 1:

Influence of waste sites on groundwater quality indicated by concentrations and statistical parameters of inorganic constituents in uncontaminated and contaminated samples

unobjectionable quality of the groundwater. This state of affairs is clearly supported by the fact that the mean concentration of toxic trace elements such as As, Pb, Cd, Ni, Cr as well as CN and NO_2 are either below the detection limit or far below (by a factor of about 10) the respective MAC of the Drinking Water Regulations. Comparatively high geogenic concentrations of iron (35*MAC) and manganese (10*MAC) do not jeopardize the statement that these groundwaters are of an excellent quality.

The extent of the impairment of an originally good groundwater quality by leachates from refuse deposits is depicted in the right-hand part of Table 1, where the concentrations downstream of 33 sites examined have been listed. There are obviously elevated mean concentrations for all inorganic parameters. The degree of the influence of waste deposit leachates on groundwater is shown even more clearly by the so-called contamination factor (CF) which can be defined as the ratio of the mean concentrations between the contaminated and the uncontaminated groundwater. It is shown in Figure 2 for the parameters discussed here and classified by its magnitude.

The CF does not provide any information on the absolute increase of the concentration of substances in groundwater, i.e. no statement can be made as to the pollutant load. Rather, it is a relative value depending on the geogenic background, the substances contained in the wastes and their mobility. The highest CF is to be expected for a substance being present in natural, non-influenced groundwater in low concentrations only, occurring in the waste in relatively high concentration and having a high mobility. Inorganic substances with CFs of >2 are defined as inorganic main contaminants. Of a total of 21 such substances (Fig. 2), boron, ammonium and arsenic are most prominent, with respective CFs of 63.5, 62.1 and 34.1. There is a variety of reasons for substances to exhibit a low CF. The most frequently found one is their high geogenic

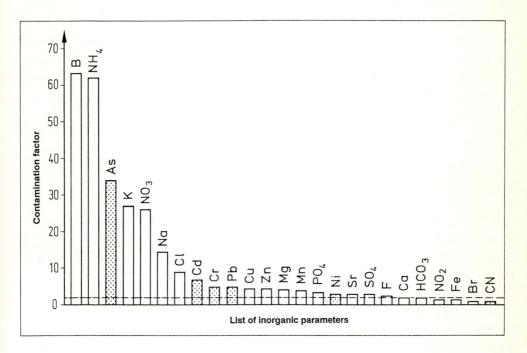

Figure 2:
Ranking of inorganic groundwater constituents based on their contamination factor
(Dotted: Toxic elements, Group D, EC-Directive)

concentration (e.g. calcium, bicarbonate and sulphate) as a considerable increase of their concentration in groundwater will be necessary to arrive at high CF values.

3. ORGANIC GROUNDWATER CONSTITUENTS

Many organic substances found in contaminated groundwater are exclusively anthropogenic in origin. Thus, on the contrary to the inorganic substances, their mere presence downstream of an abandoned waste disposal site indicates an influence of the leachate on the groundwater.

3.1 FREQUENCY OF DETECTION

So far, about 1000 organic substances have been identified in contaminated groundwater downstream of abandoned waste deposits. Their occurrence gives evidence of their migration with leachates from waste into ground-water. It is impossible in practice to routinely analyse for all these substances, so that there must be a further specific limitation of their number. Experiences made in the USA (Plumb Jr., 1985) and in the Federal Republic of Germany have shown that only a few organic substances are frequently, and most others are rarely found in groundwaters. Substances found in groundwater with a certain frequency can be considered as primary groundwater contaminants and are defined as organic main contaminants. These substances should be the first ones to be included in the list of substances to be analysed. Their number will depend on the limit to be set in terms of frequency of detection. On the one hand, most organic groundwater contaminants should ideally be accounted for, on the other, their number should be such that work on them is feasible. This can be achieved only by a suitable selection. A consideration of e.g. all substances detected in more than 1% of all groundwater samples will result in a list of about 60-70 substances. A list of about 100 organic substances will result at a limit asking for 0.2% frequency of detection. The essential factors for such a decision would be the numbers of statistical items, i.e. the number of sites examined, the number of analyses performed and the number of organic substances examined, which are considered to determine the detection frequency.

The following results are based on an evaluation of groundwater analyses for organic substances downstream of 92 waste disposal sites in the Federal Republic of Germany and Berlin (West) (Fig. 1). Only those substances have been considered for which more than 100 analytical results are available. In Table 2, these 15 organic main contaminants have been listed ranked by their detection frequency.

Owing to the different detection limits for the individual substances due to apparatus-specific and substance-specific parameters, a concentration of 1 µg/l has been defined as "positive proof". By far the most frequently detected substances are tetrachloroethene and trichloroethene with detection frequencies of 41.5% and 38.8%, respectively. Only three more substances (cis-1,2-dichloroethene, benzene, vinyl chloride) exceed the 10% limit. The other 10 substances exhibit a frequency of detection between 1 and 10 per cent.

A comparison of the detection frequency and of the statistical parameters for the concentrations found does not reveal any associations. This applies to the mean as well as the median and maximal values (Tab. 2). Thus trichloroethene, having a detection frequency of 38.8%, exhibits mean and maximal values which exceed those for tetrachloroethene

main contaminants	Federal Republic of Germany						USA			
	rank	analyt. attempts	detection frequency in %	mean	median	max	rank	detection frequency in %	mean	max
				------µg/l--------					-----µg/l-----	
tetrachloroethene	1	359	41.5	16	2.5	600	2	36.0	9700	22000000
trichloroethene	2	353	38.8	1200	3.5	130000	1	51.3	3800	800000
cis-1,2-dichloroethene	3	237	20.3	21000	75	410000	-	-	-	-
benzene	4	171	13.5	300	11	5200	14	11.2	5000	1200000
vinyl chloride	5	113	12.4	2700	475	12000	15	8.7	-	52000
trichloromethane	6	324	7.1	200	2.7	2800	4	28.4	-	220000
1,1,1-trichloroethane	7	312	5.4	40	3.0	270	7	18.9	-	620000
xylenes	8	250	5.2	*	*	20	**	8.2	-	30
trans-1,2-dichloroethene	9	113	4.4	*	*	100	3	29.1	-	75000
toluene	10	171	4.1	*	*	10	12	11.6	5200	1100000
ethylbenzene	11	136	2.9	*	*	12	**	5.8	-	25000
dichloromethane	12	180	2.8	*	*	500000	6	19.2	11000	7800000
dichlorobenzenes	13	114	2.6	*	*	70	**	1.9	-	-
chlorobenzene	14	114	1.8	*	*	300	**	5.5	-	13000
tetrachloromethane	15	309	1.3	*	*	2	**	7.0	-	20000

- no data available   * statistical evaluation not applicable   ** not represented among the 15 main contaminants

Table 2:

Ranking of 15 organic main-contaminants in the Groundwater of the Federal Republic of Germany based on detection frequency in comparison to results of the USA (Plumb Jr., 1985, 1987; Plumb Jr. & Pitchford, 1985) (detection frequency 41.5%) by two orders of magnitude. However, the similarity of the median values (2.5 and 3.5 µg/l) indicates that the high mean value for trichloroethene has to be attributed to a few extremely high concentrations.

Similar studies on organic substances in groundwater from the vicinity of abandoned waste deposits have been carried out in the USA (Plumb Jr., 1985, 1987; Plumb Jr. & Pitchford, 1985). Since these results constitute the only comparable set of data available at present, they are to be compared with the data from the Federal Republic of Germany discussed in the foregoing (Tab. 2, right side). However, one difference between both data sets has to be accounted for: The high number of analyses evaluated in the USA (in some cases > 10000) refers to a "mere" 358 abandoned sites. This reveals the fact that data from a monitoring program conducted over many years have been included in the statistical evaluation and that samples were taken repeatedly from each well. In the Federal Republic, however, samples were taken once from about 3 wells per site.

The 15 substances most frequently found in groundwater (main contaminants) are depicted separately for the Federal Republic and for the USA (Fig. 3). Chemically different groups of substances are marked by different hatching.

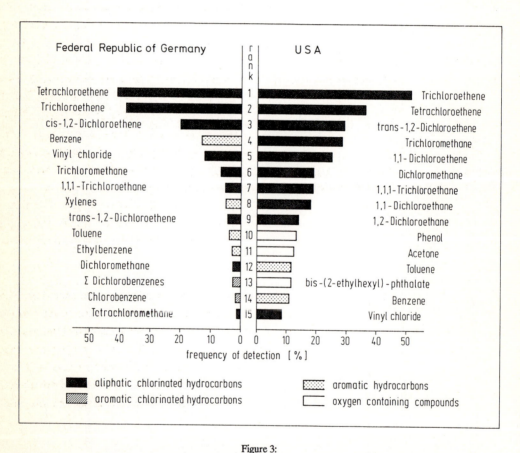

Figure 3:
The 15 most frequently detected organic contaminants in groundwaters downsteam of waste sites in the Federal Republic of Germany (92 sites) and the USA (358 sites) (Plumb Jr. & Pitchford, 1985)

The predominance of aliphatic halogenated hydrocarbons becomes evident from a comparison of both studies: There are 9 in the Federal Republic and 10 in the USA, of which 7 are identical. Another striking fact is that in both studies, although in a different sequence, tetrachloroethene and trichloroethene are the most prominent organic substances, with a frequency of detection of more than 35 per cent. Among the most frequently occurring 15 substances, the monoaromatic hydrocarbons constitute the second most frequent group of substances in groundwater in the Federal Republic with the halogen-free ones (benzene, xylenes, toluene, ethylbenzene) being more frequent than the halogenated ones (chlorobenzene, dichlorobenzenes). In the USA, this group of substances is represented by toluene and benzene only. Instead, the spectrum of substances detected in groundwater is enlarged by phenol, acetone and bis-(2-ethylhexyl)phthalate. As a result, 9 organic substances are common in both lists. It shall be pointed out that, except for tetrachloroethene, there are generally higher detection frequencies in the USA than in the Federal Republic. A reason for this may be the fact that the American results also have included repeated samplings.

Despite the difference of the data sets for both countries, a good coincidence of detection frequencies has been documented at least for the 15 most common substances and the feasibility of the method is demonstrated. By enlarging the data base for the Federal Republic, it will be possible in the future to elaborate a statistically confirmed list of main contaminants for the groundwater downstream of tipping sites in the country. These substances shall then be preferentially studied for purposes of evaluation of groundwater contaminations.

As mentioned above, such an extensive data base exists for the USA (Plumb Jr., 1985, 1987; Plumb Jr. & Pitchford, 1985), so that results beyond the 15 most frequent substances are available. The 136 substances contained in the American prio-

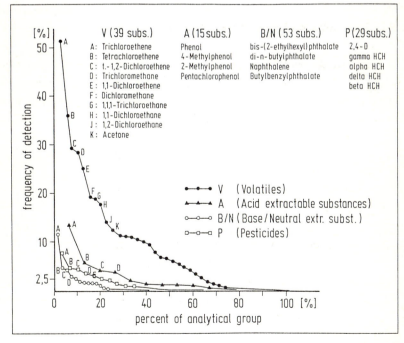

Figure 4:
Detection frequency versus percentage of substances differentiated into 4 analytical groups of organic groundwaters pollutants downstream of waste sites (modified after Plumb Jr., 1985)

rity pollutant list can be subdivided into four analytical groups:

```
- Volatiles                        n=39
- Acid extractable substances      n=15
- Base/Neutrals                    n=53
- Pesticides                       n=29
```

In Figure 4, the results are shown in a modified form, with the percentage of detection frequency of these substances in groundwater being plotted against the relative number of substances of each group. Those substances within the individual groups are named which have been detected in more than 2.5% of the investigated sites (n = 358). For better comprehension, the number of the volatiles is limited to the 10 most frequent ones.

The predominance of the volatiles as outlined above is also demonstrated by the fact that within this group, 25 substances (67%) show a detection frequency of more than 2.5% while for the other three groups, the corresponding figures are a mere 8-25%.

For a future wastes disposal in the sense of a final storage, it is of decisive importance that individual substances are found in groundwater downstream of tipping sites more frequently than others. It can be assumed that these substances are more frequently present in groundwater because of their elevated currency potential on the path waste-soil-groundwater due to their physico-chemical properties. Alternatively, their higher frequency in wastes may be an explanation.

3.1.1 Physico-chemical behaviour within the path waste-soil-groundwater

Essentially, the groundwater currency of a substance is determined by characteristics such as its mobility, persistence and accumulation potential, each of which may be described by a different number of physico-chemical properties.

Examples of parameters to describe the mobility are e.g. solubility, vapour pressure, boiling point, evaporation number, Ostwald's solubility, volatility from watery solution, adsorptive potential, density, viscosity, dissociation constant, surface tension, lipid solubility, Henry's law constant etc. In view of the fact that many of these parameters correlate with one another and may also overlap as to their meaning (e.g. vapour pressure and boiling point) and in view of the non-availability of data for many of these parameters (e.g. evaporation number) the number of parameters to describe mobility can

Table 3:
List of selected physico-chemical parameters describing the behaviour of substances on the path waste-soil-groundwater

Parameter	described by values of	dimension
Mobility	Solubility Vapour Pressure	mg/l hPa
Persistence	Hydrolysis aerobic degradation anaerobic degradation	half-life time
Accumulation	Octanol/water partition coefficient(POW) First-order molecular connectivity index (^{1}chi)	---- ----

be reduced to a few. Of these, physico-chemical properties should be known for almost all substances, as is the case for the solubility and the vapour pressure (Tab. 3).

In this way, a quantification may become possible. Likewise, also persistence and accumulation potential may be described by a small number of properties, e.g. the bioaccumulation by the octanol-water partition coefficient (P_{OW}) and the adsorption onto soils (geoaccumulation) by the first-order molecular connectivity index (^{1}chi-index) (Sabljic, 1987). In this connection, the latter one seems to be of particular interest, because the theoretical capacity of a substance for adsorption to soil particles may be calculated from its molecular properties, thus saving the necessity of expensive serial investigations. The situation is more difficult in respect of persistence, since for most substances, no data are available on hydrolysis or half-life periods for aerobic or anaerobic degradation, respectively. For this reason, this parameter has not been considered in the present case.

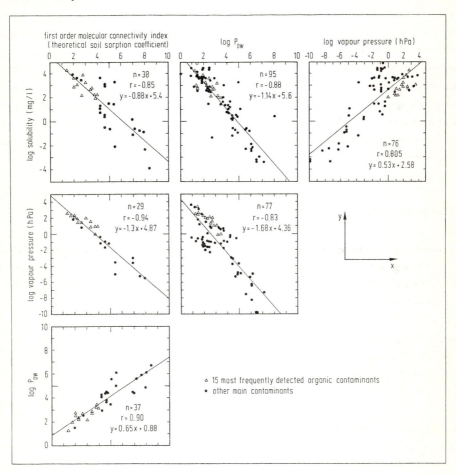

Figure 5:
Correlations between four selected physico-chemical parameters describing mobility and accumulation potential of contaminants on the path waste-soil-groundwater

From the available analytical data for 92 abandoned deposits, a list of substances preferentially found in downstream groundwater has been drawn up and completed by the properties of the four parameters, solubility, vapour pressure, P_{OW} and ^{1}chi-index for these substances. A calculation of their correlation (r > 0.8) and a comparison with the detection frequency in groundwater have shown a close association (Fig. 5).

In Fig. 5, the presently known 15 main contaminants for the Federal Republic of Germany have been marked by triangles. All of them are characterized by high solubility, high vapour pressure, low P_{OW} and low ^{1}chi-index, i.e. they are of a high mobility and have a low accumulation potential. Substances having such characteristics should be avoided for final storage. This applies in particular to substances which are also of high persistence.

3.1.2 Occurrence of organic substances in wastes

No data are available on the occurrence in point of quantity of organic substances in wastes of abandoned deposits, so that it is not possible to study for a possible relationship between the frequency of detection in the groundwater and the frequency of occurrence in the wastes, e.g. in the form of a scattergram. It is likewise impossible to reach this objective by basing calculations on the amounts of the organic substances produced because production figures and amounts in the waste must not necessarily be related. The dependency of the detection frequency and the concentration of an organic substance in groundwater on its amount within the material deposited at a site cannot be established for abandoned deposits. This problem is, however, of minor importance for future final storage as water soluble and persistent substances have an inherent high potential hazard for groundwater and should not be disposed of by tipping, for that reason.

4. TOXICITY OF GROUNDWATER CONTAMINANTS

The groundwater mobility of a substance must not be confused with its potential to endanger groundwater. The latter is a result of the additional evaluation of its toxicological and hygienic relevance. There are substances with a high groundwater currency potential which, for that reason, are found quite often and in high concentrations downstream of tipping sites but which are of minor importance toxicologically (e.g. NaCl).

In contrast to the groundwater mobility which may quite well be described by physico-chemical properties, an evaluation of the toxicity of substances is far more complex. On the one hand, there are many toxicity parameters which are not standardized in part and thus not comparable. On the other, toxicity tests often do not supply figures but non-quantifiable alternatives. Finally, results can be extrapolated to man only to a limited degree.

The guide and limit values of the Drinking Water Regulations (BMJFG, 1986) and the EC Directive (EC, 1980) which in part are based on toxicology, presently constitute an evaluation aid. However, they constitute a legal provision with regard to drinking water and can be employed in the toxicological evaluation of groundwater contaminants with reservation only. With such limitation, the maximum acceptable concentration (MAC) of a substance in its empirical groundwater concentration range may be used to estimate its toxicological groundwater hazard potential. This will be described in the following by means of examples.

Figure 6 shows such groundwater concentration ranges in form of frequency distributions and cumulative curves, for two inorganic (lead, cadmium) and two organic (trichloroethene, tetrachloroethene) substances sampled downstream of 92 tipping sites in the Federal Republic of Germany.

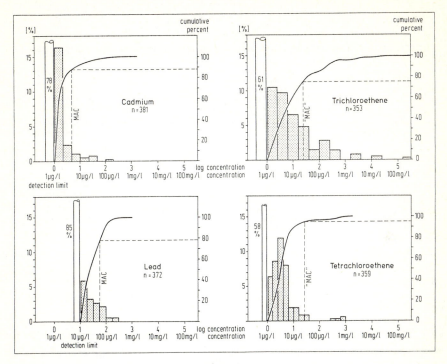

Figure 6:
Frequency distributions and cumulative curves of Lead, Cadmium, Trichloroethene and Tetrachloroethene concentrations from contaminated groundwaters downstream of waste sites in the Federal Republic of Germany (MAC = Maximum Acceptable Concentration of the German Drinking Water Regulation)

While in the histogram all measurements are considered, only "positive proofs" are reflected in the cumulative curves. For the two inorganic substances, these are the values above the detection limit (Cd: 1 μg/l, Pb: 10 μg/l), for the two organic ones, all values >1 μg/l. For all four substances, the MACs have been marked which for the two inorganic substances correspond to those of the Drinking Water Regulations (Cd: 5 μg/l, Pb: 40 μg/l) and for the organics to 25 μg/l, which is the MAC of the Drinking Water Regulations for the cumulative annual sum of chlorinated hydrocarbons (trichloroethene, tetrachloroethene, 1,1,1-trichloroethane). The percentage of "positive proofs" in the groundwater downstream of tipping sites above the marked MAC can be read from the cumulative curve. It is 12% of the measurements for Cd and 22% for Pb. This means that the ranking is inverse to that by the contamination factor (cf. Fig. 2) being >6.5 for Cd and 5.0 for Pb. Also for the two organic substances, trichloroethene and tetrachloroethene (values above MAC 25% and 5%, respectively) an inverse ranking as compared to the detection frequency can be observed. It is both the contamination factor and the detection frequency, considered together with toxicological criteria, which will permit quantitative statements on the potential groundwater hazard presented by a substance.

At present, a method for a standardized evaluation of groundwater contamination from abandoned waste deposits is being developed at the Institute for Water, Soil and Air Hygiene of the Federal Health Office (Arneth et al., 1988; Schleyer et al., 1988). Toxicological criteria are to be included in this evaluation as well.

Table 4:
List of selected parameters and tests for the characterization of the toxicity of groundwater contaminants

TOXICOLOGICAL TESTS:	(sub)acute toxicity (sub)chronic toxicity reproductive toxicity (sub)acute immunetoxicty (sub)chronic immunetoxicity	NOEL or LOEL in mg/kg bodyweight
CARCINOGENITY:		5 alternatives (A, B_1, B_2, C, D)
OTHER TOXICOLOGICAL TESTS:		
Metabolism and distribution: Biochemical toxicity: Mutagenity:	cumulation in mammalia reactive metabolites placenta accumulation hemogram/bloodserum changes enzymes/hormone/transmitter changes metal homeostasis DNA-damages point mutations chromosome mutations	positive or negative, respectively
ECOTOXICOLOGY:	fish test daphnia magna test algae test bacteria test	concentration

In Table 4, the most important criteria are listed which are to be accounted for in the evaluation of the toxicity of groundwater constituents where e.g. the primary use of this water will be for drinking purposes involving the oral intake of a pollutant.

The availability of data for these parameters with regard to substances detected in groundwater varies considerably, ranging from "well tested under all aspects" to "not tested at all". For this reason, toxicological evaluation of many substances is difficult, the more so, since until today, there have been only approaches to a linkage of the criteria mentioned in Table 4 towards the objective of establishing a toxicological "evaluation number" for each substance.

It should be also possible to determine the potential for groundwater hazards on a toxicological basis, using a "toxicity parameter" geared to the path waste-soil-groundwater, and linking it to the contamination factor, the detection frequency, the concentration range and the physico-chemical properties of all contaminants which are of relevance for groundwater (main contaminants). Substances involving the highest potential hazard may then be defined as priority pollutants.

From the group of the 21 inorganic main contaminants (contamination factor >2, Chapter 2., Fig. 2), the following 5 inorganic priority pollutants of groundwater result from an application of the EC Directive as a tool of toxicological evaluation (group D = toxic elements): Arsenic, cadmium, lead, chromium and nickel.
All the 15 organic main contaminants (Chapter 3.1, Fig. 3) are toxicologically relevant. This is documented by their presence in several substances lists (Catalogue of Substances Hazardous to Water, U.S. Priority Pollutant List, Netherlands List etc.). All of them are priority pollutants as well.

5. CRITERIA FOR THE SELECTION OF TIPPING SITES

A complete avoidance of priority pollutants would not appear likely even in the future, so that a very important objective should be an optimal sealing of future tips. Since even this would not offer a complete warranty of a protection of groundwater against pollutants from leachates over decades and centuries, a selection of suitable sites is of additional and major importance. Justifiable sites would be only those where either no contaminable groundwater is present or - if present - where a utilization of existing groundwater, especially a utilization for drinking purposes can be excluded even for the future. Further criteria may be a general protection of groundwater as an ecosystem and protection of surface waters against influents of contaminated groundwater. One should also think of possible damage to buildings by contaminated groundwaters.

These associations may be illustrated by a matrix where the objectives of protection are plotted against those portions of the groundwater downstream of a waste deposit which may have a detrimental effect on the objectives of protection (Fig. 7).

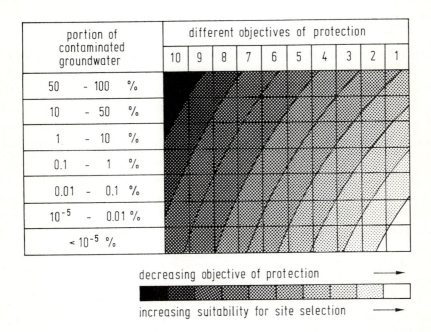

Figure 7:
Schematic procedure of site selection considering different objectives of protection (1 to 10; 10 = drinking water) and percentage of contaminated groundwater

The fields of the matrix schematically shown would correspond to the degree of protection worthiness as a function of priority in shades of grey from black to white.

6. CONCLUSIONS

Groundwater downstream of abandoned waste deposits contains a wide spectrum of substances mobilized from the wastes which contaminate the groundwater. The importance of the individual substances as groundwater contaminants can be determined only by statistical evaluation of as large as possible a number of groundwater analyses for a representative number of abandoned waste deposits. The contamination factor and/or the detection frequency serve to establish a list of main contaminants characterized by a high groundwater currency, i.e low accumulation potential, high mobility and high persistence. Taking into account toxicity and hygienic relevance, substances involving a high potential groundwater hazard can be defined as so-called priority pollutants. For future final storage of wastes, the presence of these substances should be avoided.

An evaluation of groundwater analyses from the vicinity of 92 abandoned waste deposits in the Federal Republic of Germany and Berlin (West) has revealed the following priority pollutants: arsenic, cadmium, lead, chromium and nickel as the inorganic ones and tetrachloroethene, trichloroethene, cis-1,2-dichloroethene, benzene, vinyl chloride, trichloromethane, 1,1,1-trichloroethane, xylenes, trans-1,2-dichloroethene, toluene, ethylbenzene, dichloromethane, dichlorobenzenes, chlorobenzene, and tetrachloromethane as the organic ones.

The fact cannot be accounted for in this connection that individual priority pollutants may also be degradation products (e.g. cis-1,2-dichloroethene, vinyl chloride) which do not occur at all or much less frequently in wastes than this is suggested by their prominent presence in groundwater. This is also true of other phenomena such as the complex of solubility increment which may result in the presence of some substances in groundwater in concentrations beyond their normal solubility in water. At present, there is also a lack of knowledge with regard to an evaluation of the toxicity of many substances, mixtures of such substances and metabolites that would take into account the path wastes-soil-groundwater. There is also a need for research to complete the data on the physico-chemical properties of substances, in particular those that describe persistence in soil and groundwater (e.g. hydrolysis, aerobic and anaerobic degradation). A continuously growing data base will lead to an improved and extended list of priority pollutants, so that by avoiding the presence of these substances in future wastes and simultaneous improvement of the sealing of tips and tipping sites a better protection of groundwater can be achieved.

ACKNOWLEDGEMENTS

The present study was supported by funds from the Federal Minister for Research and Technology and conducted under the auspices of the Federal Environmental Agency (FKZ.: 1440464 3). The extensive data base was created with the kind assistance of many government organs of the Federal Republic of Germany.
The authors are indebted to many colleagues and friends in our institute, especially to Monika Jung for making drawings and to Manfred Frank for translating the manuscript.

REFERENCES

Arneth, J.-D.; Kerndorff, H.; Brill, V.; Schleyer, R.; Milde, G. and Friesel, P. (1986): Leitfaden für die Aussonderung grundwassergefährdender Problemstandorte bei Altablagerungen.- WaBoLu-Hefte, 5/1986.

Arneth, J.-D.; Schleyer, R.; Kerndorff, H. and Milde, G. (1988): Standardisierte Bewertung von Grundwasserkontaminationen durch Altlasten. I. Grundlagen sowie Ermittlung von Haupt- und Prioritätskontaminanten.- Bundesgesundheitsblatt, 31, 4, p. 117-124.

BMJFG (Bundesministerium für Jugend, Familie und Gesundheit) (1986): Verordnung über Trinkwasser und über Wasser für Lebensmittelbetriebe (Trinkwasserverordnung - TrinkwV) vom 22.Mai 1986.- BGBl, Jg. 1986 (Teil I, Nr. 2).

Brill, V.; Kerndorff, H.; Schleyer, R.; Arneth, J.-D.; Milde, G. and Friesel, P. (1986): Fallbeispiele für die Erfassung grundwassergefährdender Altablagerungen aus der Bundesrepublik Deutschland.- WaBoLu-Hefte, 6/1986.

EG (Europäische Gemeinschaft) (1980): Richtlinie des Rates vom 15. Juli 1980 über die Qualität des Wassers für den menschlichen Gebrauch.- Amtsblatt der EG vom 30.8.1980, Nr. L229.

Kerndorff, H.; Brill, V.; Schleyer, R.; Friesel, P. and Milde, G. (1980): Erfassung grundwassergefährdender Altablagerungen - Ergebnisse hydrogeochemischer Untersuchungen.- WaBoLu-Hefte, 5/1985.

Plumb Jr., R. H. (1985): Disposal Site Monitoring Data: Observations and Strategy Implications.- In: Hitchon, B. and Trudell, M. (eds.): Hazardous Wastes in Groundwater: A Soluble Dilemma.- Proceedings of the second Canadian/American Conference on Hydrogeology, Banff, Alberta, Canada, June 25-29 1985, p. 69-77.

Plumb Jr., R. H. (1987): A Practical Alternative to the RCRA Organic Indicator Parameters.- In: Bursztynsky, T. P. E. (eds.): Proceedings of HAZMACON 87, Hazardous Materials Management Conference and Exhibition, 21-23 April 1987, Santa Clara, California, p. 135-150.

Plumb Jr., R. H. and Pitchford, A. M. (1985): Volatile Organic Scans: Implications for Groundwater Monitoring.- Proceedings of the National Water Well Association/American Petroleum Institute Conference on Petroleum Hydrocarbons and Organic Chemicals in Groundwater, 13-15 November 1985, Houston, Texas, p. 1-15.

Sabljic, A. (1987): On the Prediction of Soil Sorption Coefficients of Organic Pollutants from Molecular Structure: Application of Molecular Topology Model.- Environ. Sci. Technol., 21, p. 358-366.

Schleyer, R.; Arneth, J.-D.; Kerndorff, H.; Milde, G.; Dieter, H. and Kaiser U. (1988): Standardisierte Bewertung von Grundwasserkontaminationen durch Altlasten. II: Stoffbewertung, Expositionsbewertung und ihre Verküpfung.- Bundesgesundheitsblatt, 31, 5, p. 160-168.

1. What is "final storage quality" ?

A final storage is defined as a landfill which does not have a negative impact on the environment for long time periods. In order to define more precisely this time period, it is helpful to remember that the processes of erosion and sedimentation constantly change the earth surface. Therefore, for Swiss conditions, time periods of 10^4 to 10^5 years are to be observed. The term "negative impact" can be defined in various ways. As a practical working hypothesis it was considered that an increase of the natural concentrations in water, soil or air by 10-20 % has no negative impact on the ecosystem (these values of 10 - 20 % may differ considerably according to the compounds and the landfill-environment to be considered).

In Figure 1, this concept is explained in more detail: The concentration (C_4), which results from the interaction of a landfill leachate with a groundwater, is determined by the natural groundwater concentration, the groundwater and leachate flux, the leachate attenuation in the soil below the landfill and the landfill body emission concentration C_1. While for most landfill sites C_4 is much smaller than C_1 because of attenuation, degradation, dispersion and dilution, a worst case was assumed with $C_4 = C_1$. In order to avoid lengthy discussions about models to calculate C_4 from C_1 according to individual sites and hydrological regimes, this worst case situation was used for practical reasons as a base for further work by group B2. It was assumed that ecotoxicological considerations will not be necessary for the evaluation of a potential impact from a final storage landfill, if C_1 equals about C_{GW}. The question, if the requirements for "final storage" can be met by concentrations of C_1 which are higher than C_{GW} was not addressed.

Due to existing experiences with landfill envelopes, the goal of very low emissions of final storage landfills has to be mainly achieved by decreasing the leachability of the waste material itself and not by technical or natural barriers around the landfill. This means that the waste material must be transformed into a form which is stable and non mobilizable for long periods of time. Such a concept is based on the assumption, that the long term geochemical conditions around the landfill sites are biologically, chemically (pH, redox potential, complexing agents, ionic strength etc.) and physically stable, too.

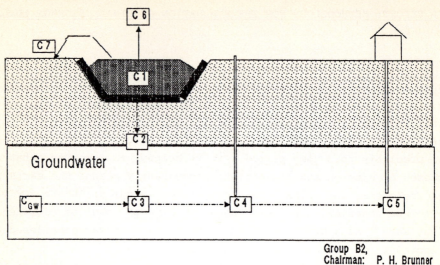

Group B2,
Chairman: P. H. Brunner

LEGEND: C_{GW} = 100%   C1 < 120%   C4 < 120%

C6 = emission concentration in the air
C7 = emission concentration in the soil
C_{GW} = concentration in the groundwater

Figure 1: Concept of final storage: The emission concentrations of a
final storage landfill to water, air or soil shall not exceed 120 % of
the natural concentrations around the landfill site. It is understood
that natural concentrations are not heavily influenced by anthropogenic
activities nor by geological anomalies.

Despite of these stringent requirements on the immobility of the waste
material, additional barriers (liners, geological environment) are necessary for monitoring and security reasons (Fig.2). It may be that final storage quality of a waste product can not be achieved due to certain highly mobile elements or compounds. Regulatory actions should be
aimed towards a reduction of problematic compounds in consumer goods
and industrial products.

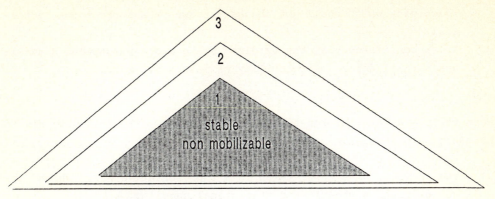

1   internal   waste barrier
2   technical   barrier
3   geological   barrier

Figure 2: The three barriers necessary for final storage landfills. Despite the fact that the waste is supposed to have final storage quality, barriers two and three are needed for monitoring and security reasons.

2. What are the main problems in developing tests for waste characterization ?

The different properties of various waste materials require specifically tailored test methods. There is no universal leaching procedure for all waste materials.
In order to reach final storage quality, the following requirements have to be fulfilled:
-    The waste contains predominantly inorganic materials.
-    Under landfilling conditions the waste has no or only a very low

reactivity which shall not change its own pH and the redox potential.

- It is considered that the oxidized state and a pH between 8 and 11 are optimal for a low solubility of most heavy metals.

When testing a material for final storage quality, the following has to be taken into account:

a) The samples have to be taken in a representative way, which requires often sophisticated statistical sampling.

b) The composition of the waste has to be known, e.g. the wastes have to be analyzed for the following parameters:
   - 95 % of the matrix elements and traces of toxic components.
   - Organic carbon content.
   - The presence of organic, biodegradable material will lead to the production of carbon dioxide and to changes in the pH and redox potential. Therefore available organic carbon in solids has to be tested. This requires the development of an incubation test using selected microorganisms.
   - The reactivity of the material, e.g. the generation of gas in contact with water; the presence of strongly reductive materials in waste. Presence of metals in a finely particulate metallic form etc.

c) Wastes have to be tested in individually tailored leaching tests. The purpose of a leaching test is to predict the concentrations of the leachates in waste landfills, and to improve the understanding of the mechanisms and kinetics of the reactions of the waste material in the landfill. Numerous leaching tests exist, each designed for a particular purpose. In a leaching test the resulting concentrations of chemical substances depend from the composition of the waste and from the test method itself, e.g. the following factors:
   - single or multiple extractions,
   - liquid to solid ratio
   - leaching media
   - leaching interval
   Regulatory action must take into account these factors when establishing concentration limits. The use of a high solid to liquid ratio and the use of acid leaching agents will increase concen-

trations. For these reasons leaching tests have also to be standardized.

d) The testing has to take into account "final storage quality" as it is defined by the actual national waste management policy. The local landfill conditions (e.g. mono landfill, ratio waste/leachate, groundwater regime) have to be observed.

e) For each quality criteria there should exist an approved testing method. Until now only a few tests have been well established. It is urgent that specific tests for important waste materials (incinerator ash, bottom ash, electroplating sludges etc.) are developed.

f) The evaluation of the leachate results has to take into account actual landfill conditions. With respect to many problems in design and interpretation of leaching tests it is proposed to compare the results from testing wastes with those from testing naturally occurring materials.

g) Until now leaching tests and transport models have not been coupled. At present predictions of concentrations in landfills are subject to error.

3. What are the relevant parameters in emissions from final storage deposits which are to be measured to meet the required safety in the ambient water (ground and surface water), air and soil (soil/dust)?

The very many substances which are in a landfill have different tendencies to be either immobile or mobile by different pathways (water, other liquid phases, gaseous phases etc.). To establish the relevant parameters for emission control and to investigate the tendency of a material to be emitted, pathway analyses are necessary. In such analyses information on the waste composition has to be combined with the existing knowledge of the physico-chemical behavior of the different substances. The tendency to be emitted by different pathways can be described by different parameters such as solubility, vapor pressure, octanol/water distribution coefficient (P_{ow}), [1]chi-index etc. All substances characterized e.g. by high solubility, high vapor pressure, low

P_{OW} and low [1]chi-index are of high mobility and have a low accumulation potential. Therefore they are relevant parameters when emissions from final storage landfills to the ground water are to be controlled. Additional parameters have to be looked for by systematic investigations into the emissions from existing landfills.

These investigations do not yet allow to assess the emissions due to the erosion of the final storage landfill.

When the relevant parameters have been established, these parameters have to be considered for the " final storage quality assessment" of the waste material in the laboratory as well in liquid and gaseous emissions of existing landfills. These values have to meet the quality requirements which have been established for "final storage".

4. How are environmental impacts of landfills evaluated in respect to final storage quality ?

Impacts on the groundwater below a landfill can be assessed by measuring either C_1 or C_4 (cf. Figure 1).

The advantages of measuring concentration C_1 are:
- Immediate control of the leachates before they leave the landfill body and enter the environment.
- Indicator for the necessity of additional measurements (C_2, C_3, C_4, gaseous emissions etc.)
- Validification of the "final storage quality" control test, helps to develop and optimize such tests.
- Yields base parameters to be applied in a model to calculate the pollutant transport to the groundwater.

Disadvantages of measuring concentration C_1:
- Gives no indication about the emissions to be expected in the groundwater (transport mechanisms, attenuation etc.)

Advantages of measuring concentration C_4:
- Gives information about the impermeability of the envelope (if C_1 is known to be $> = 0$!)
- Gives qualitative information about the emissions
- Predictions can be made for downstream pollution transport if the hydraulic regime is known.

Disadvantages of measuring concentration C_4:
- Gives no information about the quantity of emissions
- Time lag between a concentration increase at C_1 and C_4 means that remedial action may be late.
- The geological and hydrogeological regime between C_4 and C_5 must be
  known well enough to take efficient protection measurements in case
  of essential immissions at point C_4.

Complementary:
- To have the complete information concerning landfill emissions and
re sulting groundwater concentrations we need both measurements at
point C_1 and C_4.

The measurement of the concentrations C_1 and C_4 allows to assess the
possible negative impact on the groundwater of a landfill. It has to be
kept in mind, that the groundwater concentration C_4 is the result of
groundwater- and leachatefluxes, and of biochemical, chemical and
physical reactions of these material fluxes . To be able to calculate
the relationship existing between C_1 and C_4, it is necessary to know
the value of the adsorption, degradation and dilution coefficients relevant for the migration of the contaminants. In other words it is necessary to know the properties of the soil below the landfill and the
hydrology of the groundwater .
The concentration C_1 can be determined directly on the site (drainage
at the landfill bottom) or tentatively in the waste material in the
laboratory ("C_1"). Both determinations are necessary, the on site measurement for monitoring and safety reasons, and the laboratory determination for "final storage quality" assessment. In the latter case, as
the knowledge of the composition of the waste can give only a first indication on the solubility of the contaminants, it is necessary to
carry out tests to simulate the actual landfill environment. When the
laboratory "C_1" is determined, it has to be kept in mind that the
laboratory conditions are never completely identical with the field
conditions and that therefore deviations of "C_1" and C_1 are to be expected. When the laboratory "C_1" is compared with water quality standards, these possible deviations have to be taken into account (safety
margins).
In order to have a better knowledge of the potential risk of the landfill, it is necessary to monitor the leachates. When the leachate data
is combined with information on the quantities of waste deposited and
on the water household of the landfill, the expected flux of emissions

can be estimated. The size of this flux, calculated as indicated, has
to be considered in relation with the global hydrological balance of
the site. The same approach can theoretically be used for assessing the
air pollution. However, the working group B2, keeping in mind the
stringent standards imposed on the "final storage landfill" and particularly the low organic carbon content present in "final storage"
waste material, expresses the opinion that the negative impact on the
atmosphere, on the basis of today's analytical knowledge, is not quantifiable.

Summary

1. Taking a conservative and preventative approach, the following practical working hypothesis for the definition of final storage quality is
established: The emission concentration of a final storage should not
exceed environmental concentrations by more than 10 to 20%.

2. Test methods for final storage quality have to be tailor made for
each group of waste materials and must take the following into account:
- The composition of the waste must be known and the test samples must
  be representative.
- The test results are not only depending on the waste material but
  also on the test method.
- The test method must represent the proposed landfill practice as well
  as possible.

3. The reactivity of the waste material should be low, and long time
changes in the landfill environment should be a minimum.

4. It is important to select parameters specifically to evaluate the
final storage quality of a waste material. This selection should be
done by pathway analysis which is based on the chemical, physical and
biological behavior of a waste material in the specified environment.

5. In order to evaluate the environmental impact of a final storage
landfill, the concentrations C_1 and C_4 should be monitored (Fig.1). A
pathway analysis, based on mass balances and the hydrogeology of the
landfill body and site is required to evaluate the long term fluxes
from a final storage.

CONCLUSIONS AND OUTLOOK

CONCLUSIONS AND OUTLOOK

Peter Baccini and Thomas Lichtensteiger

Swiss Federal Institute for Water Resources and Water Pollution Control

1. Challenges by the chosen topic

The notions "reactor landfill" and "final storage landfill" in the workshop title have stimulated manifold and opposite reactions among the participants. For the engineer the two notions stood in opposition, at least at the beginning of the discussion. For the geologist it became soon primordial to combine the two, to see them as essential parts of a whole.

To handle the sanitary landfill as a reactor, a "biogeochemical" reactor so to say, was an acceptable procedure. The engineers were not seriously disturbed by the open questions put up by biologists and chemists. On the contrary, the extension of a geotechnically defined system, a "black box" with a combination of liners around it, to a chemical engineer's "fixed bed reactor" (Belevi 1987) or even to a biologist's small "ecosystem" (Aragno 1988) was welcomed heartily and considered to be an improvement. New horizons appeared and new technical possibilities were postulated (technical biology included) to control more efficiently the processes in the landfill and the emissions from it.

The trouble started with a new objective named "final storage". It was interesting to see that the geologists were strongly attracted by the task to predict the long-term behavior and to draw conclusions, on the basis of today's knowledge, about the type of material to be deposited in the future. From these conclusions it became obvious that the future of the technical system "reactor landfill" was endangered, a whole business branch and an established field of research were seriously questioned. Consequently some of the participants rejected principally the Swiss concept to attain final storage quality in landfills within less than 50 years for different reasons. They ranged from "unrealistic goals" to "too ambitious and utopia guidelines". It became very clear that many countries have still enough territorial resources for potential landfill sites. Furthermore it can be argued that landfills are man-made works to be handed over to the next generations, analogously to houses, streets and cathedrals. It is the merit of these opponents to have provoked very clearly the following axioms to base upon a scientific discussion on "landfill strategies":

(1) Landfills are an indispensable equipment of a national economy. Total recycling of solid residues is not possible.

(2) Quantity and quality of solid residues to be landfilled depend on the economic and cultural activities of a society.

(3) A landfill strategy is based on political guidelines. Ethical, social and economical principles are hidden behind the practical procedure in landfilling, consciously or unconsciously.

(4) Landfill strategies determine decisively the whole set of pretreatments in the waste management system of an economy. The more precise the landfills are defined with respect to their geochemical boundary conditions in space and time, the more precise wastes have to be "manufactured" in order to be landfilled.

At the end of the workshop all participants have accepted the challenge "to play Swiss" for a week or to take the role of the devil's advocate. The "final storage concept" has proven to be an efficient catalyst in a workshop on landfills. One statement from the final discussion is cited here, representing many similar comments: "If the Swiss want to achieve final storage quality in landfills within a generation, they have to exclude the classical municipal solid waste landfills from their waste management systems" (Chester Francis).

2. Main conclusions

The workshop objectives (listed in the introductory chapter on page 2) were achieved. The answers to the specific questions are well documented in the four group reports. They contain also, not explicitly stated, a critical revision of existing technical and regulatory processes. This revision may have an impact on the political guidelines in waste management. In the following, four questions with a small set of short answers are given to summarize the main conclusions of this workshop.

(1) How do we qualify a landfill?

We need to know
- the chemical composition and the phase assemblage (crystalline and noncrystalline phases) of the waste
- the geotechnical properties of the site
- the properties of the site's environment
- the methods to monitor the landfill

(2) How should the waste for a final storage look like?

It should be
- inorganic
- in an oxidized form
- solid and poorly soluble in water

(3) How can the above mentioned properties be measured?

- A uniform omnipotent test is principally not possible
- At least a half a dozen tests are required, tailor made for each waste type meeting the above mentioned criteria,

(4) How can the required properties be achieved?

- None of the available waste treatment systems can achieve these properties. A reactor landfill for municipal solid waste will produce leachates not compatible with the environment for at least hundred years.
- Liners are made to control material fluxes for decades. They are not suited to control for longer time periods.
- In a first step it is reasonable to adapt existing waste treatment processes (e.g. incineration) to improve the quality of the solid residues (e.g. bottom ash).
- In a second step it will be necessary to adapt the consumer products and their precursors. It will hardly be possible to develop omnipotent treatment processes to achieve economically final storage quality for every product.

4. Outlook

The workshop has also shown the needs for further research. Independent of the landfill strategy chosen it became clear that scientists lack conceptual models for landfilling. We have large data set from different sites to sketch simple empirical transport and transformations models with a few parameters. In other words we are aware of the fact that landfilling with contemporary wastes is a large scale and long-term experiment for which we have poor knowledge. We know also that in the field of production (supply of consumer goods) we would not run chemical and biological processes under such circumstances.

The final storage approach is one way to develop and control landfills on a conceptual basis. We have realized that the earth sciences have a great potential to build such a basis, for example with their knowledge of diagenetic processes (Lichtensteiger 1988). However this knowledge has not yet been applied in waste management, except in some specific cases, e.g. with radioactive materials. The results of this workshop may motivate more scientists, especially geologists and geochemists, to design anthropogenic sediments to be integrated in the large biogeochemical cycles.

The reactor landfill and its environmental impacts will be part of every country during the next generations, even if some regions are able to start immediately with final storage landfills. There was agreement among all workshop participants that a better control of existing reactor landfills can not be based solely on a barrier concept with liners. A better control and efficient restoration programs are only possible if we know more about the processes within the landfill, about the short-term and long-term reactions in the "black box". During the last decades engineers were asked to construct landfills and to run them. Their knowledge and their skill is still indispensable now and in the future. However the team has to be enlarged with chemists and biologists, especially microbiologists, who apply their scientific methods in the study of reactions in the landfill and develop biological and chemical concepts to run the reactor.

At first sight the two notions "reactor" and "final storage" seemed to be on opposite sides, like two knights in a medieval tournament, ready to duel. During the workshop the academic duel was sportsmanlike and very useful. It helped to screen very critically the scientific and technical concepts used in the daily practice of landfilling. At the end of the tournament it became clear that the little word "and" in the workshop title is the important link for the future research work on landfills.

References

Belevi H. and P. Baccini, "Water and Element Fluxes from Sanitary Land-fills", Proc. of the ISWA Int. Sanitary Landfill Symp., Cagliari 1987

Aragno M., "The Landfill Ecosystem: A Microbiologist's Look Inside a Black Box", this book, page ... (1988)

Lichtensteiger Th., Brunner P.H., "Transformation of Sewage Sludge in Landfills", ISWA 88 Proc. of the 5th Int. Solid Wastes Conf., Copenhagen 1988

LIST OF PARTICIPANTS

Prof. M. Aragno

Institut de Biologie
Université de Neuchâtel
CH-2000 Neuchâtel
Tel.: 038/25 64 34

Prof. P. Baccini

Abt. Abfallwirtschaft und
Stoffhaushalt
EAWAG
CH-8600 Dübendorf
Tel.: 01/823 51 21

Dr. H. Belevi

Abt. Abfallwirtschaft und
Stoffhaushalt
EAWAG
CH-8600 Dübendorf
Tel.: 01/823 51 30

H. Billard

Chef du service Recherches-
Etudes-Développement ANRED
2 square La Fayette, B.P. 406
F-49004 Angers Cedex
Tel.: 41/87 29 24

Dr. P.H. Brunner

Abt. Abfallwirtschaft und
Stoffhaushalt
EAWAG
CH-8600 Dübendorf
Tel.: 01/823 51 29

Dr. E. Bütow

Institut für wassergefährdende
Stoffe, Universität Berlin
Hardenbergplatz 2
D-1000 Berlin 12
Tel.: 030/262 70 41

Prof. F. Colin

Institut de recherches
hydrologiques
10, rue Ernest Bichat
F-54000 Nancy
Tel.: 83/96 65 10

Dipl.-Ing.-ETH M. Egli

Büro für Kies und Abfall AG
CH-3118 Uttigen
Tel.: 033/45 48 48

PD. Dr. H.-J. Ehrig

ITW Ingenieurberatung Gmbh
Friedrich-Kaiser-Str. 23
D-5860 Iserlohn
Tel.: 02371/3901-02

Dr. H.P. Fahrni

Bundesamt für Umweltschutz
Hallwylstr. 4
CH-3003 Bern
Tel.: 031/61 93 28

Prof. G. Farquhar

Faculty of Engineering
Department of Civil Engineering
University of Waterloo
Waterloo, Ontario, Canada
Tel.: 519/885 12 11

Dott. Ing. F. Filippini

Dipartimento dell'ambiente
Sezione protezione acque
CH-6500 Bellinzona
Tel.: 092/24 11 11

Prof. U. Förstner

Arbeitsb. Umweltschutztechnik
Technische Universität
Hamburg-Harburg
Eissendorferstr. 38
D-2001 Hamburg 90
Tel.: 040/7 71 70 26 09

Dr. Ch.W. Francis

Environmental Sciences Division
1505, MS 036, Oak Ridge
USA-Tennessee 37831
Tel.: 615/574 7257

Prof. D.H. Gray

Department of Civil Engineering
University of Michigan
USA-Ann Arbor, Michigan
48109-2125
Tel.: 313/764 84 95

Prof. R.K. Ham

Department of Civil and
Environmental Engineering
University of Wisconsin
USA-Madison, Wisconsin 53706
Tel.: 608/262 72 49

Dr. E. Höhn

Buchserstr. 44
CH-8157 Dielsdorf
Tel.: 01/853 30 00

Dr. P. Huggenberger

Département de géologie
Université de Genêve
13, rue des Maraîchers
CH-1211 Genêve 4
Tel.: 022/21 93 55

Dr. W. Kanz

Abt. Umweltschutz
Baudep. Kt. AG
CH-5001 Aarau
Tel.: 064/21 27 49

Dr. J. Krebs

Mühlemattstrasse 66
CH-3007 Bern
Tel.: 031/45 09 38

Dr. A. Lagerkvist

Upptek K.B.
Box 501-56
SW-951 05 Luleà
Tel.: 0920 11 825

Dipl.-Ing. P. Lechner

Institut für Wassergüte und
Landschaftswasserbau
Technische Universität Wien
Karlsplatz 13
A-1040 Wien
Tel.: 0222/588 01

Dr. Th. Lichtensteiger

Abt. Abfallwirtschaft und
Stoffhaushalt
EAWAG
CH-8600 Dübendorf
Tel.: 01/823 51 25

Prof. F.T. Madsen

Tonmineralogisches Labor
ETH-Zentrum/NO
CH-8092 Zürich
Tel.: 01/256 37 84

Dr. Ch. Matter

Ernst Basler und Partner AG
Zollikerstr. 65
CH-8702 Zollikon
Tel.: 01/395 11 11

Dipl.-Ing.-ETH J. Messmer

Büro für Kies und Abfall AG
CH-3118 Uttigen
Tel.: 033/45 48 48

Prof. G. Milde

Inst. für Wasser-, Boden-
und Lufthygiene
Corrensplatz 1
D-1 Berlin 33
Tel.: 030/8308 2337

Dr. H.P. Müller

Abt. Umweltschutz
Baudep. Kt. AG
CH-5001 Aarau
Tel.: 064/21 27 59

Dipl.-Ing.-ETH P. Oggier

Bundesamt für Umweltschutz
Hallwylstr. 4
CH-3003 Bern
Tel.: 031/61 93 76

Dr. E. Peiser

Alpenstr. 14
CH-3653 Oberhofen
Tel.: 033/43 21 03

Prof. A. Pfiffner

Geologisches Institut
Universität Bern
Baltzerstr. 1
CH-3012 Bern
Tel.: 031/65 87 57

Dr. G. Piepke   Umweltphysik
                                       EAWAG
                                       CH-8600 Dübendorf

Dipl.-Ing. W. Ryser   Rytec AG
                                       Abfalltechnologie +
                                       Energiekonzepte
                                       CH-3110 Münsingen
                                       Tel.: 031/92 62 92

Prof. M. Salkinoja-Salonen   University of Helsinki
                                       Department of General
                                       Microbiology
                                       Mannerheimintie 172
                                       SF-00280 Helsinki
                                       Tel.: 080/ 47 35 402

Dr. Ch. Schlüchter   Ingenieurgeologie
                                       ETH-Hönggerberg
                                       CH-8093 Zürich
                                       Tel.: 01/377 25 21

PD. Dr. R. Schwarzenbach   Seenforschungslaboratorium
                                       der EAWAG
                                       Seestr. 79
                                       CH-6047 Kastanienbaum
                                       Tel.: 041/47 11 75

Dipl.-Chem. D. Stämpfli   Abt. Abfallwirtschaft und
                                       Stoffhaushalt
                                       EAWAG
                                       CH-8600 Dübendorf
                                       Tel.: 01/823 51 38

Prof. R. Stegmann   Arbeitsb. Umweltschutztechnik
                                       Technische Universität
                                       Hamburg-Harburg
                                       Eissendorferstr. 38
                                       D-2100 Hamburg 90
                                       Tel.: 040/77 182 703

Dipl-Ing. K. Stief

Umweltbundesamt Berlin
Bismarckplatz 1
D-1000 Berlin 33
Tel.: 030/721 15 76

Dr. W. Strub

Büro für Kies und Abfall AG
CH-3118 Uttigen
Tel.: 033/45 48 48

Dr. J. van Stuijvenberg

CSD AG
Kirchstr. 22
CH-3097 Liebefeld
Tel.: 031/53 64 12

Dr. J. Zeyer

Seenforschungslaboratorium
der EAWAG
Seestr. 79
CH-6047 Kastanienbaum
Tel.: 041/47 11 75

Dr. A. Zingg

Geol.- Paläontolog. Institut
Universität Basel
Bernoullistr. 32
CH-4056 Basel
Tel.: 061/25 25 62

Dr. A. Zurkinden

Eidg. Verkehrs- und Energie-
Wirtschaftsdepartement (EVED)
Abt. Strahlenschutz
CH-5303 Würenlingen
Tel.: 056/99 39 30

SPONSORS

The Workshop was financially supported by Swiss enterprises and public authorities, listed below in alphabetical order:

- Amt für Gewässerschutz und Wasserbau des Kt. Zürich
- Amt für Verkehr, Energie und Wasser des Kt. Bern
- AG für Abfallverwertung (AVAG), Thun
- Baudepartement des Kt. Aargau
- Brunner Rolf, Direktor, Thun
- Bundesamt für Umweltschutz (BUS), Bern
- Büro für Kies und Abfall AG, Uttigen
- Ciba-Geigy AG, Basel
- Deponie Teufthal AG (DETAG), Frauenkappelen
- Einwohnergemeinde Köniz
- Ersparniskasse Konolfingen
- Hänsenberger Arthur, Ständerat, Oberdiessbach
- Hofstetter Karl, Kaufmann, Bern
- Kästle Theo, Dipl.Ing. ETH, Bolligen
- Kies AG Aaretal (KAGA), Uttigen
- Kieswerk Steinigand AG (KIESTAG), Wimmis
- Läderach Arthur, Firma Däpp AG, Oppligen
- Lerch Bruno, Fürsprecher, Thun
- Migros-Genossenschafts-Bund, Zürich
- Schweizerischer Bankverein, Thun
- Schweizerische Mobiliarversicherung, Bern
- Sonderabfallverwertungs-AG (SOVAG), Brügg
- Spar- und Leihkasse, Münsingen
- Verband der Betriebsleiter Schweizerischer Abfall-beseitigungsanlagen (VBSA), Luzern und Uttigen
- Verband Kehricht Worblenthal und Umgebung (KEWU), Münchenbuchsee
- Winterthur-Versicherungen, Thun